Planning Asian Cities: Risks and Resilience

In *Planning Asian Cities: Risks and Resilience*, Stephen Hamnett and Dean Forbes have brought together some of the region's most distinguished urbanists to explore the planning history and recent development of Pacific Asia's major cities.

They show how globalization, and the competition to achieve global city status, has had a profound effect on all these cities. Tokyo is an archetypal world city. Singapore, Hong Kong and Seoul have acquired world city characteristics. Taipei and Kuala Lumpur have been at the centre of expanding economies in which nationalism and global aspirations have been intertwined and expressed in the built environment. Beijing, Hong Kong and Shanghai have played key, sometimes competing, roles in China's rapid economic growth. Bangkok's amenity economy is currently threatened by political instability, while Jakarta and Manila are the core city-regions of less developed countries with sluggish economies and significant unrealized potential.

But how resilient are these cities to the risks that they face? How can they manage continuing pressures for development and growth while reducing their vulnerability to a range of potential crises? How well prepared are they for climate change? How can they build social capital, so important to a city's recovery from shocks and disasters? What forms of governance and planning are appropriate for the vast mega-regions that are emerging? And, given the tradition of top-down, centralized, state-directed planning which drove the economic growth of many of these cities in the last century, what prospects are there of them becoming more inclusive and sensitive to the diverse needs of their populations and to the importance of culture, heritage and local places in creating liveable cities?

Stephen Hamnett is Emeritus Professor of Urban and Regional Planning at the University of South Australia, Adelaide. He is also a Commissioner of the Environment, Resources and Development Court of South Australia.

Dean Forbes is Matthew Flinders Distinguished Professor at Flinders University, Adelaide. He is an urbanist with a special interest in knowledge-based development in contemporary Pacific Asian cities.

Planning, History and Environment Series

Selection of Published Titles

Planning Asian Cities: Risks and Resilience

edited by Stephen Hamnett and Dean Forbes

Routledge
Taylor & Francis Group

LONDON AND NEW YORK

First published 2011 by Routledge

This paperback edition first published 2013
by Routledge
2 Park Square, Milton Park, Abingdon, Oxon OX14 4RN

Simultaneously published in the US and Canada
by Routledge
711 Third Avenue, New York, NY 10017

Routledge is an imprint of the Taylor & Francis Group, an informa business

British Library Cataloguing in Publication Data
A catalogue record of this book is available from the British Library

Library of Congress Cataloging in Publication Data
Planning Asian cities : risks and resilience / edited by Stephen Hamnett and Dean Forbes.
  p. cm.—(Planning, history, and environment)
Includes bibliographical references and index.
  1. City planning—Asia. 2. Community development, Urban—Asia. I. Hamnett, Stephen. II. Forbes, D. K. (Dean K.)
HT169.A77P5293 2011
307.1'216095—dc22

2011009976

ISBN: 978–0–415–56335–2 (hbk)
ISBN: 978–0–203–80424–7 (ebk)
ISBN: 978–0–415–83220–5 (pbk)

Typeset in Aldine and Swiss by PNR Design, Didcot

FSC
www.fsc.org
MIX
Paper from
responsible sources
FSC® C013056

Printed and bound in Great Britain by
TJ International Ltd, Padstow, Cornwall

In memory of Ooi Giok Ling
An eminent Asian urbanist and colleague

Contents

Editors and Contributors

Editors

Stephen Hamnett is Emeritus Professor of Urban and Regional Planning at the University of South Australia. He is also a Commissioner of the Environment, Resources and Development Court of South Australia. His research interests focus these days on metropolitan strategic planning and on planning law and governance. He is a member of a number of journal editorial boards, including *Built Environment* and *Cities*. He has a long record of involvement in planning issues in Asia, including numerous projects on urban and regional development for AusAID and other international agencies in Indonesia, the Philippines, China and elsewhere.

Dean Forbes is Matthew Flinders Distinguished Professor at Flinders University in Adelaide. His previous appointments were at the Australian National University, Monash University, and the University of Papua New Guinea. His research focus in recent years has been on the connections between cities and universities in the context of changes in the global knowledge economy. He was a chief investigator in a joint international project on strengthening slum upgrading and urban governance in Southeast Asian cities funded by the Australian Research Council from 2005 to 2008. His book, *Asian Metropolis*, was published by Oxford University Press in 1996.

Chapter Authors

Ian G. Cook is Emeritus Professor of Human Geography at Liverpool John Moores University in the UK. Cook graduated from the Universities of Aberdeen (BSc Hons) and Nottingham (PhD), and was co-editor and co-founder of *Contemporary Issues in Geography and Education*. He serves on the board of two international journals, the *Journal of Geography in Higher Education* and the *International Journal of Society Systems Science*. He has co-authored and co-edited eight books to date, on China, Asia generally and geography, and has published widely in international journals and edited volumes. His main research interests are in the areas of health, environment, urbanization and the elderly in China, Asia and globally. He leads an annual field course to Beijing.

Tommy Firman is Professor of Regional Planning at the School of Architecture, Planning and Policy Development, Institute of Technology Bandung (ITB), Indonesia. He is a former Dean of the Faculty of Civil Engineering and Planning (2000–2002), Director of the Graduate Program at the Institute (2002–2004), and is currently

Chairperson of the Academic Senate of ITB. He earned his PhD degree in urban and population geography from the University of Hawaii East-West Center. Firman has written extensively on urbanization and urban and regional development in Indonesia in several international journals, including *Habitat International*, *International Development Planning Review*, *Urban Studies*, and *Cities*.

Gu Chaolin is Professor in the Department of Urban Planning, School of Architecture, Tsinghua University. He is the Vice President of the Geographical Association of China, Director of the Urban Geography Professional Commission and Councillor of the Urban Planning Society of China. He serves on the editorial boards of a number of journals including the *International Journal of Urban and Regional Research*, *Urban and Regional Planning Research*, *Acta Geographica Sinica*, *Geographical Research*, *City Planning Review*, *Economic Geography*, *Urban Planning Forum* and *Urban Studies*, and he has written extensively on Chinese cities and on Beijing in particular.

Seong-Kyu Ha is Professor of Urban Planning and Housing Policy in the Department of Urban and Regional Planning at Chung-Ang University, Korea. He received his Masters degree in City Planning from Seoul National University in 1979, a Masters from the London School of Economics in 1981, and a Doctorate in Urban and Regional Planning from University College London in 1984. In addition to his academic positions, he has served as Dean of the Graduate School of Social Development, Chung-Ang University, and Director of the Korea Center for City and Environment Research. Professor Ha also serves as President of the Korea Association for Housing Policy Studies. He has published two books in Korean, *Housing Policy* (Bakyoungsa) in 1995 and *Urban Management* (Hyungseol) in 1992. He also edited the English-language volume *Housing Policy and Practice in Asia* (Croom Helm, 1987). He has written many articles in Korean and in English on housing and urban redevelopment problems in Korea and elsewhere in Asia.

Liling Huang is an Assistant Professor in the Graduate Institute of Building and Planning at the National Taiwan University. The courses she teaches include planning theory, urban design and globalization studies. Her research interest is urban development in Asia with a focus on the cities of Taiwan and Vietnam. On the urban configuration of Taipei, she has published 'Urban Politics and Spatial Development: the Emergence of Participatory Planning in Taipei', in *Globalizing Taipei* edited by R.Y. Kwok (Routledge, 2005). Her most recent publication is 'Against the Monster of Privatization: Qing-Tien Community's Actions for Urban Livability in Taipei', published in *International Development Planning Review* in 2008.

Reginald Yin-Wang Kwok is a Professor of Asian Studies, and Professor of Urban and Regional Planning at the University of Hawaii at Manoa. He is Visiting Professor at the National Taiwan University, Tsinghua University, Tongji University, Zhongshan

University and Wuhan Academy of Urban Construction. Before joining the University of Hawaii, he was the founding Director of the Centre of Urban Planning and Urban Studies (renamed Centre of Urban Planning and Environmental Management) at the University of Hong Kong. Professionally, he has served as a consultant to professional firms, local governments and international organizations in Asia and the United States, including New York Chinatown Planning Council, Hong Kong Housing Authority, Hong Kong Town Planning Board, Shenzhen Planning Committee, Zhuhai Municipality, Taipei City Council, East-West Center, and the World Bank. His areas of specialization are political economy of Chinese development and urbanization, globalization in East Asia, urban economic and spatial development, and cultural impact on urban form. Recent publications include: *Globalizing Taipei: The Political Economy of Spatial Development* (editor)*; The Shek Kip Mei Syndrome: Economic Development and Public Housing in Hong Kong and Singapore* (with Manuel Castells and Lee Goh), and *Chinese Urban Reform: What Model Now?* (co-edited with William L. Parish, and Anthony Gar-On Yeh).

Chuthatip Maneepong is Visiting Assistant Professor at the School of Politics and Global Studies, Arizona State University. She was Assistant Professor in the School of Technology, Shinawatra University, Thailand from 2003 to 2007. Since 2004, she has been external monitor of environmental projects in eight Southeast Asia countries for the Canadian International Development Agency. She also acts as a consultant to public and private agencies on urban environmental and social issues in Southeast Asia. Dr Maneepong worked as a Senior Policy and Planning Analyst at NESDB, the Thai national planning agency, from 1991 to 1999.

Sirat Morshidi is Professor of Urban Geography in the School of Humanities, Universiti Sains Malaysia, Penang. Morshidi has researched and published several books, book chapters and journal articles on the transformation of the city of Kuala Lumpur, particularly on spatial restructuring and realignment of the urban fabric as a consequence of the globalization process and the global aspirations of the city's leaders. Morshidi is now the Director of the National Higher Education Research Institute (IPPTN) and Research Dean for the Social Transformation Research Platform, Universiti Sains Malaysia.

Asyirah Abdul Rahim is a Lecturer in the Geography Section of the School of Humanities, Universiti Sains Malaysia, Penang. She graduated with a BSc in Ecology from Universiti Malaya. She then obtained her MSc in Planning and her PhD from Universiti Sains Malaysia. Asyirah's research interests cover environmental management best practices in the construction industry and local authorities, development of sustainable indicators for environmental monitoring and auditing, and education for sustainable development.

Brian Roberts is Emeritus Professor at the University of Canberra. During the past 30 years, as a professional planner, project manager, academic and adviser, he has been involved in a wide range of urban and regional planning, urban management, institutional capacity building, land management and administration and economic development projects in more than twenty-five countries. He has held senior positions with the United Nations Centre for Human Settlements, Queensland state government, two academic institutions and the consulting industry. He has been involved in teaching and research, authoring or co-authoring 100 conference papers, journal articles, book chapters and books, including the 2006 book *Urbanization and Sustainability in Asia: Good Practice Approaches to Urban Region Development* (Manila, Asian Development Bank with T.H. Kanaley).

Wilmar Salim graduated from the Institute of Technology, Bandung in Indonesia with a Bachelor of Engineering in Regional and City Planning (1996), the University of Queensland in Australia with a Master of Regional Development (2000) and the University of Hawai'i at Manoa with a PhD in Urban and Regional Planning (2009). Before pursuing his doctoral study under the East-West Center Degree Fellowship (2003–2007), he taught regional planning and location analysis courses in the Department of Regional and City Planning at the Institute of Technology, Bandung. His dissertation is on policy implementation in a decentralized context and analyses how current central and local governments in Indonesia formulate and implement poverty reduction strategies. He has authored and co-authored several journal articles and book chapters, as well as co-editing a book in 2008 on *Sustaining a Resilient Asia Pacific Community*.

Andre Sorensen is Associate Professor of Urban Geography in the Department of Geography and Program in Planning, University of Toronto. He received his PhD from the London School of Economics in 1998. He has published widely on Japanese urbanization, urban planning, and planning history. He was awarded the 2004 International Planning History Association book prize for his book *The Making of Urban Japan: Cities and Planning from Edo to the 21st Century* (Routledge 2002). His current research project compares the role of civil society organizations in managing shared spaces in two very different cities – Toronto and Tokyo. In 2007 he was elected a Fellow of the University of Tokyo School of Engineering in recognition of his research on Japanese urbanism and urban planning. Recent publications include: 'The developmental state and the extreme narrowness of the public realm: the 20th century evolution of Japanese planning culture' in *Comparative Planning Cultures* edited by Bishwapriya Sanyal (Routledge 2005); 'Liveable cities in Japan: population ageing and decline as vectors of change', *International Planning Studies*, 2006; and *Living Cities in Japan: Citizens' Movements, Machizukuri and Local Environments*, co-edited with Carolin Funck (Routledge, 2007).

Susan Walcott is an Associate Professor in the Department of Geography at the University of North Carolina, Greensboro. Dr. Walcott's academic interests focus on regional economic development based on high technology industrial clusters such as the life science industry and urban growth coalitions. GIS as a research methodology has featured in articles on water quality in Shanghai, metropolitan growth in Atlanta, ethnic settlement patterns, and monitoring development in Bhutan. Regional study areas include the American Midwest, San Diego, Atlanta, China, India and Bhutan. She has published articles in a number of journals including *The Professional Geographer, Urban Geography, Eurasian Geography and Economics, Environment & Planning, Growth & Change, Asian Geographer, Southeastern Geographer,* and the *Journal of Cultural Geography,* Indian and Chinese specialty urban economic journals, a book on science parks in China, and chapters in eight books.

Douglas Webster is Professor at Arizona State University in the School of Global Studies, School of Sustainability and School of Urban Planning. He has been Senior Advisor to the National Economic and Social Development Board, Prime Minister's Office, Thailand (1993–2004) and a regular advisor on East Asian development to major international organizations, including the World Bank and USAID. Webster is the author of over ninety journal articles, book chapters and books, primarily on East Asian urban development processes. He has written extensively on Thai urban development including *Supporting Sustainable Development in Thailand: A Geographic Clusters Approach* (World Bank, 2006).

Antony Yeh is Dean of the Graduate School, Professor and Head of the Department of Urban Planning and Design, Director of the Centre of Urban Studies and Urban Planning, and Director of the GIS Research Centre at the University of Hong Kong. He is an Academician of the Chinese Academy of Sciences and a Fellow of the Hong Kong Institute of Planners, the Royal Town Planning Institute, the Planning Institute of Australia, the British Computer Society and the Chartered Institute of Logistics and Transport. He received the Hong Kong Croucher Foundation Senior Research Fellowship Award in 2001 and the UN-HABITAT Lecture Award in 2008. Yeh's main areas of specialization are in urban development and planning in Hong Kong, China, and South East Asia and the applications of GIS in urban and regional planning. He has carried out consultancy work related to his expertise in projects for the Hong Kong Government, World Bank, Canadian International Development Agency, UN Urban Management Programme, and Asian Development Bank. He has served on various government advisory boards of the Hong Kong Government and is at present a member of the Lands and Building Advisory Committee of the Hong Kong SAR government. He is on the editorial boards of key international and Chinese journals and has published over thirty books and monographs and over 180 academic journal papers and book chapters.

Belinda Yuen is a qualified urban planner with a PhD in Environmental Planning. During a long affiliation with the National University of Singapore, she carried out a wide range of research projects related to spatial planning and urban policy analysis. Her recent interests have included planning liveable, sustainable cities and 'vertical living'. She is on the editorial board of *Asia Pacific Planning Review*, *Cities*, *Regional Development Studies* and the *Journal of Planning History*. She has been involved in a number of local planning activities in Singapore, including the Singapore Master Plan 2003, and the Focus Group and Action Programme Working Committee for the Singapore Green Plan 2012. She has been President, Singapore Institute of Planners (2005–2008), Vice-President, Commonwealth Association of Planners (Southeast Asia) (2006–2008) and a member of the United Nations Commission on Legal Empowerment of the Poor Working Group. Belinda is also on the International Advisory Boards of several United Nations urban publications the Global Research Network on Human Settlements (2007–2011) and the Expert Group, State of Asian Cities Report 2009. Belinda has carried out consultancy work on ecocity, master planning and other planning studies for the Asian Development Bank and the World Bank.

Chapter One

Risks, Resilience and Planning in Asian Cities

Stephen Hamnett and Dean Forbes

The large metropolises strung along Asia's Pacific coasts[1] present an escalating challenge to governments and city planners due to their size, rate of growth and sheer complexity. New forms of urban development are emerging at a rapid pace, challenging the ways we think about cities and posing questions about our global urban future.

The development of Asian cities during the nineteenth and twentieth centuries was uneven and sometimes chaotic. Wars, colonialism, regime changes, periodic economic crises and natural disasters have all been influential in shaping the large and important global cities which now dominate this dynamic region. Their emergence brings to mind John Dewey's comment on Chicago in the 1890s as the place 'to make you appreciate at every turn the absolute opportunity which chaos affords' (Menand, 2001, p. 318).

This book provides a set of new perspectives on the global issues confronting eleven of the most significant Asian cities located along the western rim of the Pacific Ocean. Each of these cities is large by world standards and exerts a powerful influence on the country in which it is located. They are expanding their influence further into their rural hinterlands, to the point where they are better thought of as agglomerations and mega-urban regions rather than as individual cities, and there is a good deal of literature acknowledging this (see, for example, Laquian, 2005; Jones and Douglass, 2007; Douglass, 2010).

Tokyo is an archetypal world city. Singapore, Hong Kong and Seoul have acquired many world city characteristics. Taipei and Kuala Lumpur have been at the centre of expanding economies in which nationalism and global aspirations have been intertwined and expressed in the built environment. Beijing, Hong Kong and Shanghai have played key, and sometimes competing, roles in China's rapid economic growth. Bangkok's amenity economy is currently threatened by political instability, while Jakarta and Manila are the core city-regions of less developed countries with sluggish economies and significant unrealized potential. Each of these cities has prominence and visibility on the world stage. Many attract large numbers of tourists and visitors. Each has felt the full impact of globalization and is intent on capitalizing on its locational and other advantages. They are simultaneously dealing with new risks and threats ranging from global financial crisis to the consequences of climate change, and must come up with strategies for addressing these new challenges if they are successfully to leverage their global prominence.

Asian Pacific cities have survived wars, revolutions, economic crises and environmental catastrophes. They have also demonstrated resilience in the face of rapid unplanned growth and diverse government responses to the problems which have accompanied this. The contributors to this book are well-known urbanists and highly-regarded analysts and commentators on Asian urban matters who reflect on these experiences and discuss the challenges that lie ahead.

Urbanization and the Cities

The middle of the twentieth century was a political watershed for most Pacific Asian countries. The retreat of colonialism saw the emergence of new independent nations. It coincided with an escalation in the pace of urbanization, especially through the migration of rural residents to capital cities and growing centres of commerce and trade. By the last quarter of the twentieth century, the growing integration of many parts of Asia into the world economy meant that globalization became a highly significant factor in shaping its cities.

Around two-thirds of the world's population live in Asia. China alone has a population of 1.4 billion. Indonesia's population is 230 million and Japan has 127 million. Next is the Philippines with 92 million, then Thailand with 68 million, Korea with 48 million, Malaysia with 28 million, Taiwan with 24 million and Singapore with 5 million. In total, 1.6 billion residents of Asian countries live in towns and cities (UN-Habitat, 2010b, p. 252). Almost one in every two urban residents on the planet lives in an Asian urban area. Around 18 per cent of the world's urban residents live in a city or town in China. Paradoxically though, Asia is one of the least urbanized global regions. A little over 42 per cent of the population lived in urban areas in 2010. By comparison, almost 80 per cent of the population of Latin America and the Caribbean live in towns and cities (UN-Habitat, 2009).

Urbanization is an outcome of rural-urban migration, fertility levels in cities, and boundary changes as cities grow and encroach upon the surrounding rural areas. Asian nations have urbanized at different rates. Singapore is a city-state, and the entire population of the Hong Kong Special Administrative Region is urban. The more economically advanced economies, Japan, Korea and Taiwan, have over half their populations living in cities. So too do Indonesia, Malaysia and the Philippines. China and Thailand have fewer, but will catch up quickly.

China's socialist regime slowed urbanization from the 1950s until economic reforms commenced in earnest in the early 1990s. By requiring permits, known as *hukou*, to move to cities and by controlling urban labour and housing markets, as well as access to subsidized food, China reduced rural-urban migration rates to well below those of neighbouring countries with market economies. However, this began to change when economic reforms were implemented and China joined the global economy. Rural-urban migration has increased significantly in recent years. While 44 per cent of China's population lived in cities in 2009, this is projected to rise to 65.9 per cent in 2050. Thailand is also expected to urbanize significantly. Urban residents represented 33.6 per cent of its total population in 2009, but this figure is projected to reach 60 per cent in 2050 (United Nations, 2010).

Asia has hosted the growth of over half of the world's megacities, or cities with populations of 10 million or more. In 2010 Tokyo, Shanghai, Beijing and Manila, were in the list of the world's twenty-one megacities (see table 1.1). Several other large Asian cities expanded far beyond their formal urban boundaries into extended metropolitan regions (EMRs) and in functional terms are of megacity size. Jakarta is the hub of the Jabodetabek (Jakarta-Bogor-Depok-Tangerang-Bekasi) urban region, the population of which is in excess of 17 million. Tokyo is at the core of the Tokaido megalopolis, connected by 'bullet train' to Nagoya, Osaka, Kobe and Kyoto and including some 68 million people. An EMR wrapped around the Bohai Gulf connects Beijing, Tianjin, Tangshan and Qinhuangdao. Shanghai sits at the mouth of the massive Yangtze River Delta, and Hong Kong has a similar location at the mouth of the Pearl River Delta. Both cities are connected to hinterlands with vast populations. Manila extends its influence over the whole of the island of Luzon. Bangkok and Seoul also are central to large EMRs that dominate their national populations and economies.

UN projections of city populations in 2025 reveal contrasting patterns (United Nations, 2010). Two of the cities analyzed in this book are expected to grow very slowly or not at all. Seoul will remain much the same as in 2010 and Tokyo is expected to increase by less than 1 million. By contrast, Shanghai, Beijing and Manila will each add 3 million more residents. Yet it is risky to draw many specific conclusions from these projections. For example, Seoul is at the centre of an EMR spotted with satellite cities which are expected to continue to grow. Singapore has significant physical constraints to its growth while Hong Kong's ability to expand beyond its boundaries will depend increasingly on collaborative planning efforts with neighbouring cities and provinces in the extended Pearl River Delta. Shanghai, Beijing, Kuala Lumpur, Jakarta, Bangkok and Manila will almost certainly continue to grow both within the city boundaries

Table 1.1. City populations.

City	Population 2010 (million)	Projected population 2025 (million)
Tokyo	36.7	37.1
Shanghai	16.6	20.0
Beijing	12.4	15.0
Taipei	2.6	3.1
Seoul	9.8	9.8
Hong Kong	7.1	8.0
Singapore	4.8	5.4
Kuala Lumpur	1.5	1.9
Jakarta	9.2	10.9
Bangkok	7.0	8.5
Manila	11.6	14.9

Source: United Nations (2010) *World Urbanization Prospects: the 2009 Revision (File 12)*. New York: Department of Economic and Social Affairs, Population Division.

and in their urban hinterlands, with some further restructuring occurring between the globalizing urban cores and the rest of their EMRs. Demographic growth at the periphery of these large Asian cities is generally faster than in their central areas. In Bangkok, for instance, growth in the inner area between 1990 and 2000 was less than 1 per cent per year, whereas it was 3.3 per cent in peripheral areas. A range of factors drives the fringe development of these large cities, including increasing affluence among the urban population and the desire to move manufacturing away from high-value inner-city land.

The age profiles of Asian countries are varied. Over one-quarter of the populations of the Philippines (34 per cent), Malaysia (30 per cent) and Indonesia (27 per cent) are aged 14 or under. In contrast the 14 and under cohort comprises just 13 per cent in Hong Kong, and is also low in Japan (14 per cent), Singapore (16 per cent) and Korea (17 per cent). People of 60 years and over comprise 29 per cent of Japan's population, the highest proportion in the entire Asian Pacific region.

Asian economies span a significant range. Japan and Singapore are in the World Bank's high-income range. Taiwan, Korea, Malaysia, Thailand, and the Philippines are in the middle-income band. China and Indonesia are low-income countries. A similar pattern emerges in the country rankings in the United Nation's Human Development Index (HDI), a composite measure of human wellbeing based on life expectancy, schooling and income. Japan (11th), South Korea (12th), Hong Kong SAR (21st) and Singapore (27th) were in the top thirty nations in the 2010 survey. Malaysia (57), China (89), Thailand (92), Philippines (97) and Indonesia (108) are ranked lower. China, Indonesia and South Korea have had the most significant improvements in their HDI rank in recent years (UNDP, 2010, p. 3). The major cities have higher levels of income overall than the rural areas, higher proportions of middle- and upper-income earners, and higher HDI attributes. This results in urban-rural disparities, such as striking contrasts in GDP between the thriving cosmopolitan centre of modern Shanghai and the impoverished rural regions of western China.

Industrialization had a major impact on Tokyo as Japan's economy grew in the post World War II period (Johnson, 1982). It was followed by industrial expansion in the four 'Asian Tigers', where Singapore, Hong Kong, Taipei and Seoul led the way in expanding the productive capacity of Asia and in establishing a major export-oriented focus for their economies. However, since the latter years of the twentieth century manufacturing industries have been giving way to emerging post-industrial activities. In Tokyo, Singapore, Hong Kong, Seoul and Taipei the emphasis is now on higher value activities, with industrial capacity increasingly pushed out to countries and large cities with lower labour costs, such as Bangkok, Jakarta and Manila as well as Chinese and Vietnamese cities. Spatial restructuring of industry is also taking place within and around Asian cities such as Bangkok and Beijing, where manufacturing industry is moving out to more peripheral locations, replaced by high-tech industry and producer services in the core areas.

Despite increasing industrialization, however, not all the available labour force in cities in Asia's less economically developed countries has found work in industry. In Jakarta, Bangkok and Manila, numerous migrants to the city work in the 'informal sector', in micro-enterprises providing cheap goods and services. Metropolitan authorities have frequently sought to suppress these activities, viewing them as inefficient and disruptive to the city. There is a growing awareness, however, of the need for more positive policy interventions to provide employment for these growing numbers of city workers, and of the low cost of the services micro-enterprises provide (UN-Habitat, 2010*b*).

Migration-driven urbanization meant that cities absorbed large numbers of rural workers and their families. It produced spontaneous informal settlements and slums across many Asian cities. In 2005 it was estimated that 43.7 per cent of urban residents in the Philippines lived in slums, 32.9 per cent in China, 26.3 per cent in Indonesia and 26 per cent in Thailand (UN-Habitat, 2009, pp. 242–243). Residents of these areas have generally built their own houses, often from second-hand or scavenged materials, and seldom have secure title to their land.

There have been shifts in attitudes to slum settlements in the last few decades. Once they were considered to be fit only for demolition, but now there is greater recognition of the plight of informal settlement residents who have nowhere else to live. New strategies to increase capability and encourage self-help solutions to urban slum upgrading are being implemented, such as in Quezon City, the largest of the seventeen cities comprising Metro Manila (Veneracion, 2008) and in Taguig City (see Chapter 12 of this book). The effort required to improve the quality of housing in Asia's cities is enormous (UN-Habitat, 2010*b*; 2009), but this is just one of the major challenges facing modern Asian cities.

Writing the Asian City

An extensive and growing literature has emerged on Asian cities in recent years, laying the foundations for a comprehensive and integrated approach to understanding the dynamics of Asian cities, and their different styles of metropolitan governance and planning. Amongst the more recent contributions to this literature is *The State of Asian Cities Report 2010/11*, published by the United Nations (UN-Habitat, 2010*b*). This addresses city issues from Turkey in the west to New Zealand, a long way off the southeastern rim of the Asian continent. Key themes include consideration of the scale of urban expansion in Asia, the changing economic role of Asian cities, the problems created by poverty and economic inequality, environmental challenges, and appropriate governance structures. *The State of the World's Cities Report 2010/11*, also a product of UN-Habitat (2010*a*), and UN-Habitat's *Planning Sustainable Cities* (2009) both have a global focus but include significant commentary on Asia's cities.

In *The Rise of Mega-Urban Regions in Pacific Asia: Urban Dynamics in a Global Era* (2007), Gavin Jones and Mike Douglass address the consequences for the city of demographic changes. The book focuses on the three concentric bands of population that spread out from the core through the middle and outer suburbs to the periphery of Pacific Asian extended metropolitan regions (EMRs). Population growth rates tend to be slower in the inner band and faster in the outer band, which often provides sites for new industrial developments and industries moving out from the city core. Mike Lindfield and Royston Brockman's (2008) *Managing Asian Cities* reviews key urban challenges, with a strong emphasis on internationally-funded development project solutions to problems. Roberts and Kanaley (2006), in *Urbanization and Sustainability in Asia*, provide detailed case studies of good practice in sustainable urban development in relation to a framework which emphasizes arrangements for governance and urban management on the part of national governments. *Pacific Asian Cities: Challenges and Prospects*, edited by Stephen Hamnett and Dean Forbes (2001), includes papers on Hanoi, Manila, Kuala Lumpur, Singapore, Hong Kong and Jakarta and emphasizes, amongst other things, the disparities in welfare and income found within its selected cities, as well as the varying degrees of progress towards more open and consultative styles of governance. Inequality is also a major focus of Gugler's (2004) *World Cities Beyond the West: Globalization, Development and Inequality*, a survey which is global in its scope but which includes chapters on Shanghai, Seoul, Bangkok, Hong Kong, Singapore and Jakarta.

Aprodicio Laquian's *Beyond Metropolis: The Planning and Governance of Asia's Mega-Urban Regions* (2005) is a significant book, bringing together the collected thoughts of a distinguished Asian urbanist over many years. Laquian provides a comprehensive comparative perspective that ranges across individual city case studies from Pacific Asia, and also South Asia, focusing on housing, urban services and on broader questions of the planning and governance of emerging mega-regions. With regard to the latter, he concludes that planning has to be

geographically inclusive as well as functionally comprehensive, covering the whole urban field of influence and dealing holistically with the full range of economic and social factors that determine quality of life. (Laquian, 2005, p. xxiii)

Laquian's ideas on the governance of extended metropolitan regions are considered further later in this introductory chapter.

Another recent book, which explores governance arrangements appropriate to extended metropolitan regions, is the collection edited by Catherine Ross entitled *Megaregions: Planning for Global Competitiveness* (2009). This includes a number of case studies from Western and Asian countries. The role of the 'developmental state' in Asia and its influence on the style of planning is considered, in particular, in the chapter by Yang Jiawen (2009) on 'Spatial Planning in Asia: Planning and Developing Megacities and Megaregions'. Yang finds that China, Japan and South Korea all have '… strong central governments that have created top-down planning systems and actively promoted spatial development and planning…'. The corollary is that they are criticized for a lack of effective public participation (*Ibid*., p. 50) (see also Ooi, 2008, p. 76, writing of Singapore).

Jiang Xu and Anthony Yeh (2011) have also published a recent edited collection of international case studies entitled *Governance and Planning of Mega-City Regions: An International Comparative Perspective*. Like Yang, they note (at p. 17) that decision-making in China is, unsurprisingly, top-down but find that this has not yet led to effective planning arrangements at the mega-region scale, because these extended regions cut across existing political and administrative hierarchies, as well as the fragmented planning functions of central government ministries. Vogel *et al.* (2010) provide a third recent collection on a similar theme in their monograph, *Governing Global City Regions in China and the West*. Vogel (2010*a*, p. 67) finds that

In both East and West, globalisation is shaping patterns of urban development, undermining distinctive local cultures, and leading to greater inequalities… (F)ree trade policies and neo-conservative ideologies apparent in the West are also embraced in China. At least three factors separate the responses to globalisation. First, in China, democracy is only a recent and emerging concern and to date does not greatly impinge on the decision-making process of national, regional or local governmental actors. Second, the stage of development in China is still low compared to the West, although catching up. Third, and perhaps the most important point to make, however, is that the contexts of rescaling in the West and China are quite different. In China, there is a strong national state that is regulating and mediating the market. In the West, the nation-state is increasingly leaving this to the markets, not without criticism. China is still in a position to dictate which city regions emerge and what role they play in the world economy.

The sustainability of cities has become an increasingly important focus of scholarship on Asian cities. Fu-chen Lo and Peter Marcotullio's edited collection, *Globalization and the Sustainability of Cities in the Asia Pacific Region* (2001), highlights the impact of globalization on sustainability. *Challenging Sustainability: Urban Development and Change in Southeast Asia*, edited by Tai-Chee Wong, Brian Shaw and Goh Kim Chuan (2006), is organized around three themes: environment and infrastructure; economy and culture; society and politics. Andre Sorenson, Peter Marcotullio and Jill Grant's *Towards Sustainable Cities: East Asian, North American, and European Perspectives on Managing Urban Regions* (2004) includes chapters on Japanese and Korean cities. It provides insights into, amongst other things, the questionable relevance of Western notions of sustainable urban form to cities like Tokyo and Seoul where densities are already very high and where civil society remains fairly weak.

Most of the volumes mentioned above focus around social, economic and environmental aspects of cities. In contrast, Hans-Dieter Evers and Rudiger Korff's *Southeast Asian Urbanism: the Meaning and Power of Social Space* (2000) is a different kind of book, focusing on the cultural symbolism of urban form in Southeast Asia and Sri Lanka. More recently, Douglass, Ho and Ooi (2008), in their edited collection *Globalization, the City and Civil Society*, have explored further aspects of the nature of urban form and of public and private spaces and places in Asian cities. They start by observing the tension between the responsibility of city governments to provide better services and liveable environments for their residents while also attempting to attract transnational capital to the cities, in the hope of building up the urban economy. They suggest that, during the 'developmental state' era from the 1960s to the1990s,

the implicit social contract between government and citizens was the promise of economic growth and material benefit in exchange for curtailment of democratic rights, including freedom of speech and assembly, for the citizenry. (Douglass *et al.*, 2008, p. 1)

However, the economic success that the developmental state has given rise to has now created middle classes which demand more participation in local affairs, the conservation of remaining heritage buildings and traditional quarters and a greater attention to environmental imperatives. There is evidence in several cities of a growing propensity to form civil society associations which are the building blocks of social capital (Putnam, 1993):

Through routines of collaborative engagement in civil society, localities build capacities for problem solving and innovation. Multiple forms of association with a high degree of autonomy from state and private business are associated with local economic resilience in the new global economy of informational networks and rapid innovation in knowledge industries. (Douglass *et al*., 2008, p. 5)

Much of the literature on public spaces comes from Western sources, and this unqualified transposition to the Asian context has been criticized recently by some authors (see, for example, Hogan *et al.*, 2011). Nevertheless, Douglass *et al.* (2008, p. 6) observe that political and social struggles over the public realm are intensifying as urbanization accelerates:

In particular, the polarization of population and economic growth in a few metropolitan regions has exacerbated already very high population densities and, as a consequence, struggles over urban land use and the built environment have tended to limit expansion of public sites that might be used as civic spaces.

John Friedmann's recent work (2007 and 2010) on the importance of place-making to the global cities of Asia is discussed further in a later section of this introductory chapter.

Key Themes in Planning Asian Cities

Three main themes emerge through the chapters that follow in this book:

◆   A number of Asian cities have achieved, or aspire to, world city status. There is fierce competition between cities to increase their importance as regional and global hubs, with significant consequences for spatial restructuring and for the built environment within cities.

◆   Large Asian cities and their EMRs face a number of daunting risks that will test their economic, social and environmental resilience.

◆   There is a need to restructure governance frameworks in order to provide both clear spatial development policies for extended metropolitan regions in the environmentally challenging and highly competitive twenty-first century, while also providing for community participation in the planning of neighbourhoods and local places.

Creating World Cities

Several cities are developing in Pacific Asia with high order finance, management, law and other producer services that support the location of head offices of significant global companies. Tokyo, Hong Kong and Singapore are the prime examples of Asian world cities. Seoul is at the next level down. A third tier of cities aspiring to global status includes Shanghai, Taipei, Beijing, Kuala Lumpur and Bangkok. Jakarta and Manila are at a lower level again.

The claims of a particular city to 'world' or 'global' city status can be assessed in a number of different ways and its ranking relative to other cities may vary according to the measures adopted. Any ranking of world cities is likely to raise theoretical and methodological issues about the choice of measures included and the way these were applied. One of the most authoritative rankings to date is the Mastercard Worldwide Centers of Commerce Index (2008). Indicators for this were developed by a panel of distinguished researchers in the field of global cities, including Saskia Sassen and Peter Taylor. Seventy-five cities were compared against seven different criteria: legal and political framework; economic stability; ease of doing business; financial flow; business centre; knowledge creation and information flow; and liveability. Table 1.2 summarizes the results of this study, with the cities included in this book highlighted.

It is increasingly evident that many Asian cities are actively striving to achieve a high-ranking on one or more of the various global city indexes. The chapters on

Table 1.2. Summary ranking of seventy-five world centres of commerce, 2008.

2008 Rank	City	Index Value	2008 Rank	City	Index Value
1	London	79.17	39	Dusseldorf	50.42
2	New York	72.77	40	Geneva	50.13
3	Tokyo	66.60	41	Melbourne	49.33
4	Singapore	66.16	42	Bangkok	48.23
5	Chicago	65.24	43	Edinburgh	47.79
6	Hong Kong	63.94	44	Dubai	47.23
7	Paris	63.87	45	Tel Aviv	46.50
8	Frankfurt	62.34	46	Lisbon	46.46
9	Seoul	61.83	47	Rome	45.99
10	Amsterdam	60.06	48	Mumbai	45.71
11	Madrid	58.34	49	Prague	45.50
12	Sydney	58.33	50	Kuala Lumpur	45.28
13	Toronto	58.16	51	Moscow	44.99
14	Copenhagen	57.99	52	Budapest	44.52
15	Zurich	56.86	53	Santiago	44.49
16	Stockholm	56.67	54	Mexico City	43.33
17	Los Angeles	55.73	55	Athens	43.25
18	Philadelphia	55.55	56	Sao Paulo	42.70
19	Osaka	54.94	57	Beijing	42.52
20	Milan	54.73	58	Johannesburg	42.04
21	Boston	54.00	59	Warsaw	41.26
22	Taipei	53.32	60	Shenzhen	40.04
23	Berlin	53.22	61	New Delhi	39.22
24	Shanghai	52.89	62	Bogota	38.27
25	Atlanta	52.86	63	Buenos Aires	37.76
26	Vienna	52.52	64	Istanbul	36.14
27	Munich	52.52	65	Rio de Janeiro	35.91
28	San Francisco	52.39	66	Bangalore	35.78
29	Miami	52.33	67	St Petersburg	35.55
30	Brussels	52.16	68	Jakarta	35.40
31	Dublin	51.77	69	Riyadh	35.37
32	Montreal	51.60	70	Cairo	35.29
33	Hamburg	51.53	71	Manila	35.15
34	Houston	51.30	72	Chengdu	33.84
35	Dallas	51.25	73	Chongqing	33.13
36	Washington DC	51.19	74	Beirut	31.81
37	Vancouver	51.10	75	Caracas	26.11
38	Barcelona	51.90			

Source: Mastercard Worldwide, 2008, pp. 20–21

Kuala Lumpur, Seoul, Hong Kong, Shanghai, Beijing, Taipei and Singapore make this abundantly clear. Douglass has referred to this as 'intentional world city formation' through capital investments in urban services, infrastructure, architecture and urban design with emphasis on both functional and symbolic elements. In the aftermath of the 1997–1998 economic crisis in East and Southeast Asia,

the competitive drive to support urban mega-projects re-emerged at a grander level of ambition and intensity… The more determined neoliberal shift in public policy greatly affected the built environment and form of the city by relaxing or eliminating many of the regulations on land use and, more generally, public oversight of development projects… In many projects, however, government involvement was needed. Evictions of people from occupied housing and land to be assembled for very large projects required government action. Initial investment in projects that would not have streams of profit for many years ahead would not be readily taken by the private sector without government subsidies. (Douglass, 2010, pp. 51–52)

Douglass notes that competition to be the site of the world's tallest structure (see the Kuala Lumpur, Shanghai, Jakarta, Taipei and other chapters in this collection) is emblematic of megaprojects being adopted for symbolic status as well as for economic gain. In central cities new middle- and high-income housing developments are often intrinsic parts of new complexes that also feature offices catering for the financial, producer services and hi-tech industries, restaurants and other services in demand from knowledge workers. In addition, global visibility is enhanced by attracting major events. The FIFA Football World Cup was held in Seoul and Tokyo in 2002. Beijing hosted the Olympics in 2008, and Seoul was the host in 1988. The Shanghai Expo was held from May to October in 2010. This event attracted 73 million visitors to its main 5.3 square kilometre site at Pudong on the eastern bank of the Huangpu River.

A theme in several chapters is the importance of investment in super-fast rail systems and mass rapid transit systems (MRTs). Japan has had a world-leading high-speed rail system for 30 years. China is rapidly establishing a high-speed rail network, including an ambitious plan for a fast train to Tibet. South Korea is also expanding its network of high-speed trains, although its ability to connect with the Chinese network will remain limited until political relations with the North improve. Pudong Airport is connected to the middle suburbs of Shanghai by a magnetically levitating train that travels at speeds of around 4 40 kilometres per hour.

At the city level, Singapore, Hong Kong, Beijing, Shanghai and Bangkok now have modern and expanding MRT systems. Kuala Lumpur, despite recent investments in light rail and monorail systems, has now acknowledged the importance of investing in a modern high-speed MRT to underpin the further growth and integration of its metropolitan region. Jakarta and Manila, by contrast, have failed to match their public transport systems to the needs of their growing metropolitan regions and pay a heavy price in traffic congestion.

Six Asian airports were ranked in the Skytrax top ten world airports in 2010 (Skytrax, 2010). Singapore's Changi Airport has been a pacesetter, being ranked first globally, out-competing rivals Kuala Lumpur International Airport (5th) and Bangkok's Suvarnabhumi Airport (10th). Seoul's Incheon International Airport was ranked 2nd, followed in 3rd place by the Norman Foster designed Hong Kong International Airport. Beijing Capital Airport has undergone a massive upgrade, with giant new terminals, and was ranked 8th. In Shanghai, in addition to the huge new Pudong Airport, the older Hongqiao Airport has been substantially expanded. These airports, with their dramatic architecture, make symbolic statements about their cities. They also incorporate large shopping malls, retailing global designer brands, and offer increasingly luxurious lounges to business travellers as well as restaurants, gymnasiums and cinemas. Around the airports are concentrations of businesses connected to travel and international commerce, including hotels, warehouses and freight-forwarding enterprises, as well as residential areas for airport workers.

In summary, since around 1990, the large cities of Asia have been experiencing ambitious and spectacular megaprojects:

the world's tallest structures, mega-malls with places for religious worship in them, world business triangles, cyber-cities, mega-airports, and private new towns for hundreds of thousands of people. Together these projects are creating a new layer of the built environment in core and suburban areas that make the era of rapid urban-industrial growth pale in comparison. (Douglass, 2010, p. 52)

The reference to 'cyber-cities' is a shorthand term for a particular set of strategies which Asian cities have been pursuing to strengthen their engagement in information and knowledge industries and the associated infrastructure of the digital revolution, the internet and the development of broadband. High profile manifestations of the enhancement of knowledge city capacities include the development of specialist knowledge-based industry parks and an enhanced emphasis on the contribution of universities to economic development. Thus, Kuala Lumpur is now connected to the Multimedia Supercorridor that extends south of the city to Kuala Lumpur International Airport at Sepang. Examples from other cities noted in chapters of this book include the Zhongguancun technology hub in Beijing's Haidian District, Shanghai's university-based new towns, Seoul's Digital Media City, Hsinchu near Taipei and the Singapore and Hong Kong Science Parks.

Continued expansion of higher education as city populations grow is also part of the strategy of developing the knowledge economy. However in some Asian countries, the opportunities for progression through school and on to university study have increased more slowly than populations have grown. As a result, a small but significant proportion of students from the large Asian cities go abroad to America, the UK and Australia to attend university and acquire skills. The Chinese and South

Korean governments have led the way in expanding domestic higher education by strengthening local universities. Singapore has also attracted significant foreign universities to establish small campuses and research centres in the city in order to build local skills, and attract foreign students to Singapore as part of its aspiration to be a 'global schoolhouse'.[2]

An important theme in several chapters of the book is the trend through globalization towards standardization or 'homogenization' of urban environments – the Las Vegas-based Sands Hotel in Singapore, for example and Starbucks everywhere. There is also a simultaneous striving for difference as policy-makers realize that the potential of their city to compete may be enhanced by emphasizing the diverse cultural qualities of particular parts of the city. Thus, several chapters describe belated attempts to save remaining heritage buildings in pursuit of 'urban quality'. Sassen (2011, p. 105) has observed that

The specialized economic histories of major cities and regions matter in today's global economy because there is a globally networked division of functions. This fact is easily obscured by the common emphasis on competition and by the standardization (no matter how good the architecture) of state-of-the-art built environments, from offices to airports … the common notion of the homogenization of the urban landscape in today's economy misses a critical point. It neglects, or obscures, the fact of the diversity of economic trajectories through which cities and regions become globalized, even when the final visual outcomes may look similar… Similar landscapes may contain very different economies.

Likewise, Vogel (2010b, p. 9) notes that while there is sometimes an appearance of convergence, cities retain elements of distinctiveness brought about through the actions of local communities. However, this sets up a tension between the actions of local groups and the intentions of investors and developers to build and promote megaprojects attractive to corporate and government leaders aiming to secure a city's niche in the competitive world economy.

John Friedmann has explored this tension in a number of recent articles. In particular, in a 2010 paper titled 'Place and Place-Making in Cities: A Global Perspective', he deplores the fact that

the art of place-making has not informed planners of the swaths of the urban in the newly industrializing global regions of Asia and elsewhere. Their principal preoccupation has been with the branding of cities and the advanced infrastructure required by global capital. In the process, millions of ordinary folks have been displaced and their neighborhoods erased, as speed, movement, and power have been valued more than the fragile social infrastructure of place-based communities. (Friedmann, 2010, p. 149)

Friedmann, drawing on Meyer (2009, p. 40) describes how, between 1998 and 2001, more than half a million people were officially displaced from their old neighbourhoods in the centre of Beijing and moved into apartments on the city's periphery beyond the fourth ring road. In the run-up to the Summer Olympics of 2008, several hundred thousands more followed the first contingent, as entire *hutong* quarters were earmarked for demolition to make way for shopping malls, office buildings and high-rise luxury condominiums.

Friedmann (2010, p. 149) rejects this approach to urban planning, arguing that 'place-making is everyone's job, local residents as well as official planners, and that old places can be "taken back" neighborhood by neighborhood, through collaborative people-centered planning'. In a tradition of urban advocacy which can be readily traced back at least as far as Jane Jacobs in *The Death and Life of Great American Cities* (1961), Friedmann emphasizes the importance of the local and the pedestrian scale 'which allows people to interact in a variety of mostly unplanned ways, on the street or in business establishments among other spaces of habitual encounter' (*Ibid.*, p. 158). He sees hope in the experience of Japan's local citizen groups – *machizukuri* – described in Sorensen's chapter on Tokyo (and see also Sorensen and Funck, 2007) which challenge Japan's long-established style of planning through remote central government ministries and represents a much more active role for local communities in environmental management and neighbourhood planning. Friedmann also notes how, following Japan's devastating Kobe earthquake, it was the better-organized neighbourhoods with established *machizukuri* that were able to recover quickest. The importance of social capital at the local level is considered further in the following section, which examines the second major theme of the book, risk and resilience.

Risk and Resilience

Accounts of risks and resilience are threaded through the chapters. Cities can be impacted by many kinds of catastrophic events, with vastly different consequences (Vale and Campanella, 2005). Catastrophes can be single events. The earthquake that devastated the Chinese city of Tangshan in 1976, the tsunami in 2004 that caused such destruction to coastal cities bordering on the Indian Ocean and the more recent earthquake and tsunami which, in March 2011, caused enormous damage and loss of life in Japan's Fukushima Prefecture are three dramatic examples**.** Many Asian cities are situated in low-lying areas along the coast. 'Tsunami' is, of course, a Japanese word, and Tokyo is vulnerable to the impact of tsunamis and to flooding. Typhoons sweeping in from the Pacific are a constant danger for coastal communities in Taiwan, China and the Philippines. Earthquakes are a hazard in several Asian cities, with Tokyo and Beijing sitting on seismic faults. The Indonesian island of Java is subject to regular volcanic eruptions, and Jakarta is not far from areas exposed to devastating damage from this 'ring of fire', although its greater vulnerability is to flooding and the potential impacts of sea level rise, as Salim and Firman make clear in Chapter 10.

Terrorist attacks, such as the 1995 sarin gas attack on the Tokyo subway and the terrorist bombing of the Australian Embassy in Jakarta in 2004, are also capable of causing major disruptions to Asian cities. Jakarta is not the only Asian city exposed to terrorist activity, but inequities among people make this a greater risk for the poorer and less well-managed cities. The management of post-conflict reconstruction provides another challenge.

Not all catastrophic episodes, however, have such an immediate impact. Just as significant, but without the headline-grabbing coverage in the media, are the catastrophes that are the result of a slow build-up of problems, or the confluence of multiple disruptive events. Globalization itself has introduced a new set of risks to Asian cities as the integration of economies has exposed them to the shock of regional and global economic crises. The Asian economic crisis reversed economic growth in several countries in 1997–1998. In Chapter 11 Webster and Maneepong describe the profound effects of this crisis on Bangkok. The global financial crisis (GFC), which peaked in 2008–2009 but continues to have an impact, had more serious consequences in America and Europe than in Asia. Nevertheless the shock waves have been felt throughout the region. The UN-Habitat report on *The State of Asian Cities 2010/2011* (UN-Habitat, 2010*b*) suggests that the size of the domestic market in many Asian countries, plus fiscal stimulus by governments, helped Asian cities recover relatively quickly from the GFC. Manila's experience was distinctive (see Chapter 12 of this book by Roberts) because of the importance of overseas remittances for the economy. Most Asian exporters, however, have been affected by the decline in demand for goods from western countries and ongoing volatility in the global economy.

Poor environmental quality in urban areas is a significant concern throughout most of the Asian region. In the most disadvantaged Asian cities, the problems centre on water quality and management. The most significant risk to human wellbeing is the lack of access to clean water for human consumption. Cities expanding into vast hinterlands create additional environmental pressures and expose their populations to new environmental disasters. Often it is the poor who live in the most vulnerable sites, on floodplains, close to rubbish dumps, or adjacent to polluted waterways. Slum housing is attracted to hazardous areas and there is no will to stop the settlements expanding. Global warming and climate change will increase the vulnerability of city populations, especially those located on low-lying areas such as those of Jakarta, Bangkok and Tokyo.

Communities will always face natural hazards, but today's disasters are often generated by, or at least exacerbated by, human activities. At the most dramatic level, human activities are changing the natural balance of the earth, interfering as never before with the atmosphere, the oceans, the polar ice caps, the forest cover and the natural pillars that make our world a livable home. But we are also putting ourselves in harm's way in less visible ways. At no time in human history have so many people lived in cities clustered around seismically active areas. Destitution and demographic pressure have led more people than ever before to live in flood plains or in areas prone to landslides. Poor land-use planning; environmental mismanagement; and a lack of regulatory mechanisms both increase the risk and exacerbate the effects of disasters. (Annan, 2003, p. 3)

Climate change adds to the already substantial sources of risk for Asia's cities and confronts them with problems never before encountered, or possibly even imagined. Nevertheless, the history of urban resilience, evidenced by the continuing functioning of major cities despite doomsday concerns about their excessive size or congestion or fragile environments, provides some comfort. Asian cities are proof of the ability of human societies to adapt to changing conditions. But can we afford to rely on this record of adaptability in future? The weight of argument seems to be that we need to beware of complacency and to take active steps to mitigate the risks of future disasters and crises. We need to adapt our cities as best we can to reduce the likely damage of extreme events. Adaptive capacity is mediated through the social and economic circumstances of a city and its residents, which is why poorer cities, and the poorer areas of rich cities, are most at risk. Poverty increases vulnerability and limits the opportunities to recover, while higher levels of public and private wealth provide cities with greater resilience and more scope to engage in effective planning. Very few Asian cities have significant plans in place to manage the consequences of climate change, but wealthy Singapore is an exception to this, as Belinda Yuen shows in Chapter 8, and a city such as Hong Kong has more adaptive capacity than Jakarta or Manila. In the latter cities, corruption and inept urban governments further cramp the actions of planners and their ability to prepare for future disasters, as Chapters 10 and 12 make clear.

Vulnerability varies with the social resilience of the population – its ability to deal with catastrophes and to bounce back – which, as the evidence from Kobe and elsewhere, referred to briefly above suggests, seems related to existing levels of social organization and capital (Olsson *et al.*, 2004; Putnam, 2000; Resilience Alliance, 2007; and Healey, 2009). Berke (2010) advocates a four point framework for enhancing resilience. The key elements are hazard mitigation plans; moral rather than actuarial planning; citizen participation in planning for resilience; and recovery planning from the bottom up. Fundamentally, it is the combination of effective urban governments together with well-organized community groups and structures that seems essential:

Resilience … takes into account the economic, social, psychological, physical and environmental factors that are necessary for humans to survive and to thrive. Locally rooted strategies to build resilience need to incorporate a strong focus on both disaster reduction and climate change adaptation and need to be embedded within a city's institutional and organizational framework. Many aspects of resilience are closely associated with a holistic approach to development. Individuals and households that have access to adequate food, clean water, healthcare and education will inevitably be better prepared to deal with a variety of shocks and stresses. (International Federation of Red Cross and Red Crescent Societies, 2010, p. 131)

Metropolitan Governance and Planning

There is a strong emphasis throughout this book on the role of government in planning cities. The chapters reveal a number of perspectives on appropriate forms of government and governance for metropolitan regions. The chapters on Manila and Jakarta illustrate the problems of weak metropolitan governance. In Manila the principal issue appears to be the reluctance of powerful municipal governments, which Roberts describes as 'fiefdoms', to give up any of their power to allow the Metropolitan Manila Development Authority to operate effectively. In Jakarta the post-Soeharto emphasis on local autonomy and the World Bank's championing of 'good governance' and subsidiarity have transferred more and more decision-making powers to the local government level. However challenges like climate change are not best tackled only at the *kota* or *kabupaten* level, and there are problems of multiple levels of government and a lack of fit between the extent of the metropolitan region and the boundaries of the responsible planning authorities.

At the other extreme is Singapore where there is a close fit between the territorial extent of the unitary authority and not much encouragement to alternative views (the recent official discovery of public participation in the planning and policy process notwithstanding). Tokyo also stands out for its strongly centralized planning arrangements, focused on economic development and infrastructure and much less on the quality of urban environments, albeit with the recent promise of the *machizukuri*, referred to earlier. Other cities such as Seoul and Taipei also emerge as strongly centralized and primarily focused on economic development and the maintenance of world city status. As noted earlier, several recent works have also examined the 'top-down' nature of Chinese planning and its appropriateness as a model for the planning of extended metropolitan regions or mega-regions.

It can be argued that polycentric cities lend themselves to more effective management structures and have a more resilient settlement structure. Laquian (2005), whose work was referred to earlier, suggests that the more viable mega-urban settlements in Asia are those that have been planned and managed as 'polynucleated regions'. Around Shanghai, a number of compact and densely populated urban nodes, which are more or less self-sufficient in employment, housing, and urban services, surround the central city. In the Pearl River Delta no one urban centre dominates, but 'symbiotic relationships exist among a number of urban nodes to create a coherent city-region'. In terms which disciples of Ebenezer Howard would recognize, Laquian observes that residents of the urban nodes lead the greater part of their daily lives within their communities. However, when they need to go to the central core or to other nodes, they can avail themselves of efficient transport, including mass transit systems. (Laquian, 2005, pp. 382–383)

The areas between the central city and the densely settled urban nodes can be used for open space, green areas and peri-urban agriculture. The merits of the polycentric city as a more sustainable urban form have many advocates amongst Western scholars (for example, Newman *et al.*, 2009; and see also Hall, 2011). However, Manila's polycentric structure is less successful, as explained in Chapter 12, mainly because of the lack of an effective rail system to knit it together.

Several of the chapters in this book address the relationships between core cities and their extended metropolitan regions – Tokyo and the Tokaido, Jakarta and Jabodetabek, Hong Kong and the PRD, Shanghai and the Yangtze River Delta, Bangkok and the BEUR and so on. There seems a clear need in many of the cities examined in the book for planning arrangements to be extended to match the spatial scale of the mega-urban regions which are emerging. Whatever models emerge are likely to require a strategic planning authority at the level of the extended metropolitan region and lower level planning authorities capable of engaging with neighbourhood issues and organizations.

Laquian (2005) comments that, in general, attempts to plan Asian mega-urban regions are frustrated by jurisdictional fragmentation, the decentralization of authority and power to autonomous local government units, and the uneven distribution of economic and financial resources among various local units.

The main challenge … is how developments in the mega-city, the extended metropolitan region, and the megapolitan region can be effectively planned and governed in such a way that these agglomerations can continue to be economically productive, provide gainful employment, meet ever-rising levels of demand for key urban infrastructure and basic urban services, protect and conserve the physical and cultural environment, foster civic involvement and participation of citizens in public affairs, achieve equity and social justice, and ensure the sustained livability of these human settlements. (Laquian, 2005, p. 6)

Mega-regions provide opportunities to balance the global pressures of core cities with the needs of lower cost peripheral areas within the same extended region. According to Sassen (2011, p. 104) a mega-region is an

… internally diverse, economic territory that can contain diverse spatial logics – particularly, agglomeration and dispersal logics – which might translate into high-cost high-density areas and low-cost low-density areas. We know that large integrated firms need both … for their operations. Thus, the megaregional scale could enable the exploring of novel development strategies predicated on this diversity of spatial logics, hopefully to the advantage of both the more advanced and the least advanced areas within that megaregion.

Sassen's notion of a mega-region, which incorporates both high-end, high-tech economic activities and lower-level economic activities, might assist in framing policies to redress regional imbalance within these extended regions, of the sort described in Chapter 6 on Seoul, for example. Critical is the need to support the least advanced sectors in these regions because of their economic role in general, but significantly also because they are integral to the success of the more advanced industries as well.

Vogel (2010*b*, p. 5) identifies a number of approaches to effective metropolitan strategic planning along a continuum from 'regional government' to 'regional governance'. The former represents 'old-style regionalism' with a formal hierarchical structure created to set strategy and provide services directly, whereas the latter involves structuring intergovernmental relations among existing institutions and private actors to achieve a form of co-ordination that is more fluid and flexible. Public-private partnerships enable the financing of innovative pilot projects in infrastructure and services for sustainable development, allowing governments to do more while spreading the risk. Yet they are sometimes criticized because they neglect aspects such as social inclusion, everyday service delivery or high-quality urban design (UN-Habitat, 2009, p. 16). Laquian notes the diversity of existing metropolitan planning arrangements in Asia, ranging from autonomous local authorities to unified regional governments. He also emphasizes the tension that often exists between effective and powerful regional governments and local community participation. Clearly, it is no easy task to retain a focus on the local as regional bodies becomes spatially more extensive and remote. Nevertheless, stronger regional authorities, which are balanced by increasing opportunities for greater and more effective public participation in local planning processes and place-making, seem essential if cities are to cope with future growth challenges and to develop the social resilience that they will certainly need.

The Chapters

The authors represented in this book each tell a story about a city. All have considerable knowledge and experience of their chosen city. Some write from the inside looking out, casting a critical and informative eye over a city in which they are engaged as both citizens and analysts. Others look at the city from the outside, combining their expertise as urbanists with the knowledge gained from regular visits to a city with which they have a strong and personal relationship. Each chapter covers some common ground but they were not written to a template. Each also reflects the style and interests of its authors and their perceptions of the factors significant in the shaping of their city and critical to its future.

The individual chapters explore three main themes in relation to each city. First, city planning histories provide a baseline for understanding the evolution of the current urban morphology and the main social, political, economic and environmental events that have shaped the city to date. Second, each city's 'world city' role is described, the way that this is evolving and the consequences for different groups and areas within the city and its EMR. Third, each chapter analyzes the contemporary challenges and risks facing the cities and the varying strategies that governments and urban managers are developing to balance continuing growth and enhanced world city status with environmental protection, an increased ability to adapt to environmental and economic risks and the enhancement of urban liveability, including a higher level of involvement by citizens in urban affairs.

André Sorensen's chapter on Tokyo provides a rich insight into the history of the world's largest city and its distinctive set of sustainability and planning issues. He traces the development of Tokyo from its role as the imperial capital of the Edo period to becoming one of the world's three leading global cities in the 1980s and its stalling economy since then. Sorensen emphasizes the continuing influence of historical urban development patterns on present-day Tokyo – in particular, the spatial division of the Edo period between the High City of the samurai and the Low City of the merchants and artisans. This division was perpetuated in the first zoning plans of the early twentieth century and remains evident in current socio-spatial patterns. The Low City today is one of the poorest and most vulnerable areas of Tokyo with higher levels of low-wage employment, cheap rental housing and concentrations of the homeless. Substantial parts of the Low City are also below sea level and at serious risk, therefore, from sea level rise and flooding.

Tokyo was almost completely destroyed twice in the twentieth century, first by the Great Kanto earthquake of 1923 and then by the fire-bombings of World War II. Following the earthquake, rapid development of new areas outside the city occurred, creating a belt of unplanned sprawl comprising many areas of high-density wooden houses and narrow streets. Some of these areas remain at high risk of devastating fires in the event of earthquakes, as few houses are fireproofed and many stand within inches of each other. Tokyo's next major earthquake has been expected since the mid-1990s. Despite the damage caused by fire following the 1923 earthquake and during the final stages of World War II, no opportunities have been taken to provide these areas with the wide roads and public parks that could also serve as firebreaks.

The neglect of redevelopment of these areas, Sorensen explains, is a consequence of a Japanese approach to urban development which has consistently been characterized by a high degree of centralization of policy and administrative power and which has had an enduring focus on economic growth, at the expense of urban liveability. Japan's 'developmental state' has shaped Tokyo's growth and form through large-scale infrastructure projects – railways, expressways, airports, bridges – but with limited investment in social facilities and few constraints on private urban investment or land development, allowing the spread of unplanned and sprawling residential areas. Following the crash of Japan's 'bubble economy' at the beginning of the 1990s, investment in large-scale infrastructure and construction projects was one of the central government's principal means of counteracting recession and stimulating economic activity. This was accompanied by deregulation of planning and building control systems in order to encourage private sector investment in urban projects. A consequence has been the redevelopment of low- and medium-rise areas in central Tokyo with high-rise, high-density condominiums but it is not at all clear that this makes sense in a city like Tokyo where densities are already very high and where the disbenefits of overcrowding may already outweigh the advantages of compactness in urban form. There is no overall planning framework to guide intensification, and few areas are protected from indiscriminate high-rise redevelopment. A major risk of this approach, according to Sorensen, is that overbuilding will create more vulnerable populations in central areas, requiring ever larger infrastructure investments to keep them supplied with water, wastewater treatment facilities and energy. Protests have been made against these forms of redevelopment, without much success, although Sorensen does see, in the growth over the past few decades of the *machizukuri* social movements at the neighbourhood level, an important increase in citizen engagement in place-making and community development which may lead to a growth in resistance to pressures for redevelopment.

Tokyo has shown great resilience in recovering from the destruction of the twentieth century, but Sorensen's main conclusion is that recent and current planning approaches, with their failure to address Tokyo's vulnerability to flood and fire, may have reduced the city's resilience and exacerbated the risks of future disasters.

Chapters 3 and 4 provide perspectives on Shanghai and Beijing respectively, China's world cities. Susan Walcott begins with Shanghai's colonial history, still evident in the architecture and urban form of parts of the inner city, followed by an account of the urban and economic development of the communist era. She then examines Shanghai's emergence, after China's adoption of an 'open door' policy, as the country's model city and face to the world – the 'Dragon's Head'. Her chapter provides detailed case studies of the recent transformation of the old inner core of the city and of the new urban area of Pudong, across the Huangpu River to the east. Broad spatial rearrangements have occurred as Shanghai's planners have attempted to alleviate the problems of very high settlement densities and industrial pollution in the inner city by promoting the growth of surrounding suburban areas while simultaneously redeveloping the central core to accommodate and attract global city functions – most recently, the Shanghai World Expo of 2010.

Walcott provides a thorough account of Shanghai's recent uneven experience in the establishment of self-contained satellite towns at some distance around the city. Shanghai is the hub of the emerging Yangtze Delta mega-region that also includes the cities of Hangzhou, Suzhou, Ningbo, Nanjing and Wuxi. Combined, the region represents 6 per cent of China's population and nearly 20 per cent of its GDP. The astonishing growth of Shanghai and its region over the past three decades has not occurred without environmental impacts, including poor air and water quality. Attempts to improve the city's environmental image have been hampered by the failure of the much-publicized Dongtan eco-city project and by allegations of corruption on the part of environmental officials. Walcott is optimistic, nevertheless, that Shanghai's role as one of China's main windows to the world will ensure that central and local governments will continue to strive to demonstrate their green credentials to the global community, but significant challenges remain in accommodating continuing economic and population growth, providing affordable housing for the workforce and reducing pollution.

Chapter 4 on Beijing, by Gu Chaolin and Ian Cook, traces the breathless transformation of China's capital over the last 30 or so years from a dull, drab and austere communist city into a modern, outward-looking metropolis. There are parallels, of course, with Shanghai, especially in its recent experience of economic development. There are also differences resulting from the symbolism expressed through Beijing's physical form over time as an imperial capital, with the Forbidden City at its heart; as Mao Zedong's city which sought to be the political, cultural, and industrial heart of a socialist nation, celebrating the 'working people's labour and production'; and, more recently, as one of China's major globalizing cities. Beijing's global role is now expressed through its new high-rise buildings and international architecture, including the remarkable structures built as part of the preparations for the 2008 Olympic Games. This latest phase of development has also seen the relocation of manufacturing industry to the outer suburbs and beyond, as well as the destruction of much of the city's historic urban fabric, including many of the *hutong* areas of courtyard houses. Belated attempts are now being made to save what is left as part of the desire to make Beijing a more 'humanistic and habitable city' and also to capitalize on the tourist potential of these historic areas. This has been accompanied by modest reforms to extend the legal rights of citizens to be consulted about development decisions and to object to these in certain circumstances.

Gu and Cook trace Beijing's development through an analysis of its successive plans. The city's emergence as a modern city with spectacular architecture is also very well illustrated through an extensive set of images and area studies. Beijing's transformation has been accompanied by environmental problems of water shortages, pressure on energy supplies, occasional severe dust storms and poor air quality, with the city being described as 'the most polluted on earth' in 2005. Social polarization is also pronounced, with much of Beijing's migrant workforce living in overcrowded 'villages' in the city, lacking security of tenure and without access to basic health or education facilities. Environmental and social injustices are, of course, interlinked. Richer residents can afford to move away from polluted areas, whereas poorer people are more likely to have to put up with them, with consequences for their health. Gu and Cook conclude that Beijing's rapid pace of development is likely to be maintained for the foreseeable future, backed as it is by the great wealth of the People's Republic of China. In the longer term, as with the other cities discussed in this book, they see its future trajectory depending on international co-operation between megacities and their national governments in meeting the difficult challenges and risky conditions of the twenty-first century.

Co-operation between cities across borders is also a theme of Chapter 5, by Liling Huang and Reginald Yin-Wang Kwok, which discusses Taipei's metropolitan history and growth. Collaboration across the Taiwan Strait is now clearly occurring in both the environmental and economic spheres, but this is a fairly recent development. Huang and Kwok provide a fascinating account of the way in which Taiwan's evolving relationship with China has shaped Taipei's development, intertwined over the last half century or so with its aspirations to global city status, as the capital of an 'Asian Tiger' economy, and with the shifting politics of national identity.

Huang and Kwok describe Taipei's growth under earlier Chinese dynasties, Japanese colonization, the rule of the Kuomintang after 1949 and in the more recent era of better relations with China, following the re-opening of the Taiwan Strait to travel and commerce. Successive ruling regimes have used urban landmarks and sites to underpin their legitimacy and to project particular cultural values. For example, after the Japanese colonial period commenced in 1895, traditional Chinese buildings in the old walled city area were demolished and replaced with official buildings in Japanese style. In the 1960s the Kuomintang under Chiang Kai-shek adopted a Classical Chinese style for important buildings to symbolize their claim to be the rightful heirs to Chinese history and culture. More recently, in 2007, the Democratic Progressive Party sought to rename the Chiang Kai-shek Memorial Hall as the 'National Taiwan Democracy Hall' and the forecourt as 'Freedom Square' to symbolize 'the end of the authoritarian rule of the Kuomintang'. Huang and Kwok examine the spatial implications of these and other political and ideological changes, both for the physical development of the capital itself and for regional development in other parts of the country.

When Taiwan became cut-off from China in 1949, it was obliged to develop its own economy. It succeeded, especially from the 1960s, by pursuing an export-focused industrialization approach under strong state leadership that led to the profitable production of food, clothing and electronic goods for the international market. This period of economic success was accompanied by a quadrupling of Taipei's size as rural migrants flocked to the city. The capital became overcrowded, inadequately provided with infrastructure and afflicted by serious environmental pollution. By the mid-1980s, changing global conditions and increasing labour costs led Taiwan to move to a new economic approach which involved a shift to higher-technology products and the relocation of Taiwanese firms to lower cost countries in Southeast Asia. Some moved to coastal China, which was now open to Taiwanese investment and goods following China's adoption of a socialist market economy in the late 1970s. Closer trading links with China were not universally popular in Taiwan but Huang and Kwok show how, over the past 20 years or so, the growing economic power of China has made it increasingly unlikely that Taiwan can pursue a separate development path. Since the early 1990s, many high-tech manufacturing firms have been relocating from Taiwan to China, clustering particularly in the Yangtze Delta Region, and by 2006 it was estimated that more than 5,000 Taiwan citizens were living in and around Shanghai.

Taipei has become the core of a larger metropolitan region that incorporates the Taoyuan International Airport and the nearby Hsinchu County, now linked to Taipei by a high-speed rail network which also extends to southern Taiwan. Taoyuan airport is an important international transport node with strong connections to China. There is now a Greater Taipei Plan for this extended metropolitan region, although tensions between national and local governments, and between adjoining municipalities themselves, seem likely to complicate the implementation of this plan in the short term. The core of Taipei has been intensively redeveloped since the 1990s, in part with the (unfulfilled) intention of turning Taipei into an international financial centre to rival Hong Kong and Singapore. As part of this process, Taipei became yet another city able to claim to have the tallest building in the world for a while, following the opening of Taipei 101 in 2004.

Taipei's move towards more high-technology industries and producer services has allowed a shift also towards a more environmentally responsible mode of economic development and the city has made progress in recent years in establishing parks, dealing more effectively with solid waste and cleaning up its rivers. Collaboration on environmental initiatives with Chinese cities is developing and Taipei was given the opportunity to showcase its environmental achievements at the 2010 Shanghai World Expo. The overall conclusion of Huang and Kwok's chapter is that, while issues of national identity are deep-seated in Taiwan, there is a new global dynamic that will inevitably continue to strengthen the Cross-Strait relationship.

In Chapter 6 Seong-Kyu Ha describes the recent development of the capital of another of the former Asian Tiger economies – Seoul in South Korea. Seoul was a devastated city in the early 1950s following the Korean War which did great damage to the physical fabric of the city and gave rise also to a substantial influx of refugees and returning expatriates. Growth in Seoul was rapid in the 1960s and 1970s, fuelled by the rural-urban migration of workers coming to participate in the huge programme of industrial development which was driven by the military government of Park Chung-hee. Since the 1980s the growth of the city itself has slowed and the population is ageing. Seoul's population in the mid-1990s was over 10 million but it has now fallen below that figure and the latest UN forecasts are that it will be only 9.8 million in 2025 (see table 1.1). However, the extended Seoul Metropolitan Region now has a population of more than 24 million – about half of South Korea's total population, encompassing the separate city of Incheon with about 2 million people and neighbouring parts of

Gyeonggi Province. As well as being one of the largest cities in the world, Seoul is also amongst the most crowded, with some areas in its inner city having a higher net density than Tokyo.

As Seoul spread outwards during its rapid industrialization, a green belt was proposed to limit its growth. Seoul's green belt, unlike many others, remains largely intact as an extensive area separating the city from its satellite suburbs and now providing a valuable recreational resource. However, the green belt has helped drive up densities in the redevelopment areas within it, turning Seoul into a city of high-rise apartments. Suburban development leapt over the green belt in the 1980s and was boosted towards the end of that decade by a programme of government-supported new towns, which also took the form of high-rise, high-density apartments. A second round of new towns was begun in 2001 some 30 to 40 kilometres from the city centre. These have lower densities and more attention has been paid to principles of sustainability in their design (although most of these more recent towns will not be served by rail lines in the foreseeable future).

Ha examines also the attempts – largely unsuccessful – over several decades to redress the imbalance in growth between Seoul and the rest of the country, including a proposal in 2003 by the President of the time to relocate government functions to a new administrative capital in Chungcheong province. Strong political opposition to the proposal prevented it proceeding. Ha suggests a tension persists between policies of balanced regional development and the continuing tendency for high-level financial and producer services to want to concentrate in Seoul as it acquires more global city functions. The chapter includes a useful discussion of the 'global city' or 'world city' concept, including a review of the arguments (after Hill and Kim, 2000) of the difference between 'market-centred and bourgeois' Western global cities, such as New York, and 'state-centred and bureaucratic' Asian global cities, such as Seoul. Ha acknowledges that this distinction may be less clear-cut than it was some years ago, but stresses, nevertheless, that Seoul's development for much of the latter part of the twentieth century was led predominantly by government agencies under authoritarian regimes in a style associated with a 'developmental state'.

Much of Seoul's recent development has been driven by the explicit desire on the part of national and metropolitan policy-makers to increase its standing as a world city. Ha charts Seoul's progress in the international comparative rankings of cities and notes also its success in attracting international sporting events. Major national infrastructure programmes in recent times have also been aimed at improving Seoul's competitive position in relation to other major cities in the region – the Incheon International Airport, for example, and high-speed rail links to Busan and Mokpo.

Ha notes remarkable recent growth in a new range of industries. Electronic games, animation and software industries have become important contributors to the Seoul economy and it has also carved out a niche in film production and distribution. Korean pop music, television dramas, movies, fashion, food and celebrities are popular in China, especially amongst young people, and that popularity has expanded to Hong Kong, Taiwan, Vietnam and Japan in what has been termed 'the Korean Wave'. Seoul's cultural policy is strengthening the city's image through the promotion of these new forms of culture. However, much of the urban area is low in amenity and in the quality of its public realm. The chapter describes recent strengthening of the city's image through an explicit policy of 'place-making' and the protection of heritage sites.

Seoul's environmental problems are inextricably linked to its growth and the concentration of population and activities. During the first two decades of South Korea's economic boom, there was little attention paid to the damaging effects of rapid industrialization. In the 1980s South Korea began to show some concern for the environment, though Seoul continues to suffer from air and water pollution, solid waste disposal problems and a shortage of green space. Recently the South Korean government has taken a major initiative to position the country at the forefront of 'green growth', with a massive increase in expenditure on new environmental technologies and materials, renewable energy initiatives, sustainable transport, green buildings and ecosystem restoration. At the city level, Ha also describes the project which some see as the turning point in Seoul's attitudes to both environmental issues and the quality of urban design, the Cheonggyecheon project. This saw the removal of a major highway through the city and the restoration of a former stream, which had been enclosed in a concrete culvert in the 1950s, to create an attractive area of public space extending through the heart of the city. Ha sees in this project evidence of a paradigm shift in Seoul, away from an emphasis on the quantitative aspects of economic development and towards a more participatory approach concerned with the quality of the urban environment.

Anthony Yeh uses the image of the 'dragon head' in reference to Hong Kong in Chapter 7, as Walcott did in relation to Shanghai in Chapter 3. Yeh explores the risks to Hong Kong's long-established role as a centre for producer and financial services for China as a result of the rise of competitor cities in the Pearl River Delta (PRD) and elsewhere in China. He describes Hong Kong's importance as 'the factory of the world' in the decades after World War II and then traces how, following China's adoption of economic reforms in the late 1970s, Hong Kong gradually moved its manufacturing to nearby areas of mainland China while continuing to provide the financial, logistical and marketing functions for these industries. This worked well initially but is now under threat as cities like Guangzhou and Shenzhen are developing their own cheaper producer and financial services, as well as investing in major new transportation infrastructure and freight-handling capacity, such as the new Baiyun International Airport and the Nansha Port in Guangzhou. Hong Kong is showing signs of decline with job losses, deflation, a rising unemployment rate, declining relative income and a slowing down of infrastructure development.

Yeh notes that, during the lead up to the end of colonial rule and the return of Hong Kong to China in 1997, there was considerable uncertainty about what this would mean for Hong Kong's future development, but he suggests that Hong Kong's relationship with China has become much clearer now. Tourism and retailing remain important parts of Hong Kong's economy and have been boosted significantly by China's economic growth and by the easing of travel restrictions between China and Hong Kong. Under the 'One country, two systems' arrangement which is in place until 2047, Hong Kong still retains some clear competitive advantages over the rest of China in its open and fair market system, an efficient government with minimal corruption and a good legal system. Nevertheless, there is a perceived need for Hong Kong to reposition its economy to differentiate it from the increasingly sophisticated PRD. An economic strategy adopted in 2008 seeks to develop Hong Kong's strengths in areas like educational and medical services, and in environmental, creative and cultural industries as points of difference.

The West Kowloon Cultural District is an illustration of Hong Kong's aspiration to become a more important centre of global culture. The development on 40 hectares of reclaimed land on the Kowloon waterfront was first announced in 1999 as a deliberate attempt to bring together cultural and entertainment facilities likely to attract tourists. Yeh describes this as part of a conscious strategy of 'place-making', although this is not, it seems, the sort of place-making advocated by Friedmann (see above) which focuses on the small spaces of the city and the needs of the people who inhabit them. Rather, it is urban remodelling on a grand scale, intended to reposition Hong Kong as a more interesting and attractive destination, relative to its regional competitors. However, whatever its aspirations, a decade or so later, the project is a long way from completion and Yeh suggests that, in fact, it has come to represent an area of competitive weakness for Hong Kong. Cities in China seem able to bring major projects to fruition much more quickly than Hong Kong can and are making very effective use of 'mega-events' and associated urban developments for marketing purposes.

Hong Kong's producer services are still well regarded by the international business community for the time being and continue to operate in one the world's best business environments. However, their longer-term future, and the future of Hong Kong more generally, seems inevitably to lie in closer links with the cities of the PRD, as the infrastructural and policy barriers to integration continue to be removed. A step in this direction was provided by the 2009 *Study on the Co-ordinated Development of the Greater Pearl River Delta Townships*. Jointly prepared by the government of Guangdong province and the Special Administrative Regions of Hong Kong and Macao, it proposed collaborative efforts to develop the Greater PRD as a centre for world-class advanced manufacturing, supported by modern services, major transportation hubs, a cultural centre of global influence and a residential environment that is 'affluent, civilized, harmonious and liveable'. A major piece of infrastructure which will be highly significant in the further integration of the component parts of this mega-region is the Hong Kong-Zhuhai-Macao bridge, planned for completion in about 5 years time.

Singapore is the subject of Chapter 8 by Belinda Yuen. Singapore's credentials as a world city include being a major logistics hub and the easiest place to do business according to the World Bank. The city has reached this global eminence in a fairly short time, transforming itself from a low-rise British colonial trading port with slums and squatter settlements in the 1960s to a high-rise, modern, post-industrial garden city in which nearly 90 per cent of residents own their own homes. The public transport system, including a rapid transit rail line, feeder light rail networks and frequent buses, is a model of efficiency, as is the management of the private car, through an advanced system of electronic road pricing. A comprehensive and integrated approach to land-use and transport planning is facilitated by Singapore's single level of government, which assists with the policy co-ordination process.

These elements of Singapore's urban development are highly regarded, but this regard has often been tempered by comments that Singapore is rather too ordered and, as a consequence, a little too sanitized and dull. Yuen describes how, since the early years of the twenty-first century, Singapore's planners have been attempting to change this perception through deliberate policies to improve the quality of the public environment – to make it not just efficient and ordered, but also exciting. This is seen as an important part of a strategy to discourage Singaporeans from emigrating and to attract 'creative class' migrants to the city. The typically ambitious aim is to reposition Singapore as a 'Renaissance City', an innovative global city for the arts and culture.

As part of the process of selling a new and positive image of the city to both residents and visitors, Singapore's city centre has been replanned recently, with the intention of adding more opportunities for leisure and consumption – theatres, museums, a casino and restaurants – to the already excellent business facilities. The preferred approach to this is through 'post-industrial mega-flagship projects' with stunning architecture, an example of which is the Marina Bay redevelopment with its landmark resort designed by Moshe Safdie. Singapore's heritage, treated with little respect in the early post-Independence days, is also now seen as important as globalization reinforces an interest in locality and place.

The shift in Singapore's planning goals, towards a greater acceptance of diversity and an emphasis on quality as well as efficiency, is also accompanied by a change in the style of planning to include public participation, allowing local people to have more say in what happens to their local environments. What this will actually mean in practice in the city-state of Singapore, where the same political party, the People's Action Party, has won every election since Independence, remains to be seen, but it clearly forms part of an agenda to present Singapore as a more open and cosmopolitan society than in earlier years when decision-making was unambiguously 'top-down'. As Yuen observes, new partnerships between government, business and civil society are the embodiment of enlightened governance.

Yuen's chapter also sets out Singapore's recent initiatives in relation to sustainable development. Proposals announced as part of the 'Sustainable Singapore Blueprint' include programmes to reduce both carbon dioxide emissions and energy intensity while increasing dramatically the provision of green roofs and gardens on the upper floors of Singapore's many high-rise buildings. Reductions in the volume of waste are proposed through behaviour change programmes and by the conversion of waste to energy. Until fairly recently, Singapore was heavily dependent on water imported from Malaysia but it has now reduced its imports by half and makes more use of both desalinated and recycled water. A particularly ambitious plan involves the construction of an underground network of tunnels for the collection and treatment of sewage and the reclamation of water. This also has the benefit of freeing up a large area of land which, in the past, was used for more conventional sewage treatment. Yuen explains that Singapore's ambition is to become a centre of excellence for water research and investment.

Singapore has been a model of effective planning since the early years following its Independence. The centralized planning system, the continuity of political direction that allows a long-term perspective, the small land area and the single level of government all provide a context for planning that is unique. These are certainly attributes which will assist in pursuing strong environmental programmes and the prospects for successful 'behaviour change' polices are probably as good in Singapore as they are anywhere. Whether these same attributes can readily produce the qualitative changes in Singapore's urban environment, which are now sought, and the diversity, which its latest plans hope to foster, is not so obvious.

Chapter 9, by Sirat Morshidi and Asyirah Abdul Rahim, focuses on Kuala Lumpur and the recently established administrative capital of Putrajaya, about 30 kilometres to the south, on the high-speed rail line linking Kuala Lumpur with its new international airport. Much recent development of Kuala Lumpur is explained by the authors as part of the attempt to elevate the city's status as a global city by investment in megaprojects, including a new international airport and the Petronas Twin Towers, the tallest buildings in the world at the time of their completion in 1998. The transformation of Kuala Lumpur into a modern, globally-networked city in the last decade of the twentieth century was driven to a significant extent by the desire of the then prime minister of Malaysia, Mahathir Mohamad, that Malaysia should be a developed country by 2020 with a world city at its heart. Putrajaya, the new federal capital, had the stated aims of reducing congestion in Kuala Lumpur, as well as freeing up land in the city centre for new commercial development. But Putrajaya was also a symbolic development, an attempt to discard old colonial legacies and to create a city to embody the spirit of Malaysia and its aspirations in the twenty-first century. Its plan was appropriate for a capital city, with a grand ceremonial axis, plazas and large areas of parkland; its architecture was grand and expressive of a 'Pan-Islamic modernity' (King, 2007); and, as a city conceived at the end of the twentieth century, it was also to be a model of environmental sustainability, as a consequence, in particular, of its extensive areas of artificial wetland.

The authors consider the prospects of a more sustainable future for Kuala Lumpur and Putrajaya in relation to a set of particular environmental risks to the Greater Kuala Lumpur region. They identify the growing problem of a 'heat island' effect in the central city as a consequence of the continuing concentration of high-rise buildings in an area which is already highly developed and has limited amounts of green or open space. Secondly, vulnerability to flash flooding is a particular risk to which Kuala Lumpur is exposed by its climate and topography. The incidence of flooding appears to have become more frequent in recent years, despite some innovative flood control projects, and the chapter suggests that risks from both flooding and the heat island effect are likely to be further exacerbated by predicted changes to the climate. Thirdly, the authors note the increased traffic congestion in Kuala Lumpur in recent years as the use of private vehicles has increased. There have been several investments in new public transport infrastructure, including two light rail networks and a monorail, but these are not well integrated with the older commuter rail system and many people do not have good access to suitable public transport. As a consequence only about 12 per cent of daily trips are made by public transport. Putrajaya has a planned monorail system but the completion of this has been much delayed and there are doubts about its viability. The new capital's claim to being a model of sustainability is threatened by its current levels of automobile dependency which are related to its fairly low density.

Morshidi and Asyirah observe that, since the late 1990s, both the federal government and the Kuala Lumpur city council have begun to pay more attention to the imperatives of sustainable development and to the importance of urban design and the public realm in their policy statements. However, they also suggest that the development approach of the Mahathir period remains largely in place under the current national government. The present draft Kuala Lumpur City Plan contains proposals for an additional 100,000 people to be living in the city centre by 2020, at inner-city densities approaching those in Tokyo and Seoul. These proposals may not be easy to reconcile with the desire to improve the quality of Kuala Lumpur's urban spaces while reducing the city centre's traffic congestion and its vulnerability to flash floods and temperature rises. The authors point to the significance of current redevelopment proposals for Kampung Baru, a surviving low-rise area retaining some of the qualities of a traditional Malay settlement that provides affordable housing within the shadow of the Twin Towers. The federal cabinet has approved in principle the comprehensive redevelopment of this area and, given the value of the land, this is likely to mean modern high-rise commercial and residential buildings for at least part of the area.

This would mean, in turn, the displacement of many of the existing small lot owners in the area to the outer fringes of the city, adding to the numbers of daily commuters and reducing the already limited stock of housing for low- and middle-income earners close to the city.

Morshidi and Asyirah anticipate that Kuala Lumpur's further development will continue to be driven by the aspirations of its policy-makers to become a 'world class city', attractive to global companies and investors. The authors believe the city's future growth must be planned more effectively under a strategic planning framework which encompasses the whole of the emerging metropolitan region, including Putrajaya, Petaling Jaya and Shah Alam. They also observe that megaprojects, aimed at increasing the global importance of a city, can be empowering for local people, but they can also be harmful and disruptive. There is a need to pay careful attention to the 'local imprints of globalization' and to ensure that the environmental risks of such projects are mitigated. A failure to do so will undermine Kuala Lumpur's aspiration to 'world class status'.

Shortly before this chapter was finalized, the Malaysian federal government announced a new economic development strategy for Greater Kuala Lumpur and the Klang Valley. This foreshadows the sort of strategic planning approach for this polycentric metropolitan region that Morshidi and Asyirah advocate, as well as making a commitment to an immediate start on a mass rapid transit system as the core of much-needed improvements to the public transport system. The government has also acknowledged that Kuala Lumpur's competitive position is threatened because its 'liveability lags compared to many other Asian cities' and this leads to further proposals to safeguard heritage areas and to develop 'iconic places and attractions', including the waterfront of the flood-prone Klang river. It remains to be seen what balance is eventually struck in practice between urban liveability, the continuing desire for Kuala Lumpur to be amongst the world's twenty leading cities by 2020 and the reduction of environmental risks.

Vulnerability to flooding is also an important theme in Chapter 10, by Wilmar Salim and Tommy Firman, which examines current issues in planning and governing Jabodetabek, the extended Jakarta metropolitan region. The authors review current demographic, traffic, environmental and economic challenges facing Jakarta and discuss emerging strategic responses to date. A central concern of the chapter is the continuing inadequacy of governance and planning arrangements for the metropolitan region and the need to address this as a matter of great urgency if Jakarta's high level of vulnerability to flooding and other environmental risks is to be reduced.

Jakarta is at the core of the archetypal *desa kota* region, an extended region with some 25 million people that also includes substantial areas of rural land. This region is now administered by a number of separate authorities. In addition to Jakarta, other parts of the Jabodetabek extended metropolitan region fall within the separate districts of Bekasi, Bogor and Tangerang and the cities of Bekasi, Bogor, Tangerang and Depok, as well as within the provinces of West Java and Banten. As the national capital, Jakarta is also subject to central government influence and another factor relevant to its governance is that, for the past decade or so, Indonesia has been pursuing a decentralization policy that gives more autonomy to the municipal and district levels. The institutional arrangements for managing the Jabodetabek mega-region are thus fairly complex and this complexity leads to ineffectiveness.

Salim and Firman review Jakarta's history and conclude that there has been little respect for formal plans at any stage of its development. It is a city shaped more by its rulers than its rules and, in particular, by its first two post-Independence Presidents, Soekarno and Soeharto. Both pursued large-scale urban development projects intended to symbolize Indonesia's growing affluence and modernity. Soeharto was also assisted by his alliance with a number of large real estate developers who, in the 1980s and 1990s, undertook major new town and shopping mall developments across the metropolitan region, often with scant regard for the metropolitan plans which were adopted from time to time. Several of these plans, dating back as far as the 1960s, advocated the adoption of a rail-based rapid transit system connecting Jakarta's main business districts, but these proposals were never implemented and the city is now chronically congested.

An important thrust of metropolitan plans since 1983 has been to direct urban growth along a central east-west axis, thereby avoiding low-lying coastal areas to the north and the steep mountain areas to the south, which are subject to rapid run-off. However, substantial amounts of urban development have occurred in both the northern and the southern areas. Recent studies, based on the modelling work of the International Panel on Climate Change, have predicted increased risks of inundation by sea level rise along the northern coast, with the effects exacerbated by land subsidence, attributable in part to the depletion of groundwater supplies by the growing population. Jakarta has always been somewhat flood-prone, situated as it is on a coastal plain crossed by several rivers, but the severity of flooding has been increasing. Meanwhile, urban development permits continue to be issued for the southern upland areas around Puncak, despite regular warnings over several decades of the consequences for runoff and flooding downstream. Flood risks are increased as rivers are clogged with garbage and lined with the illegal structures of industry and informal settlements. Salim and Firman tell a depressingly familiar tale of urban authorities which appear not to understand the relationship between poorly planned settlements and flooding or which, in their zeal to promote development, neglect its environmental consequences and fail to complete flood prevention measures in a timely manner, or at all.

Most metropolitan plans for the Jakarta region to date have been weak. It has been difficult to reach agreement on their main provisions between the various levels of government involved. Such strategic plans as have been released have been advisory and not binding on either government agencies or private developers. As a consequence, they have been largely ignored. Proposals have been made to establish a unitary planning authority with some teeth, but these proposals have also foundered in the face of the wide range of conflicting interests and constituencies. Salim and Firman see the major obstacle to a more resilient Jakarta in these ineffective governance arrangements and call for a fundamental reform to change the relationship between governments and civil society. Their proposals include strengthening the long established, but hitherto ineffectual, Co-operation Agency for the Development of the Jakarta Metropolitan Area so as to provide it with real authority for planning, watershed management and the co-ordination of major infrastructure projects. An enhanced role for national government ministries in the work of this agency is foreshadowed. They also argue for better land titling, followed by the use of property taxes to raise funds to implement spatial plans; and for a more open and transparent planning process generally, with closer scrutiny, in particular, of the issuing of development permits.

Salim and Firman's chapter shows clearly that Jakarta faces great risks at present to its economic, environmental and social sustainability. Some result from forces beyond the control of citizens and governments, but improvements to governance and planning of the sort sketched are seen to have the potential to help prepare the city for the major challenges which undoubtedly lie ahead. Without such reforms, Jakarta's prospects seem bleak.

Bangkok, the subject of Chapter 11 by Douglas Webster and Chuthatip Maneepong, has a population of over 10 million and is at the heart of an extended urban region with about 22 million people. The chapter focuses on key examples of three types of risk affecting Bangkok at present – economic, political and 'natural' risks. A key theme of the chapter is the interaction between different types of risk and the consequences, difficult to forecast, which result. For example, the recent street conflicts between contending political groups in Bangkok have obvious potential to impact negatively on the increasingly important tourism sector which is now a major plank of Thailand's economy.

Webster and Maneepong describe how Bangkok was transformed by the 1997 Asian financial crisis that affected Thailand severely. It led, in particular, to the expansion of the amenity economy which, in turn, has led to restructuring of Thailand's spatial economy, concentrating wealth and opportunity along the Gulf of Thailand and marginalizing to some extent smaller cities in the south and northeast. The perceived spatial imbalance between more and less dynamic regions of the country was a root cause of the growth of the rural-based 'red-shirt' political movement, which supports former Prime Minister Thaksin Shinawatra, and of Thailand's recent political unrest. Political instability in turn can have a negative effect on tourist numbers, as well as high-end property sales. Over the past 5 years, Thailand's economy has significantly underperformed against competing Southeast Asian and East Asian states such as Malaysia, Singapore and China, and political instability is certainly part of the explanation for this. The authors observe that people with money are footloose. They also note other risks that have impacted on tourism in recent times, including fluctuating exchange rates and the rising cost of energy which is reflected in increased prices for air travel.

Webster and Maneepong also explore the links between Thailand's amenity economy and environmental deterioration. Tourist numbers and foreign investment in property are likely to fall if the environmental quality of the cosmopolitan urban core and the beach resort areas declines. Paradoxically, the more attractive that a location becomes, the more likely that carrying capacity pressures will degrade the very environmental amenity responsible for that attractiveness. Other threats to the amenity economy identified include natural hazards (Phuket was badly affected by the Indian Ocean tsunami in 2004); terrorism which remains a threat in Thailand's southern provinces; and a shortage of workers with the necessary skills for the new economy.

Bangkok is vulnerable to climate change. Webster and Maneepong draw on a good deal of research carried out by Thai government agencies and institutions to demonstrate this. The main potential impacts identified in recent studies are that a 1 in 30 year flood in 2050 will inundate an additional 180 square kilometres of the Bangkok Metropolitan Area and the adjacent Samut Prakarn Province, as compared with an equivalent event in 2008. The western area of the metropolis is at greatest risk. The east and city core are better protected as a result of works undertaken in relation to recent city centre developments, the new Bangkok subway and the recently opened Suvarnabhumi international airport.

A major flood event affecting the Bangkok Extended Urban Region (BEUR) would have serious consequences because the BEUR dominates the national urban system and economy to a greater extent than the major metropolitan region of any other country in East Asia, accounting for over half the country's economic output. Thailand would have a greater percentage of its GDP at risk from sea level rise than any other major country in East Asia except for Vietnam. Webster and Maneepong suggest that the emphasis on assessing the likely impacts of increased flood risks in Bangkok has been on physical impacts and more research is needed into the impacts on the region's dynamic production and housing sectors. Responses to flood risks have concentrated on engineering works, such as dyke building and landfill, to raise the height of important infrastructure. The authors call for more emphasis also on raising community awareness and involving local groups and NGOs in the process of preparing for future flood events. They also identify a need for a more integrated approach to adaptation along with institutional change to overcome the 'silo' approach that still characterizes Thai national government agencies.

Webster and Maneepong conclude that, while a good deal of official risk mitigation and adaptation planning and action is under way in Bangkok in regard to climate change, most Bangkok residents would view political risk as the greatest threat to the well-being of the city at present. Handling this risk has, to a large extent, been regarded as a 'law and order' problem, but increasingly it is also viewed popularly as a spatial equity issue. The least awareness, and least explicit action in regard to risk, has been shown in relation to significant, ongoing changes to the national economy and spatial system. What is clear, however, is that all of these risk areas are closely inter-related and their permutations are inherently unpredictable. The authors argue that, where risk cannot be fully mitigated, as in the case of climate change or political and social instability, the challenge is to facilitate adaptive processes that minimize the impacts of shocks on the city's overall wellbeing and performance. Grounds for some optimism can be found in the way in which Bangkok recovered from the Asian financial crisis and from other devastating events in recent times. Webster and Maneepong see this as indicative of Bangkok's high levels of social capital, a key requirement for a resilient city.

In the final chapter of the book, Brian Roberts looks at the present and future risks facing Metropolitan Manila. Manila is a city with a population in its extended metropolitan region of about 18 million. It faces enormous development problems associated with congestion, pollution, inadequate infrastructure, poverty and weak governance. It is also a city that is becoming increasingly exposed to the impacts of globalization, terrorism and climate change.

Roberts establishes a framework for an assessment of Manila's risks and resilience in relation to potential natural disasters and its economic and social vulnerabilities. He notes that some of the more encouraging efforts in developing risk management strategies have come from the local government level and he uses three detailed case studies to demonstrate this. These focus on the revival of the shoe industry in Marikina city; the recovery from a disaster at a large rubbish dump in Quezon city; and the transformation of Taguig city to a model of local government and business efficiency, with innovative programmes for providing housing for the poor. Like Webster and Maneepong in the previous chapter, Roberts stresses the interrelationships between different types of risk and the tendency for disasters and shocks to have multiple causes.

Roberts identifies the absence of effective metropolitan governance and planning arrangements as one of the major obstacles to Manila's ability to manage its urban development and to prepare for future environmental, economic and social challenges. In terms rather similar to those used by Salim and Firman to describe Jakarta, Roberts concludes that planning has been ineffective in Manila for over half a century and the unwillingness of national, metropolitan and local governments to implement plans or enforce regulations has led to an urban development pattern which is not sustainable and which has increased the risk profile of the city. The Metropolitan Manila Development Authority, established in 1995, emerges from Roberts' account as a fairly weak body and, as a result, opportunities have been lost for efficiency gains by, for instance, promoting greater consolidation of development, better integrating logistics and infrastructure systems, clustering employment and co-ordinating the provision of community services. Metropolitan Manila's municipal fiefdoms each seeks to secure whatever it can extract from central government or the business sector, rather than working collaboratively to overcome their many common development problems.

Corrupt practices and nepotism are widespread in the urban development process in Manila, according to Roberts. He also finds the Philippines to be one of the most litigious and bureaucratic democracies in Asia, with multi-layered central and local government authorities which can take years to make development decisions. Consequently, development tends to occur without formal endorsement or approval, especially on land where tenure and ownership is unclear. The more successful cities have mega-malls with the rich living nearby in secure gated communities. Meanwhile the city's 4 million urban poor are confined to older inner-city housing, squatter settlements on public land, and flood-prone and land-slip areas. The competition between cities, together with ineffective planning at the metropolitan level, has resulted in a polycentric city poorly linked by its transportation system. The light-rail network, for which planning commenced in 1980, lacks capacity and is incomplete. As a result, Manila is one of the most congested cities in Asia. This adds to the transaction costs for business and undermines the city's attempts to attract overseas investment. It also condemns many of Manila's residents, rich and poor alike, to long and frustrating daily journeys.

Roberts sets out the principal risks that Manila faces. He notes its vulnerability to typhoons and to related storm surges which create dangerous conditions for the citizens who live in low-lying areas. Roberts also discusses earthquake risks, as Manila is built on two major fault lines. The metropolitan area's failing infrastructure poses particular risks. Electric power supplies are more reliable than they were 20 years ago, but poorly managed privatization arrangements mean that they are expensive. Water supplies are extremely vulnerable. Several times in the early 1990s, the city almost ran out of water when the Angat Dam reached dangerously low levels. Chronic supply and theft problems meant that only a few areas of the city had access to a continuous water supply. Factories were forced to cut production at that time and high-rise buildings had to bring in water by truck, not only because of limited supplies but also due to low water pressure. Health problems emerged in poor areas of the city as people began using contaminated water for drinking and ablution purposes. The water supply management authority at the time was found to be corrupt, highly inefficient and in debt. Privatization in 1997 was seen as the answer but the two companies awarded contracts were hit by the Asian financial crisis later that year and failed to meet their obligations. Then, in 2004, the main tunnel servicing the Angat dam, collapsed, leaving many areas with inadequate water supplies. The position has improved little since then and indicates, among other things, the risks inherent in a megacity having too much reliance on a single water source.

Air pollution is a major problem in Metro Manila, as are solid waste management, water pollution, including large amounts of untreated waste entering the river systems, and the depletion of groundwater supplies, leading to subsidence. Roberts also discusses Manila's economic risks, the reasons why it has tended to lag behind other Asian cities and the different character of the Filipino economy as a result of its very high level of reliance on remittances from overseas workers. Over 20 per cent of Manila's population falls below the poverty line. Roberts notes the resilience of Manila's urban poor at times of economic crisis and their remarkable level of co-operation and trust in sharing resources and labour to support individuals during economic shocks, and at times of post-disaster reconstruction. To Roberts, this underlines the important role that social capital plays in rebuilding communities.

Metro Manila is highly vulnerable to natural disasters but the vulnerability of many of its citizens is further increased by the city's inadequate and failing infrastructure, and by the large numbers who are obliged to live in informal settlements, with no security of tenure, on flood plains, low-lying coastal land or sites close to sources of pollution. Roberts documents some encouraging examples of local resilience and the lessons that these provide. His main theme, however, is the importance of Manila adopting effective strategic planning arrangements for the extended metropolitan region, to increase the capacity to address and respond in a co-ordinated way to the many desperate risks to the city's future liveability and sustainability.

Notes

1   The focus of this book is on Pacific Asia. In using this term we are following the usage adopted by Fuchs *et al.*, 1987. In essence Pacific Asia comprises the countries of East Asia and Southeast Asia. See also Douglass *et al.*, 2008, p. 23.
2.   See http://www.edb.gov.sg/edb/sg/en_uk/index/industry_sectors/education/global_schoolhouse. html. Accessed 16 February 2011.

References

Annan, Kofi (2003) Foreword to *Disaster Reduction and Sustainable Development: Understanding the links between vulnerability and risk to disasters related to development and the environment*. Background paper developed for the World Summit on Sustainable Development. Geneva: UN Inter-Agency Secretariat of the International Strategy for Disaster Reduction. Available at www.unisdr.org/eng/risk-reduction/sustainable-development/DR-and-SD-English.pdf. Accessed 15 December 2010.

Berke, Philip (2010) Catastrophe, in Hutchison, Ray (ed.) *Encyclopedia of Urban Studies*. Los Angeles, CA: Sage, pp. 119–122.

Douglass, Mike, Ho, K.C. and Ooi, Giok Ling (eds.) (2008) *Globalization, the City and Civil Society in Pacific Asia*. London: Routledge.

Douglass, Mike (2010) Globalization, mega-projects and the environment: urban form and water in Jakarta. *Environment and Urbanization Asia*, **1**(1), pp. 45–65.

Evers, Hans-Dieter and Rudiger Korff (2000) *Southeast Asian Urbanism: The Meaning and Power of Social Space*. Singapore: Institute of Southeast Asian Studies.

Forbes, Dean (2010) Asian cities, in Hutchison, Ray (ed.) *Encyclopedia of Urban Studies*. Los Angeles,

CA: Sage, pp. 41–44.

Friedmann, John (2007) Place and place-making in the cities of China. *International Journal for Urban and Regional Research*, **31**(2), pp. 257–279.

Friedmann, John (2010) Place and place-making in cities: a global perspective. *Planning Theory and Practice*, **11**(2), pp. 149–165.

Fuchs, Roland, Jones, Gavin and Perina, Ernesto (eds.) (1987) *Urbanization and Urban Policies in Pacific Asia*. Boulder, CO: Westview Press.

Gugler, Josef (ed.) 2004 *World Cities Beyond the West: Globalization, Development and Inequality*. Cambridge: Cambridge University Press.

Hall, Peter (2011) The polycentric metropolis: a Western European perspective on mega-city regions, in Xu, Jiang and Yeh, Anthony (eds.) *Governance and Planning of Mega-City Regions: An International Comparative Perspective*. London: Routledge.

Hamnett, Stephen and Forbes, Dean (eds.) (2001) Pacific Asian Cities: Challenges and Prospects. *Built Environment*, **27**(2).

Healey, Patsy (2009) Developing Neighbourhood Management Capacity in Kobe, Japan: Interactions between civil society and formal planning institutions. Case study prepared for UN-Habitat *Planning Sustainable Cities: Global Report on Human Settlements 2009*. London: United Nations Human Settlements Programme/Earthscan. Available on-line at http://www.unhabitat.org/downloads/docs/GRHS2009CaseStudyChapter04Kobe.pdf. Accessed 14 December 2010.

Hill, R.C. and Kim, J.W. (2000) Global cities and developmental states: New York, Tokyo and Seoul. *Urban Studies*, **37**(12), pp. 2167–2195.

Hogan, Trevor, Bunnell, Tim, Pow, Choon-Piew, Permanasari, Eka and Morshidi, Sirat (2011) Asian urbanisms and the privatization of cities. *Cities* – article in press: doi:10.1016/j.cities 2011.01.001.

International Federation of Red Cross and Red Crescent Societies (2010) *World Disasters Report: Focus on Urban Risk*. Geneva, Switzerland: International Federation of Red Cross and Red Crescent Societies.

Jacobs, Jane (1961) *The Death and Life of Great American Cities*. New York: Random House.

Johnson, Chalmers (1982) *MITI and the Japanese Miracle: The Growth of Industrial Policy 1925–1975*. Stanford, CA: Stanford University Press.

Jones, G. and Douglass, M. (2007) *The Rise of Mega-Urban Regions in Pacific Asia: Urban Dynamics in a Global Era*. Singapore: Singapore University Press

King, R. (2007) Rewriting the city: Putrajaya as representation. *Journal of Urban Design*, **12**(1), pp 117–138.

Laquian, Aprodicio (2005) *Beyond Metropolis: The Planning and Governance of Asia's Mega-Urban Regions*. Washington DC: Woodrow Wilson Center Press/ Baltimore, MD: Johns Hopkins University Press.

Lindfield, Mike and Brockman, Royston (2008) *Managing Asian Cities*. Manila: Asian Development Bank, Manila.

Lo, Fu-chen and Marcotullio, Peter (eds.) (2001) *Globalization and the Sustainability of Cities in the Asia Pacific Region*. Tokyo: United Nations University Press.

Mastercard Worldwide (2008) *Worldwide Centers of Commerce Index 2008*. New York: Mastercard Worldwide.

Menand, Louis 2001 *The Metaphysical Club: A Story of Ideas in America*. New York: Farrar, Straus and Giroux.

Meyer, M.J. (2009) *The Last Days of Old Beijing: Life in the Vanishing Backstreets of a City Transformed* New York: Walker.

Newman, Peter, Beatley, Tim and Boyer, Heather (2008) *Resilient Cities: Responding to Peak Oil and Climate Change*. Washington DC: Island Press

Olsson, P., Folke, C. and Berkes, F. (2004) Adaptive co-management for building resilience in socio-ecological systems. *Environmental Management*, **34,** pp. 75–90.

Ooi, Giok Ling (2008) State-society relations, the city and civic space, in Douglass, Mike, Ho, K.C. and Ooi, Giok Ling (eds.) *Globalization, the City and Civil Society in Pacific Asia*. London: Routledge.

Putnam, R. (1993) *Making Democracy Work*. Princeton NJ: Princeton University Press.

Putnam, R. (2000) *Bowling Alone: The Collapse and Revival of American Community*. New York: Simon and Schuster.

Resilience Alliance (2007) *Research Prospectus: A Resilience Alliance Initiative for Transitioning Urban Systems towards Sustainable Futures.* Canberra: CSIRO in collaboration with Arizona State University and Stockholm University.

Roberts, Brian and Kanaley, Trevor (eds.) (2006) *Urbanization and Sustainability in Asia: Case Studies of Good Practice.* Manila: Asian Development Bank/Cities Alliance.

Ross, Catherine L. (ed.) (2009) *Megaregions: Planning for Global Competitiveness.* Washington DC: Island Press.

Sassen, Saskia (2011) Novel spatial formats: megaregions and global cities, in Xu, Jiang and Yeh, Anthony *Governance and Planning of Mega-City Regions: An International Comparative Perspective.* London: Routledge.

Skytrax (2010) *World Airport Awards 2010* Available at http://www.worldairportawards.com/ Awards_2010/Airport2010.htm. Accessed 13 January 2011.

Sorenson, A., Marcotullio, P. and Grant, J. (2004) *Towards Sustainable Cities: East Asian, North American, and European Perspectives on Managing Urban Regions.* Farnham: Ashgate.

Sorensen, A. and Funck, C. (eds.) (2007) *Living Cities in Japan: Citizens' Movements, Machizukuri and Local Environments.* London: Routledge.

United Nations (2010) *World Urbanization Prospects. The 2009 Revision.* New York: Department of Economic and Social Affairs, Population Division.

UNDP (2010) *Human Development Report 2010. The Real Wealth of Nations: Pathways to Human Development.* New York: United Nations Development Programme

UN-Habitat (2010a) *The State of the World's Cities 2010/2011. Bridging the Urban Divide.* London: United Nations Human Settlements Programme/Earthscan.

UN-Habitat (2010b) *The State of Asian Cities 2010/2011.* Fukuoka, Japan: United Nations Human Settlements Programme.

UN-Habitat (2009) *Planning Sustainable Cities: Global Report on Human Settlements 2009.* London: United Nations Human Settlements Programme/Earthscan.

Vale, Lawrence J. and Campanella, Thomas J. (eds.) (2005) *The Resilient City: How Modern Cities Recover from Disaster.* Oxford: Oxford University Press.

Veneracion, Cynthia C. (2008) *Capability Building for Urban Slum Upgrading: Views from Five Communities in Quezon City.* Quezon City: Ateneo de Manila University.

Vogel R.K. (2010a) The city region as a new state space (Chapter 12), in Vogel, R.K., Savitch, H.V., Xu, Jiang and Yeh, Anthony, Wu, Weiping, Sancton, A., Kantor, P., Newman, P., Tsukamoto, T., Cheung, P., Shen, J., Wu, Fulong and Zhang, F., Governing global city regions in China and the West. *Progress in Planning*, **73**, pp. 64–66.

Vogel R.K. (2010b) Governing global city regions in China and the West (Chapter 1) in Vogel, R.K., Savitch, H.V., Xu, Jiang and Yeh, Anthony, Wu, Weiping, Sancton, A., Kantor, P., Newman, P., Tsukamoto, T., Cheung, P., Shen, J., Wu, Fulong and Zhang, F., Governing global city regions in China and the West. *Progress in Planning*, **73**, pp. 4–10.

Vogel, R.K., Savitch, H.V., Xu, Jiang and Yeh, Anthony, Wu, Weiping, Sancton, A., Kantor, P., Newman, P., Tsukamoto, T., Cheung, P., Shen, J., Wu, Fulong and Zhang, F. (2010) Governing global city regions in China and the West. *Progress in Planning*, **73**, pp. 1–75.

Wong, Tai-Chee, Shaw, Brian J. and Goh, Kim Chuan (eds.) (2006) *Challenging Sustainability: Urban Development and Change in Southeast Asia.* Singapore: Marshall Cavendish Academic.

Xu, Jiang and Yeh, Anthony (eds.) (2011) *Governance and Planning of Mega-City Regions: An International Comparative Perspective.* London: Routledge.

Yang, Jiawen (2009) Spatial planning in Asia: planning and developing megacities and megaregions, in Ross, Catherine (ed.) *Megaregions: Planning for Global Competitiveness.* Washington DC: Island Press.

Chapter Two

Uneven Geographies of Vulnerability: Tokyo in the Twenty-First Century

André Sorensen

Tokyo is at the heart of the largest city-region in the world, with about 35 million people, or 28 per cent of the Japanese population in 2010, living in the Greater Tokyo Area. This comprises the cities, towns and wards of the Tokyo Metropolitan Government area, with a population of about 13 million, and the three adjoining prefectures of Saitama, Kanagawa and Chiba. The Greater Tokyo Area also forms the core of the Tokaidô Megalopolis, which extends west for about 600 kilometres through Nagoya to Osaka and Kobe in an almost continuous belt of urban development and is home to 68 million people.

The focus of this chapter is on the Greater Tokyo Area (see figures 2.1 and 2.2), which presents a distinctive set of sustainability, urban form and planning issues, in part because of its great size and high population density. Tokyo has one of the best heavy rail commuter systems in the world and its dominant central employment zone attracts over 3 million commuters each day from an area within a radius of about 70 kilometres. Although there is a high rate of car ownership and congestion, Tokyo is not an automobile-dependent city. Excellent public transit, high density and very mixed land uses mean that the 'Smart Growth' strategies typically prescribed to make cities more sustainable in Europe and North America seem of limited relevance here (Sorensen, 2010). Tokyo also compares well with other developed cities in the energy efficiency of its vehicles, buildings and industries (Fujita and Hill, 2007), and in the progress that has been made to reduce vehicle emissions and improve air quality (Okata and Murayama, 2010).

Figure 2.1. Japan.

Figure 2.2. The prefectures of Greater Tokyo.

Tokyo thus faces some different challenges from other major global city-regions. To understand these challenges, it is necessary to know something of Tokyo's history and of the processes of modernization experienced during the twentieth century. The first part of this chapter therefore describes Tokyo's historical urban development and the ways in which earlier settlement patterns and building technologies were adapted during the creation of the modern city in the late nineteenth and early twentieth century. The second section examines the development of the metropolis from the rapid economic growth period of the 1950s and 1960s, through the glory days of Tokyo's emergence as a global city matched in stature only by London and New York (Sassen, 2001) to the loss of that exalted status following Japan's two 'lost decades' of financial crisis, and economic and population decline. The focus in this section is on the ways in which successive planning approaches maintained and even exacerbated the risks and vulnerabilities of certain parts of the city-region.

The third section centres on three major challenges facing the Tokyo city-region today. The first is the enduring division between the upland High City, west of the Imperial Palace, and the Low City in the floodplains of the Sumida, Edogawa and Nakagawa rivers, most of which is now at or below the mean high-tide level. Second is the continued existence and even growth of highly vulnerable areas of substandard housing throughout the twentieth and into the twenty-first century, a situation that contributes to the heightened disaster risks facing major segments of the Tokyo population. Third is the huge pressure to redevelop and intensify land use in the central areas of the city. This is in large part a product of changes in planning regulations, which are designed more to create profitable opportunities for redevelopment than to address existing urban challenges. Tokyo has been highly resilient to past disasters, such as the massive destruction of the Great Kanto Earthquake of 1923 and the firebombing of World War II, both of which are discussed below. A central argument of this chapter, however, is that recent and current planning approaches may have reduced this resilience and made Tokyo more vulnerable to future disasters.

The Influence of the Past

Unlike virtually all the other megacities in the world today, with the exception of Beijing, Tokyo was a giant city of over a million inhabitants prior to industrialization. Europe's two giants, London and Paris, were quite small cities at the beginning of the eighteenth century and other contemporary giants like São Paulo, Bangkok, Jakarta, Lagos and Delhi were still tiny at the beginning of the twentieth century. So history is particularly important in this analysis because Japan had a fully developed urban system by the end of the Edo period (1600–1867) and Tokyo (then called Edo) was probably one of the world's largest cities in the seventeenth and eighteenth centuries (see figure 2.3). The urban culture and patterns of development of that period are still influential. After centuries of self-imposed isolation, Japan only opened to the world in 1867, casting off its feudal governance system and beginning to modernize and industrialize. As a result, Tokyo's pre-modern urban patterns have had a huge impact

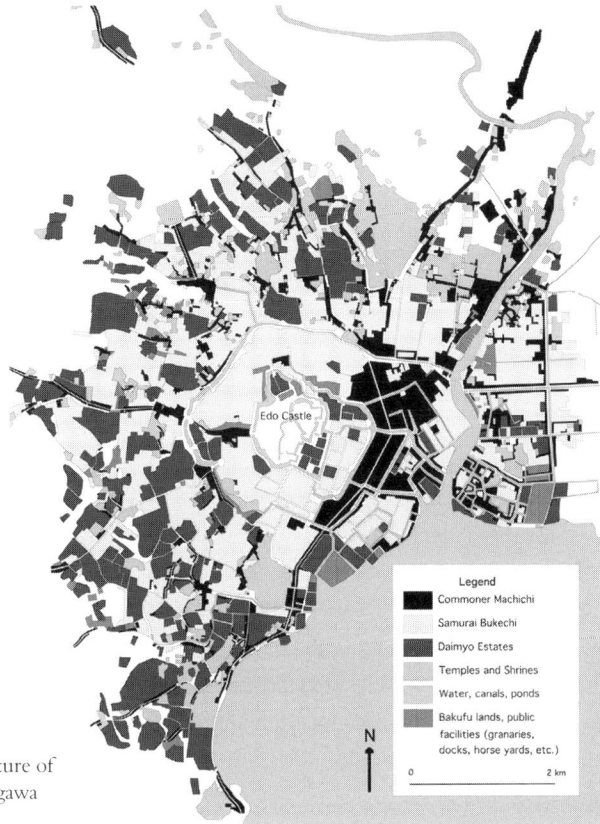

Figure 2.3. The spatial structure of Edo near the end of the Tokugawa period in 1859.

on the modern city, and much city planning in support of modernization during the first half of the twentieth century was oriented to dealing with the legacy of the pre-modern city. So understanding both the positive and negative aspects of that legacy is essential to understanding Tokyo in the twentieth and twenty-first centuries.

Two characteristics of the early development of Tokyo are essential to understanding its urban form: the fact that it was a planned city, dominated by a military government and designed primarily for effective defence against attack; and the planned spatial division of population by class and rank, instead of by wealth through market processes. Edo was created as a castle town at the end of a long period of civil war, in which castles were strategic emplacements. The country was unified after 1600 under the Tokugawa Shogunate, ushering in the long period of peace and development known as the Edo period. Especially in the early part of this period, the military kept strict control over space and deliberately allocated land to different land uses. Apart from various military infrastructures, such as the central castle and its moats and fortifications, most of the area of the city was reserved for the warrior samurai class, who were allocated space and location based primarily on rank. Smaller areas were also allocated for commoners and for temples, which were also necessary parts of the feudal political and space economy.

The main spatial division in Edo, which continues to be reflected in modern Tokyo, was that between the High City of the samurai and the Low City of the commoners. The Low City was mostly built on the marshy estuary of the Sumida River, and on land reclaimed from Tokyo Bay by landfill. The commoner city was laid out in a grid pattern of square blocks, following the ancient Chinese measurement system, and was deeply penetrated by a dense network of canals (Jinnai, 1990; Sorensen, 2002). A system of canals was important because virtually all movement of goods was by boat. The canals were lined by warehouses and markets, and extended throughout the commercial areas. In the late eighteenth century the commoner area of Edo measured about 13 square kilometres and had a total population of about 500,000, with population densities reaching as much as 58,000 per square kilometre (Rozman, 1973). This was a very high density for a low-rise city. The streets were generally lined with 2-storey merchant houses (*machiya*)*,* while the interiors of the blocks were occupied by rows of single-storey wooden shacks fronting on to narrow lanes (*ura-nagaya*), built as rental housing for poor artisans, labourers and servants.

The samurai High City was a much larger area, built on the ridges and plateaus to the west of Tokyo Castle and structured by the hilly geography of the Yamanote area. The dominant land use in the High City was housing for the various classes of samurai, including the huge compounds of the major feudal lords who were required by the Shogun to live in Edo every other year, and to leave their families there permanently as hostages. It was this requirement of residence in Edo that caused the Shogun's capital to grow so rapidly to the enormous size of a million people by the end of the seventeenth century. Consumption by that large population stimulated the economic integration of the whole of Japan during the Edo period. About two-thirds of the area of Edo was reserved for the ·samurai residences, even though samurai were only about half the population (and about 15 per cent of the total population nationally), so population densities in the samurai areas were less than a quarter of those of the commoner areas (Rozman, 1973; Smith, 1978).

In contrast to the planned grid of the commoner Low City, most of the High City was relatively unplanned with main roads following the ridgelines and valleys in a relatively organic pattern (Jinnai, 1990). So, whereas most commoners lived on the floodplain in planned mixed-use areas that were busy, noisy, dense, economically vibrant and culturally dynamic, most samurai lived in the hilly uplands in unplanned, lower density areas, which were green and quiet, but economically and culturally sterile. Most houses were set in gardens and surrounded by perimeter walls and, in the case of the estates of the hundreds of feudal lords (*daimyo*), the gardens were often very large. This was the source of one enduring legacy of the Edo period, an indigenous Japanese version of the suburban housing ideal of detached family homes set in walled gardens which became highly influential with the growth of the middle class after the First World War, when suburban growth towards the west of the city started in earnest (Smith, 1979; Jinnai, 1994).

The vibrant culture of Edo that is remembered today, including the iconic woodblock prints, geisha and the teahouses where they entertained, theatres, Sumo wrestling and Rakugo storytelling, were all based in the commoner Low City. This 'floating world' of brothels, actors, wrestlers and beautiful tea-house girls, integrated by the canals, rivers, restaurant barges and riverside palaces, was at the heart of Edo's urban culture during the feudal period (Nishiyama, 1997). The political, economic and geographic shift in the early twentieth century away from the lowland areas to the western uplands served by rail and road, the filling in of canals, industrialization, and the pollution of the waterways, were all lamented as emblematic of the slow death of Edo culture (Seidensticker, 1991).

Three other major legacies of the feudal period were the enduring problem of urban fires; the limited road system; and an excellent water supply and waste management system. In the Edo period the problem of fireproofing cities was never solved. Repeated orders to build with fireproof materials after major fires suggest that the orders were not followed or effectively enforced (Kelly, 1994). To be fair, enforcing fireproof building standards was difficult everywhere and was usually accomplished only after a major fire, as was the case for London in 1666, and Chicago in 1871. Japan also had the particular challenge of frequent large earthquakes, which made building in stone or brick unwise. Preventing urban fires became one of the major planning challenges in the modern period.

Second, roads were designed for pedestrians, not vehicles. In the Edo period virtually all travel was by foot while, as noted above, most goods transport was by boat. Streets tended to be very narrow, and were often laid out primarily with military considerations in mind, with narrow bridges, and many sharp turns and T-intersections to aid defence. Wheeled vehicles were almost non-existent and, apart from the main highways and castle moats which had military significance, urban infrastructure maintenance was not considered to be a government responsibility and was delegated to local residents. Streets were unpaved, creating clouds of dust in the dry season and turning into quagmires in the rainy season. In consequence, when street railways were being built in the early twentieth century, road widening and straightening and modern wider bridges were major priorities. Improving transportation infrastructure has been a priority of the central government ever since, a task it has carried out with unflagging zeal.

Third were the excellent water supply and waste management systems. The giant city of Edo had simple and effective systems of water supply and waste removal. A water supply system built in the seventeenth century drew water from the Tama river in the west and brought it in canals above ground to the edge of the city, then in stone main conduits underground, with wood and bamboo pipes supplying shallow local wells in the centre of each block (Hatano, 1994). Early European visitors to Edo tested the water supply and found it of very high quality, far better than in most contemporary European cities (Hanley, 1997). But major cholera epidemics in the nineteenth century meant that an early concern for planners was to ensure safe drinking water supplies, and improving the drinking water system was an important early investment, starting in the 1890s (Hayami, 1986). Sewerage was not such a priority, as the traditional system of entrepreneurs collecting human waste and selling it to farmers outside the city continued to be very effective. Sewers were gradually built, starting in central Tokyo, but in suburban areas the 'honey bucket men' continued to collect most human waste until well into the 1960s, when petrochemical-based fertilizers became more available and cheaper, putting them out of business.

The relative success of traditional urban technologies, and the order, cleanliness and discipline of Japanese urban life, lessened the need for new technologies and approaches to urbanization, delaying the adoption of many modern urban infrastructures. As suburban growth began in earnest in the 1920s, traditional urban infrastructure and social systems were employed to create very liveable and desirable (even though unplanned) suburbs in the uplands west of the central city. Municipalities were able to rely on neighbourhood-scale civic organizations for the delivery of essential social services (Bestor, 1989; Sorensen, 2007). The problematic aspects of unplanned and unserviced suburban development only really manifested themselves much later, as discussed below.

The Beginning of Modern Planning

After the Meiji revolution of 1867 the Japanese state focused on the project of industrialization and the importation of Western technologies and institutions, including constitutional government, military technologies, railways, a postal system, universal public education and private property rights. The primary motivation was to grow strong quickly enough to be able to prevent colonization by the European powers that were then dividing up the globe. Creating a modern state that could convince the great powers that Japan was a civilized nation was a top priority, as were industrialization and military development.

The initial attempts to modernize Tokyo followed typical patterns of colonial borrowing by hiring foreign advisors to create direct copies of Western town forms and planning ideas. The first major chance came after a fire destroyed a huge area of the merchant district in the Low City in 1872. The central government hired a British engineer-architect named Thomas Waters to plan and supervise the building of a European-style commercial centre in the Ginza area, later called the Ginza Renga-Gai (Ginza Brick Town) and built to a design inspired by London's Regent Street (Fujimori,

1982; Noguchi, 1988). The rebuilding stalled when it was about two-thirds complete because the buildings were expensive, damp and ill-suited to the Tokyo climate, but the project did have the long-term impact of shifting the area of highest retail rents from Nihonbashi southward towards Ginza 4-chome, an area that, by the 1990s, could boast the highest rents in the world (Okamoto, 2000). A second such attempt was a plan commissioned by the Foreign Ministry from the German architects Bockmann and Ende to restructure the government quarter in the grand neoclassical style. That plan, however, was abandoned before construction began when the minister was disgraced by his failure to renegotiate the 'unequal treaties' that had been signed under duress in 1858 when the US navy threatened Edo (Beasley, 1995).

In the end, the major early planning achievements were the results of a much more practical urban improvement plan. In 1889 the central government passed the Tokyo City Improvement Ordinance (TCIO, *Tôkyô Shiku Kaisei Jôrei*), which created the legal authority to carry out an ambitious project to improve roads and to build parks, markets and schools. Instead of a plan based on the architectural fads of European advisors, this was basically a large-scale co-ordinated municipal public works project. The major priorities of this project, which lasted from 1888 to 1918, followed fairly directly from the legacies of the giant city of Edo, including fireproofing, road-widening and water supply. This project also helped to create a new electric streetcar system after 1890, which developed into a comprehensive network by 1920. The financial contributions from the private streetcar company for the use of roads to lay their tracks paid for most of the road-widening projects. Other major initiatives were the building of a new water supply system and the beginning of a mains sewer system for central Tokyo.

Several of the enduring characteristics of Japanese city planning are first evident during this period. These include a strong centralization of power in the Home Ministry, with city mayors and prefectural governors appointed by the Ministry. This concentration of experts and plan-making authority in the central government led to a relatively quick attainment of high levels of planning expertise in the Home Ministry City Planning Bureau, but limited the ability of municipalities to devise or implement locally specific solutions to particular issues. A further problem was the weak financial base for planning, as the Finance Ministry always retained tight control over the purse strings and local governments had few of their own resources available for locally important projects (see Sorensen, 2002, p. 110).

The early Meiji period had seen a drastic halving of Tokyo's population, with the departure of most samurai to their home provinces and the collapse of the former economic system, but Tokyo started to grow once more in the 1890s. It had only reached its 1800 population again by 1900 and was still contained within its feudal boundaries in 1910, but thereafter Tokyo experienced a period of rapid population and physical growth. By the time of the First World War, Tokyo began to see major changes, as rapid industrialization resulted from a combination of the blockade of Germany and sales of industrial goods to the allied powers. This presented a new set of problems of how to manage growth and land development on the urban fringe, and of how to alleviate the steadily worsening housing conditions of the working class in the inner city. Rising rents led to increasing population densities, worsening housing conditions, and serious epidemics of cholera and tuberculosis, while the growing middle class sought new housing outside the existing urban areas.

Changes in transportation technology were also significant factors shaping urbanization in this period. The building of the national railway system was a top priority of the government during the second half of the Meiji period, and the famous Yamanote loop line, linking Tokyo Station and Ueno Station, was completed in 1919. The private railways, which proliferated after 1910, were prohibited from establishing terminals within the Yamanote loop, so their terminals were located at Ikebukuro, Shinjuku, Shibuya and Ueno. This later structured urban growth by creating important sub-centres where commuters transferred from the private commuter railways to the Yamanote Line or the Tokyo streetcars.

The streetcar system also had profound effects on patterns of urban development. As in other major cities such as London, Berlin, Paris and New York, the vast majority of the population travelled on foot before the development of public transport systems. That meant a relatively compact form of growth, usually within a radius of about 5 kilometres. The development of public transit systems allowed an enormous spread of population, with the better off who could afford daily fares able to travel the furthest. The system also encouraged the development of the great department stores in Ginza, and the central business area in Marunouchi, to serve the whole city. The Ginza subway, begun in 1920, was originally part of a system with seven routes, of which it was the only line completed before the World War II.

The 1919 City Planning Act provided the basis for Japan's first city planning system. The main elements were: land-use zoning; an urban buildings law; the building line system; excess condemnation; land readjustment; and the designation of public facilities. Unfortunately, funding measures included in the proposed legislation were opposed by the Ministry of Finance. Provisions for a betterment tax, a land expropriation law, and public financing of city planning projects all had to be dropped before the bill could become law.

Early by international standards, the new zoning system was simple, with only four land-use zones: commercial, residential, industrial and quasi-industrial. The law also allowed for undesignated zones and non-zoned areas within the city planning area. Each zone allowed a wide variety of land uses, with the main restriction being that large-scale factories could only be located in industrial areas, and noisy uses such as theatres and cabarets were only allowed in commercial zones. Small factories and retail and office uses were allowed in residential zones, however, and housing was still allowed in heavy industrial areas. It was imagined that zoning would not act primarily to restrict land uses, but rather as an indicator of the appropriate scale and design of public facilities such as roads in the different zones, with narrow roads in residential areas and super-blocks in industrial zones. This was quite different as a conceptual and political basis of zoning from, for example, New York, where the early goal of zoning was to prevent the encroachment of industrial lofts into upper-class residential areas.

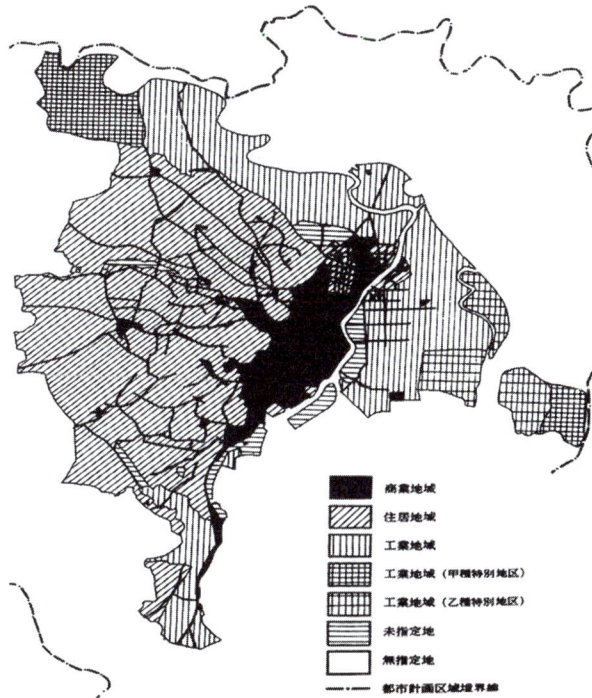

Figure 2.4. Tokyo's first zoning plan, 1923.

At first glance, the first Tokyo zoning plan (figure 2.4) can be taken at face value, as a plan to shape the structure of land use and built form of the city. It is better understood, however, as a schematic representation of land uses existing in Tokyo when it was approved in 1923. This was a new regulation for an existing fully built-up area, and it was merely confirming the broad land use pattern then existing. A few main aspects are worth noting: the entire Low City west of the Sumida River is designated as a commercial area (solid black), and the transition to the residential zone (diagonal stripes) on its western edge closely follows the change in elevation to the uplands of the High City. Throughout the residential areas of the High City, commercial areas are designated as linear corridors along main highways and arterial roads. Many, but not all, of these follow the lines of the streetcars. This pattern of commercial areas in long strips along main roads is still common in Japan and, where the building of a parallel highway or elevated expressway has allowed it, many of these have been

Figure 2.5. A pedestrian shopping street in central Tokyo in 2007.

pedestrianized (see figure 2.5). Finally, virtually all the Low City floodplain areas that are not designated commercial are zoned either industrial (vertical stripes and cross-hatched) or 'undesignated' (horizontal stripes). In 1919, as today, these were areas with a fine-grained mix of factories, housing and retail land uses.

As is very clear, the initial zoning divided the city of Tokyo into two halves, a residential zone in the upland High City, and a commercial/industrial zone in the floodplain Low City. The samurai High City and areas towards the west were to remain residential, while the commoner areas of the Low City were to be the business and industrial areas, without excluding the possibility that people would also continue to live in those areas. This division is still apparent in land uses even today, with the exception that three large commercial sub-centres have developed at Shibuya, Shinjuku and Ikebukuro, the three main points in the west at which suburban commuter trains connect to the Yamanote line. Although most Japanese cities do not show the dramatic socio-spatial divisions of, for example, US cities, and have a mixture of different classes in virtually all areas of the city, the Low City of Tokyo continues to be one of the poorest areas, with much higher levels of employment in low-wage sectors, inexpensive rental housing, and concentrations of day-labourers and the homeless (Kurasawa, 1986).

Tokyo's Two Great Twentieth-Century Disasters

Tokyo was almost completely destroyed twice during the twentieth century: in the Great Kanto Earthquake of 1923, and during the fire-bombings of World War II. These two episodes are crucial for understanding the development of Tokyo, in part because in each case the city was transformed in the process of rebuilding, and in part because these disasters provide insights into major contemporary risks and challenges for the city.

The Great Kanto Earthquake struck just before noon on 1 September 1923. Between 100,000 and 140,000 people were crushed or incinerated, and some 44 per cent of the urban area of Tokyo was completely destroyed. As Seidensticker (1991) laments, this marked the final disappearance of the built form of the feudal city of Edo, with the destruction of most remaining pre-modern buildings and the transformation of the cityscape during reconstruction. The areas worst affected were the Low City commercial and industrial areas, primarily because they were built at much higher densities than the residential areas of the High City and burned more readily. A vast area of the capital was utterly devastated and the major investment in reconstruction made over the next 8 years put a strain on the entire national budget and financial system. As anyone who has studied urban disaster recovery knows, it is usually very difficult to accomplish a significantly different urban plan during reconstruction, as the priority is often to rebuild in the same pattern in order to provide new homes and business spaces quickly. In Tokyo, however, a major redesign of the central area was undertaken, creating a new hierarchy of broad avenues, substantial commercial streets, and smaller streets and lanes. New bridges, parks and schools were also built, with a number of innovative features including fireproof concrete schools with emergency shelter areas in the playgrounds and pocket parks at each of the hundreds of new bridges.

Two problematic outcomes of reconstruction are important here. First, to allow rapid rebuilding of housing and businesses during reconstruction, the requirements for fireproof buildings were relaxed, initially temporarily but then with repeated extensions. This resulted in huge areas of these flammable buildings being consumed by fire during World War II bombing. Second, all the carefully prepared plans for suburban land development projects and development regulation schemes proposed under the 1919 Planning Act had been lost in the fires which destroyed Tokyo City Hall and the Home Ministry in the aftermath of the earthquake. Worse, in the months after the disaster, suburban development occurred too rapidly, resulting in many people made homeless by the earthquake moving to new areas outside the city, creating another vast belt of unplanned sprawl. This area of high-density wooden housing, with very narrow streets, few parks and few firebreaks, has been a key concern again since the 1980s, as Tokyo anticipates its next great earthquake (see figure 2.8). Historically the cycle of major quakes in the Tokyo region has been about every 70 years, so a major disaster has been expected since the mid 1990s.

Tokyo suffered massive destruction again during the incendiary bombing raids of 1945, with most of the built-up area burned and some 750,000 houses destroyed.

Once more, the commercial and industrial areas of the Low City were the most heavily damaged. The Tokyo Metropolitan Government prepared a very idealistic reconstruction plan to transform central Tokyo, especially those areas that had not been modernized after the Great Kanto earthquake. Redevelopment projects were planned for 20,000 hectares, while the burned area was 'only' 16,000 hectares. The intention was to create broad avenues, large parks and green corridors. These extensive open spaces could also serve as firebreaks, a proposal that made great sense given that, within the space of 25 years, central Tokyo had been burned to the ground twice.

But compared with the reconstruction effort after the earthquake, little urban restructuring was accomplished. Of the 20,000 hectares planned, only 1,380, or 6.8 per cent of the reconstruction projects, were completed (Sorensen, 2002). Most of these were at main stations along the Yamanote Line such as Shinjuku, Shibuya and Ikebukuro, and were designed to improve the railway station plazas and create better connections between Japan Rail (JR) lines, private railway lines, streetcars and buses. Most of Tokyo, however, was rebuilt quickly along the old pattern, and many areas so redeveloped are now considered at high risk from earthquake and fire.

By the 1950s the key characteristics of Japanese planning were firmly established, namely a high degree of centralization of policy and administrative power in central government, and an enduring focus of government policy on economic growth, often at the expense of urban liveability. The pervasive role of the state is a distinctive feature of Japanese urbanization, especially in Tokyo, which may seem odd for a city that appears to most first-time visitors to be profoundly unplanned and chaotic (see Shelton, 1999). As noted above, Tokyo was originally planned as a military city, with the priority placed on defence. During the Edo period the Shogun's government had extraordinary powers over patterns of activity and development (Kato, 1994; Sorensen, 2002). In the Meiji period the urban project was to modernize, industrialize, and create a prestigious capital city. After the Great Kanto Earthquake in 1923, and again after World War II, the central government took direct control of planning and reconstruction. In the rapid growth period and since, the state has focused on the planning and building of the large-scale infrastructure that has shaped overall patterns of growth – particularly the Tokyo Bay landfills for the port and industrial complexes, and the railways, expressways, bridges and airports that have facilitated economic and physical growth. These vast projects and investments established Japan's reputation as the archetypal 'developmental state' or 'construction state' (*doken kokka*), and also decisively shaped the growth of the capital city. At the same time relatively few constraints were placed on private urban investment or land development, so even with vast state infrastructure investments, Tokyo continues to display a rather haphazard and unplanned look. More important, most residential areas were developed in an unplanned manner, with little government investment or even regulation. These areas pose some of the main challenges facing the Tokyo region today.

From Rapid Economic Growth to Post-industrial Tokyo

Contemporary Tokyo, although built on the foundations of the pre-war period, and shaped by the institutions developed then, is mostly a product of two decades of rapid post-war economic growth from 1954 to 1973, another two decades of moderate growth from about 1973 to 1990, and the following two decades of recession and low growth. Throughout this period the construction of new urban areas, of public and private infrastructure, and of buildings on redevelopment sites, has been a major industry, especially during the low economic growth period of recent decades. As a result, virtually all buildings in Tokyo have been constructed during the last half-century, with many being replaced on one or more occasions. There is no space to trace in detail all the urban changes in the Tokyo region in this period, so the focus here is limited to two major aspects of urbanization and growth. The first is the continued dominance of the 'developmental state', with its politicized public works and infrastructure spending, which has skewed urban priorities and planning solutions towards large-scale engineering approaches and grandiose schemes, rather than towards systematic planning and careful regulation of development. The second is the distinctive pattern of suburban sprawl and substandard suburban development that is characteristic of Tokyo.

The political economy of the Japanese 'developmental state' model has been extensively studied, and it is unnecessary here to do more than summarize the main features. The core of the model is: a highly centralized government structure and weak local governments; weak influence of electoral politics on government policy formation, which is instead entrusted to a relatively autonomous bureaucracy that is to some degree insulated from political pressures; the prioritization of economic growth to the detriment of other goals; and a reliance on economic success as the primary basis of state legitimacy (Johnson, 1982; Cumings, 1987; Deyo, 1987; Gao, 1997; Woo-Cumings, 1999). The enduring links between the ruling Liberal Democratic Party, central government bureaucracy and big business are referred to as 'the iron triangle' and continue to be influential (McCormack, 2002; Feldhoff, 2007). This configuration was particularly powerful in the post-war 'rapid growth' decades – a 'conservatives' paradise', in which a powerful consensus prevailed that post-war rebuilding, economic growth, and particularly industrial expansion, were necessary and desirable (Samuels, 1983, p. 168).

The state played a key role in facilitating industrial expansion by allocating foreign exchange, arranging loans, promoting capital formation and plant investment, and ensuring an adequate supply of industrial land, electricity and water (Johnson, 1982; Deyo, 1987). In the Tokyo region, the major investments were in creating new integrated industrial areas on landfills in Tokyo Bay, with associated road and rail links, port facilities, highways, river engineering, storm-water management, and electricity and water supply. Samuels (1983), in his case study of the development plans for the Tokyo region, details the multiple ties linking the major ministries (Construction, Transport, and International Trade and Industry) with business lobbyists, major industrial groups, and prefectural and municipal governments. Ambitious plans were formulated to fill in two-thirds of the 12,000 square kilometre Tokyo Bay for a new airport, new Tokyo Station, industrial, residential and commercial areas, and with expressways to link everything together. Although many of the major projects were uncontroversial, some, like Narita Airport, elevated expressways and the new Shinkansen high-speed rail lines, attracted heated protests when they were being built (McKean, 1981; Samuels, 1983; Apter and Sawa, 1984; Hood, 2006). Such protests seldom had much influence, however, as most major decisions were made among elite actors long before plans were made public.

It is during the rapid growth period that the dualistic aspects of urbanization in Japan became strongly apparent. On the one hand, the central government poured money into ambitious large-scale infrastructure projects, and on the other hand, as mentioned above, there was a vast amount of uncontrolled, unplanned residential land development with few public services such as local roads, community centres, libraries or parks (Tsuru, 1999; Hasegawa, 2004). While 41 per cent of the public works budget was allocated to roads, harbours and airports in 1960, and 49.9 per cent in 1970, the percentage devoted to housing and sewer systems was 5.7 per cent in 1960 and 11.2 per cent in 1970 (Yamamura, 1992, p. 48). Careful planning and huge investments in infrastructure to aid economic growth occurred at the same time as extremely tight budgets for social infrastructure and a *laissez faire* attitude to living conditions in residential areas, with few regulations put in place to limit pollution of air, water and soil.

The results of this approach were mixed in their effects on urban structure. On the one hand, the huge investment in growth-oriented infrastructure projects undoubtedly contributed to economic development and mobility in the Tokyo region. On the other hand, lack of investment in 'social overhead capital' contributed to the 'rich Japan, poor Japan' problem of low living standards, degraded urban environments and a severe pollution crisis in the 1960s. There were major investments in transportation systems, including the construction of a large-scale elevated expressway system, new high-speed railway links to Osaka and later Tohoku and the Japan Sea, new subways and railway lines, and the major new industrial areas on landfills in Tokyo Bay. The water supply and sewer systems were rapidly extended, although until the 1990s the sewers were still expanding more slowly than the urban area. While overseas visitors often admire the lavish investment in infrastructure, and especially transportation systems, to aid economic development (Cervero, 1998; Mosk, 2001), Japanese commentators are more likely to notice the associated accumulation of government debt, the failure to invest in social infrastructure, and the spread of low-quality living environments (Hayakawa and Hirayama, 1991; Tsuru, 1999; Hasegawa, 2004).

Thus, the focus on growth-oriented infrastructure in the Tokyo region has resulted in a neglect of the infrastructure that supports liveability. The problem was not just a lack of spending, however, but also a failure to regulate private land and housing development, allowing the continuing spread of unplanned residential areas.[1] Weak regulation of private development, high housing prices and a failure by the government to invest in residential areas has meant that huge areas were developed in an entirely haphazard manner with tiny lots, without adequate basic infrastructure of roads and sewers, and without minimum social facilities, such as parks, libraries and community centres. As mentioned above, these unplanned residential areas have long been seen as some of the most vulnerable, dangerous and problematic parts of Tokyo. In the area shown in figure 2.6, the parts shaded in black have substandard housing plots of 80 square metres, which front primarily on to small dead-end streets with few connections. Virtually all streets are extremely narrow. Failure to plan or

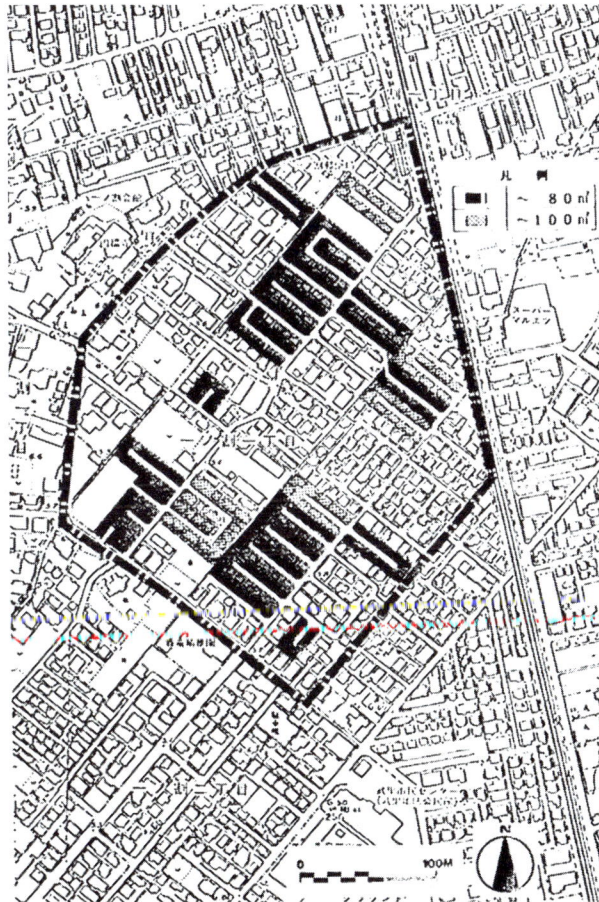

Figure 2.6. Unplanned sprawl areas north of Tokyo.

regulate suburban development has also meant that other major infrastructure, such as sewers, has had to be built retroactively at great cost, while storm-water systems must be oversized because hard surfaces are ubiquitous.

This approach to urban planning and urban infrastructure investment was a major cause of the environmental crisis in Japan that came to a head in the 1960s, with hundreds of deaths and tens of thousands of people made seriously ill, as pollution emissions expanded rapidly and housing was still being built in industrial zones (Ui, 1992). By the second half of the decade worsening pollution contributed to the growth of environmental movements which blamed the central government for the pollution crisis and lack of investment in community infrastructure (George, 2001). This in turn contributed to major gains by the opposition socialist and communist parties, particularly in local government elections, where progressive parties were able to elect mayors and governors in most of the cities and prefectures in the metropolitan areas (Krauss and Simcock, 1980; MacDougall, 1980; McKean, 1981). Of all environmental complaints, those about vibrations, bad smells and noise consistently topped the list, and all were a result of poor land-use planning. Although many of the new progressive local administrations worked hard on environmental concerns, their land-use planning powers were relatively weak, and they were largely unable to limit the continued spread of new areas of unplanned sprawl.

Three developments since the time of the environmental crisis are important here. The first is the bubble economy of the 1980s, the second has to do with the long period of economic stagnation following its collapse, and the third has been the emergence of large numbers of small-scale, citizen-initiated projects to improve local living environments, referred to as '*machizukuri*' (which literally translates as 'community development' or 'town-making'). During the 1980s Japan experienced an extraordinary asset inflation bubble, fuelled in part by the huge exporting success of Japanese corporations, in part by easy credit designed to stimulate domestic demand in response to US pressure over the trade imbalance between the two countries, and in part by speculative investment in urban land and real estate. The results were equally diverse and included a vast inflation of land values, an accelerated spread of development ever further from central Tokyo in efforts to find affordable housing land, and a severe economic crash in the early 1990s followed by stagnation which persists to the present. The Japanese financial system was pushed near to collapse by large numbers of bad real estate loans made during the bubble economy period. Since then central government has sought to encourage construction investment as a stimulus to counteract recession. Such efforts have been two pronged, including both massive government investment in infrastructure and construction projects, and a series of deregulations of the planning system, designed to stimulate private sector investment in redevelopment.

The government's economic stimulus investments have been widely criticized (Woodall, 1996; McGill, 1998; Feldhoff, 2007) as reinforcing the 'iron triangle' while creating a deepening dependency on public works. As McCormack (2002, p. 11) argues,

Japan's public works sector has grown to be three times the size of that of Britain, the US or Germany, employing 7 million people, or 10 per cent of the workforce, and spending between 40 and 50 trillion yen a year – around $350 billion, 8 per cent of GDP or two to three times that of other industrial countries. Naturally there have been short-term benefits for many, not least in terms of soaking up unemployment during the long 1990s recession. Gradually, however, public-works infrastructure has been replaced by 'extrastructure' – developments undertaken for their own sake, while the collusive alliance at the system's core has corrupted both politics and society.

Just as problematic have been the central government's moves to encourage private development investment by selectively weakening the national building code, and by removing normal municipal planning controls in specified areas of central Tokyo, in order to speed up the process of approval of large-scale redevelopment projects. As examined in detail in earlier studies (Fujii, 2005; Fujii *et al.*, 2007; Sorensen *et al.*, 2009), during the 1990s the Ministry of Construction (now the Ministry of Land, Infrastructure and Transport) weakened regulations controlling the permitted floor area, height and permissible building envelope of buildings in cities in order to permit much larger buildings on existing sites These changes were accomplished not by introducing legislation, but by changing the way the Building Standards Law was interpreted and administered. For example, a major change was the decision not to count the area used for elevators, lobbies and corridors as a part of the total floor area of large condominium buildings. Another major series of changes was to the way the permitted building envelope was calculated, in order to allow much taller buildings on sites that front on to narrow roads. These changes have had major impacts on built form. In one famous case, a 10-storey building, built in the early 1990s to the maximum size then permitted, was demolished in 2002 and replaced with a 30-storey building that took advantage of the new regulations (Sorensen *et al.*, 2009).

As can be imagined, the sudden appearance of 30-storey buildings in low-rise residential areas has provoked bitter conflicts and vigorous opposition movements (Fujii *et al.*, 2007). Few of these have been successful in preventing redevelopment, however, and during the last 15 years Tokyo has seen an extraordinary transformation, from a largely low-rise and mid-rise built form to a predominance of high-rise buildings over large expanses of the central city. This has reversed the long decline of the central city population, as intended, but increasing density in the central areas of Tokyo may not be good policy. Unlike many other developed country cities, Tokyo already has extremely high central area densities and increased burdens on an already overworked infrastructure may be very costly and may exacerbate problems in the event of a major disaster (Onishi, 1994).

More widespread and much more important than the movements in opposition to high-rise intensification are the many groups, projects and activities that seek to improve local environments and liveability in Japanese settlements. The *machizukuri* (mentioned above) represent a self-help approach to working on neighbourhood-scale liveability issues. Drawing on the famous examples of the Maruyama and Mano neighbourhoods in Kobe (Watanabe, 2007) and Taishido in Tokyo (Sorensen *et al.*, 2009), where sophisticated community movements worked to improve their neighbourhoods, *machizukuri* activisim spread throughout the country in the 1990s. *Machizukuri* includes a very wide range of activities and organizational structures, including both bottom-up citizens' organizations and more top-down projects initiated by local governments, but the essential characteristic is the voluntary participation of residents in managing and ameliorating neighbourhood scale environmental quality. Typical activities engaged in by *machizukuri* groups include the building and management of local parks and community gardens, children's play areas and small green spaces, historical preservation campaigns, community centre building, environmental remediation efforts and local disaster response preparations (Watanabe, 1999; Hein, 2001; Murayama, 2007; Sorensen *et al.,* 2009).

Although the devolution of planning powers and responsibilities under the revised City Planning Law of 1992 and the major transfer of legal authority for planning to local governments in 2000 allowed increased scope for local initiatives and greatly raised expectations, *machizukuri* processes are still important because the massive local borrowing to help finance central government infrastructure projects during the 1990s left local governments deeply in debt and little able to finance new initiatives and projects (Schebath, 2006).

In a recent paper John Friedmann (2010) argues for the central importance of place and place-making in cities today, and advocates a more positive approach by planners in supporting efforts to enhance neighbourhood-scale shared places, and to protect them from the powerful forces of erasure and dislocation. Friedmann provides examples of three cases of place-making in Japan, China, and Canada and notes the importance of the institutional frameworks that make successful citizen engagement in place-making possible. Although a wide variety of neighbourhood organizations exists in cities around the world, their capacity to influence patterns of urban change is highly uneven. In most cases, developers, governments and planners are happy to disregard the priorities of neighbourhood organizations in pursuit of other interests, and institutional links between the local state and neighbourhood organizations have very often been designed to ensure that municipal priorities and interests are communicated to the neighbourhood, not the reverse.

The crucial social innovation of *machizukuri* was the adaptation of existing institutional structures for neighbourhood self-governance, the traditional neigh-bourhood associations of *chonaikai* and *jichikai* (see Dore, 1958; Bestor, 1989), to create more democratic, inclusive and autonomous groups. This was a complex process, lasting several decades. The environmental crisis and the collapse of the economy were key moments of de-legitimation of existing institutions and, over time, new institutional forms of place-based organizing have become established and widely disseminated (see Sorensen, 2007). Two key innovations were a move from the traditional system of household membership in neighbourhood associations to individual membership in *machizukuri* groups, allowing routine membership of women and young adults instead of household representation by the eldest male, and a formalization of majority voting instead of the traditional 'consensus' decision making that had allowed most decisions to be made by small leadership groups. These changes have facilitated the emergence of more independent groups which do not automatically follow the instructions of local government, and have fought against place erasure, as well as for place values. In many cities in Asia, it is precisely the lack of institutional models and frameworks for autonomous local groups with both organizational and political capacity that is the greatest hindrance to place-making (Sorensen and Sagaris, 2010). *Machizukuri* in Tokyo are important in providing a setting and sense of legitimacy for citizen engagement in place-making and community development, which would otherwise be almost completely lacking.

Conclusions: Major Contemporary Challenges for Tokyo

The Tokyo region faces many major challenges today. Of these, probably the most prominent are earthquake risks, economic problems related to the lack of growth, huge government debts, an ageing and declining population, and growing problems of homelessness and poverty. This concluding section focuses on three issues that have a distinctly spatial manifestation, and that are deeply linked to historic patterns of planning and development. The first is the enduring divide between the upland High City and the Low City in the floodplains of the Sumida, Edogawa and Nakagawa rivers, most of which is now at or below the mean high-tide level. Second is the continued existence and even growth of highly vulnerable areas of substandard housing, a situation that contributes to the serious risks facing large segments of the Tokyo population. Third is the huge pressure to redevelop and intensify land use in the central areas of the city. This is, in large part, a product of the recent changes in planning regulations noted above, which are designed more to create profitable opportunities for redevelopment than to address existing urban challenges and risks.

The continuing spatial divide between the Low City in the east and the High City to the west of the Imperial Palace is a surprisingly enduring characteristic of Tokyo. Dating back to the feudal era policy of class-based separation of the city into different districts for samurai and commoners, and powerfully reinforced by the early zoning of the city into a 'residential' zone in the salubrious former samurai areas, and commercial and industrial zones in the areas formerly inhabited by commoners, these two halves of Tokyo continue to demonstrate very different urban form characteristics and problems. Apart from the zoning, a major factor has been that it was primarily the low-lying commoner areas that were destroyed in the great earthquake of 1923, and that were rebuilt with a grid of wide roads. This has permitted much higher built densities in these areas than in the upland areas with their very narrow roads.

The industrial areas continue to have an extraordinary mixture of factories, housing, office, retail and warehouse uses, with almost every urban block containing all of those uses. During the environmental crisis of the post-war period, these were the areas that suffered the worst effects of uncontrolled pollution emissions, but much stricter pollution controls since 1970 have greatly reduced such problems. In many ways these are extremely successful examples of the sort of low-rise, high-density, mixed-use areas promoted by advocates of urban sustainability.

They are, however, still vulnerable to a major ongoing risk that is a by-product of the unregulated industrialization of the rapid growth period: most of the lowland area of Tokyo, with a population of 2.5 million, is built on land that is up to 2 metres below the mean high tide level (see figure 2.7). Excessive groundwater pumping by factories from the aquifer below the delta from the 1940s to the mid 1960s resulted in widespread subsidence of ground levels. Although strictly enforced regulations have prohibited the removal of groundwater since the mid 1960s, and subsidence has stopped, huge areas are still vulnerable to flooding. Dykes, water gates and locks protect these areas, but Tokyo's heavy rains must be pumped out over the dykes, so these areas are always at risk of floods from a failure of the emergency pumps, from a surge of river or sea water levels during a storm, or from the failure of a dyke (as happened in New Orleans). As sea levels are expected to rise considerably over the next century, this presents a significant ongoing risk for a major part of the Tokyo population. The unstable alluvial soils of the delta area are also considered to present greater risks in the event of earthquakes, a factor that may amplify the risk of floods.

The second major vulnerability of Tokyo today is a consequence primarily of the weak planning regulations in place during the first half of the twentieth century, and the hasty rebuilding after World War II. As shown in figure 2.8, just beyond the Yamanote railway line there is a ring of suburban residential areas with very high population densities, and a high proportion of dwellings that are constructed of wood. These are almost all areas that were initially suburbanized after the 1923 earthquake, in a wholly unplanned fashion, and tend to have a substandard road infrastructure, with many houses on very narrow lanes and paths that even the tiniest car cannot enter. Some of these areas are charming living environments, with lots of small houses, potted plants in the alleyways, and largely pedestrian and bicycle traffic in the streets. The problem is that they are also at extremely high risk of unstoppable fires in the event of earthquakes, as few houses are fireproofed and they stand within inches of each other in most cases. In the Kobe earthquake of 1995 virtually all fatalities were in

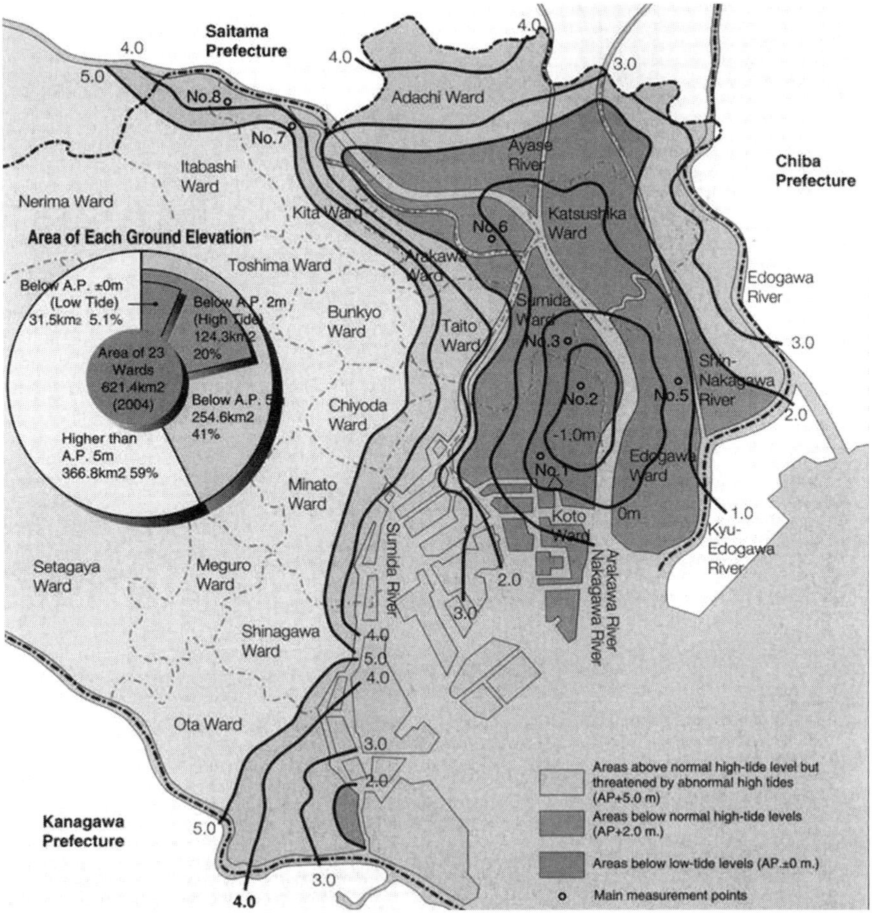

Saitama Prefecture

4.0
5.0
4.0
4.0
3.0

No.8

Adachi Ward

No.7

Ayase River

Chiba Prefecture

Itabashi Ward

Nerima Ward

Kita Ward

Katsushika Ward

Area of Each Ground Elevation

Edogawa River

Toshima Ward

No.6

Arakawa Ward

Below A.P. ±0m (Low Tide) 31.5km2 5.1%

Below A.P. 2m (High Tide) 124.3km2 20%

Bunkyo Ward

Sumida Ward

Taito Ward

Area of 23 Wards 621.4km2 (2004)

No.3

3.0

Below A.P. 5 254.6km2 41%

Chiyoda Ward

No.2

No.5

Shin-Nakagawa River

Higher than A.P. 5m 366.8km2 59%

-1.0m

2.0

No.1

Edogawa Ward

Minato Ward

Edogawa River

Setagaya Ward

Meguro Ward

Koto Ward

0m

1.0

Kyu-Edogawa River

Sumida River

Arakawa River Nakagawa River

2.0

Shinagawa Ward

4.0
5.0
4.0

3.0

Ota Ward

3.0

2.0

Areas above normal high-tide level but threatened by abnormal high tides (AP+5.0 m)

Kanagawa Prefecture

5.0

Areas below normal high-tide levels (AP+2.0 m.)

Areas below low-tide levels (AP.±0 m.)

3.0

4.0

o Main measurement points

Figure 2.7. Tokyo below sea level.

areas just like these, as collapsed buildings blocked escape routes and residents were trapped, unable to flee the advancing flames. While the risk of another great earthquake is well known, the very uneven geography of vulnerability in the city is less frequently mentioned.

The Tokyo Metropolitan Government has, since the early 1980s, identified these Low City areas as a major target for disaster prevention investment, but progress has been slow. Initial plans to bulldoze the areas and replace them with wide boulevards and high-rise towers in the modernist style were vigorously opposed by residents, so an incremental approach has been pursued instead, with investment put into building emergency escape routes, fireproofing buildings and installing water cisterns in case of disaster.

The third major risk and vulnerability identified here is rather different, as it is not associated with potential earthquakes or other natural disasters. Rather, it is one

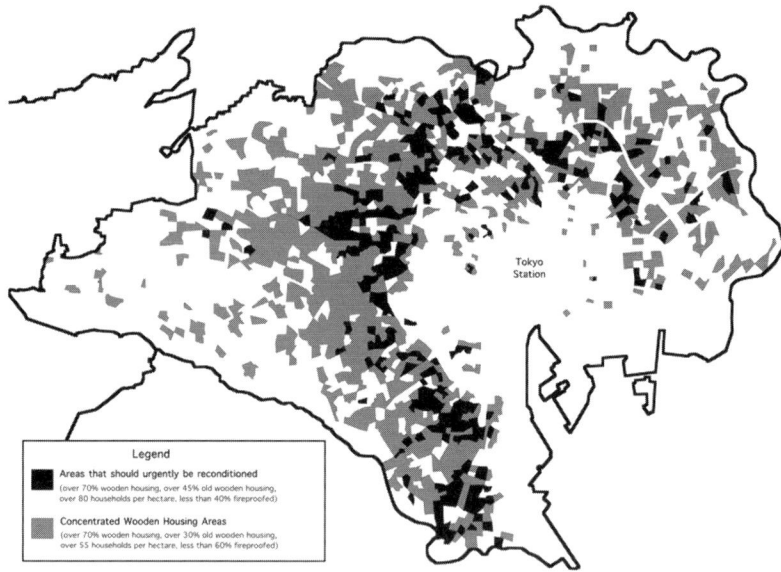

Figure 2.8. High-density wooden buildings areas, 1997.

that is entirely linked to weaknesses of the planning system, and the tendency to use urban regulation as a tool of economic policy. While the prevailing wisdom today is that higher population densities are associated with greater urban sustainability, it is not at all clear that this makes sense for Tokyo, with its already very high densities. It is not known precisely at what density the benefits of greater concentration give way to the disbenefits of overcrowding, and that density is likely to be very different in different locations, different cultures and even for different individuals. But the current policy of encouraging indiscriminate intensification through the redevelopment of low- and mid-rise neighbourhoods into super high-rise condominium buildings of over thirty storeys seems very likely to create more problems than it solves (see figure 2.10). The problem is not associated with vulnerability to earthquakes – most new high-rises are probably safer in the event of earthquakes than the buildings they are replacing. But these super high-rises tend to be much more energy intensive than low- and mid-

Figure 2.9. Shimokitazawa in Tokyo's wooden housing belt.

rise buildings, so the wholesale replacement of Tokyo's low- and mid-rise buildings with super high-rises seems more likely to reduce local resilience in the face of disaster rather than to increase it. Also, many of these new high-rises are built within a metre or two of 2-storey residences, so the immediate neighbours are understandably concerned about loss of sunlight, wind funnel effects, falling objects and increased traffic. Protests have been widespread but, as noted earlier, largely ineffective (Fujii *et al.*, 2007).

The current incentives to rebuild in the form of high-rise buildings have sparked a wave of redevelopment and intensification throughout the Tokyo region, but there is no overall planning framework to guide intensification, and few areas are protected from such redevelopment. A major risk of this approach is that overbuilding will create more vulnerable populations in central areas, requiring ever larger infrastructure investments to keep them supplied with water, wastewater treatment and energy, while the characteristic resilience and flexibility of the low- and mid-rise urban forms they replace are diminished.

Tokyo is such a vast and complex city-region that a brief survey cannot do justice to its multiple strengths and weaknesses. What is clear, however, is that the institutional structure of government has consistently worked to leverage Tokyo's centrality as a vehicle for land development profits, with less priority given to dealing with the risks that the city faces from earthquakes, floods and other catastrophic events. Until this dynamic changes, it is reasonable to expect that Tokyo will remain highly vulnerable to natural disasters.

Figure 2.10. Recent high-rise tower development in Tokyo.

Postscript

As this book went to press, the tragic events of March 2011 – the huge earthquake, the resulting tsunami which caused appalling destruction and loss of life, and the explosions at the Fukushima Daiichi nuclear power plant north of Tokyo – focused the attention of the world on the vulnerability of Japan's cities. It is premature for any detailed assessment of this tragedy, but a few points already seem obvious. First, the large investments of recent decades in careful earthquake engineering of buildings in cities have paid off. Despite the strongest earthquake in Japanese history, there were very few building failures, and almost none of newer buildings. Second, the terrifying images of the tsunami showed the vulnerability of low-lying coastal communities to sudden huge rises in sea levels and gigantic waves. In areas prone to tsunamis, settlements must be sited on higher ground. Third, the failure of multiple backup systems at the nuclear power plants suggests that, if such plants are to be located on the coast, as virtually all in Japan are, they must be much better protected. Finally, one can only hope that the 2011 Sendai disaster will prompt early and significant measures to better protect the extensive areas of Tokyo which are currently at or below mean high tide levels, as discussed in this chapter. In Japan major earthquakes and tsunamis in future are inevitable.

Note

1.   The term 'sprawl' (*supurôru*) is applied to these areas, but they are not the same as North America's sprawling suburbs. Tokyo's unplanned suburbs have been the subject of a number of studies (see Hanayama, 1986; Hebbert, 1994; Mori, 1998; Sorensen, 2001).

References

Apter, D.E. and Sawa, N. (1984) *Against the State*. Cambridge, MA: Harvard University Press.

Beasley, W. G. (1995) *The Rise of Modern Japan*. New York: St. Martin's Press.

Bestor, T.C. (1989) *Neighborhood Tokyo*. Stanford, CA: Stanford University Press.

Cervero, R. (1998) *The Transit Metropolis: A Global Inquiry*. Washington DC: Island Press.

Cumings, B. (1987) The origins and development of the Northeast Asian political economy: industrial sectors, product cycles, and political consequences, in Deyo, F.C. (ed.) *The Political Economy of the New Asian Industrialism*. Ithaca, NY: Cornell University Press, pp. 44–83.

Deyo, F.C. (ed.) (1987) *The Political Economy of the New Asian Industrialism*. Ithaca, NY: Cornell University Press.

Dore, R.P. (1958) *City Life in Japan – A Study of a Tokyo Ward*. London: Routledge and Kegan Paul.

Feldhoff, T. (2007) Japan's construction lobby and the privatization of highway-related public corporations, in Sorensen, A. and Funck, C. (eds.) *Living Cities in Japan: Citizens' Movements, Machizukuri and Local Environments*. London: Routledge.

Friedmann, J. (2010) Place and place-making in cities: a global perspective. *Planning Theory and Practice*, **11**(2), pp. 149–165.

Fujii, S. (2005) The mechanism of Manshon Conflicts and the View for a New Planning System *(Manshon funsô no kôzô to kiseishigaichi kôshin kontorôru shuhô ni kansuru kenkyû)*. Tokyo: Unpublished PhD thesis, Department of Urban Engineering, University of Tokyo.

Fujii, S., Okata, J. and Sorensen, A. (2007) Inner-city redevelopment in Tokyo: conflicts over urban place, planning governance, and neighborhoods, in Sorensen, A. and Funck, C. (eds.) *Living Cities in Japan: Citizens' Movements, Machizukuri and Local Environments.* London: Routledge.

Fujimori, T. (1982) *Tokyo Planning in the Meiji Period(Meiji no Tokyo Keikaku).* Tokyo: Iwanami Shoten.

Fujita, K. and Hill, R.C. (2007) The zero waste city: Tokyo's quest for a sustainable environment. *Journal of Comparative Policy Analysis,* **9**(4), pp. 405–425.

Gao, B. (1997) *Economic Ideology and Japanese Industrial Policy: Developmentalism from 1931 to 1965.* Cambridge: Cambridge University Press.

George, T.S. (2001) *Minamata: Pollution and the Struggle for Democracy.* Cambridge, MA: Harvard University Press.

Hanayama, Y. (1986) *Land Markets and Land Policy in a Metropolitan Area: A Case Study of Tokyo.* Boston, MA: Oelgeschlager, Gunn and Hain.

Hanley, S. (1997) *Everyday Things in Premodern Japan: The Hidden Legacy of Material Culture.* Berkeley, CA: University of California Press.

Hasegawa, K. (2004) *Constructing Civil Society in Japan: Voices of Environmental Movements.* Melbourne, Vic: Trans-Pacific.

Hatano, J. (1994) Edo's water supply, in McClain, J.L., Merriman, J.M. and Ugawa, K. (eds.) *Edo and Paris: Urban Life and the State in the Early Modern Era.* Ithaca, NY: Cornell University Press.

Hayakawa, K. and Hirayama, Y. (1991) The impact of the minkatsu policy on Japanese housing and land use. *Environment and Planning D: Society and Space,* **9**, pp.151–164.

Hayami, A. (1986) Population changes, in Jansen, M.B. and Rozman, G. (eds.) *Japan in Transition: From Tokugawa to Meiji.* Princeton, NJ: Princeton University Press.

Hebbert, M. (1994) Sen-biki amidst Desakota: Urban sprawl and urban planning in Japan, in Shapira, P., Masser, I. and Edgington, D.W. (eds.) *Planning for Cities and Regions in Japan.* Liverpool: Liverpool University Press.

Hein, C. (2001) Toshikeikaku and Machizukuri in Japanese urban planning: the reconstruction of inner city neighborhoods in Kobe. *Japanstudien:Jahrbuch des Deutschen Instituts fur Japanstudien der Philipp Franz von Siebold Stiftung,* **13**, pp. 221–252.

Hood, C. (2006) From polling station to political station? Politics and the shinkansen. *Japan Forum,* **18**(1), pp. 45–63.

Jinnai, H. (1990) The spatial structure of Edo, in Nakane, C. and and Oishi, S. (eds.) *Tokugawa Japan: Social and Economic Antecedents of Modern Japan.* Tokyo: University of Tokyo Press.

Jinnai, H. (1994) Tokyo, a Model for the 21st Century? Paper presented to the European Association of Japanese Studies conference, August.

Johnson, C.A. (1982) *MITI and the Japanese Miracle, the Growth of Industrial Policy, 1925–1975.* Stanford, CA: Stanford University Press.

Kato, T. (1994). Governing Edo, in McClain, J. L., Merriman, J. M. and Ugawa, K. (eds.) *Edo and Paris: Urban Life and the State in the Early Modern Era.* Ithaca, NY: Cornell University Press.

Kelly, W.W. (1994) Incendiary actions: fires and firefighting in the shogun's capital and the people's city, in McClain, J.L., Merriman, J.M. and Ugawa, K. (eds.) *Edo and Paris: Urban Life and the State in the Early Modern Era.* Ithaca, NY: Cornell University Press.

Krauss, E.S. and Simcock, B. (1980) Citizens' movements: the growth and impact of environmental protest in Japan, in Steiner, K., Kraus, K.E. and Flanagan, S. (eds.) *Political Opposition and Local Politics in Japan.* Princeton, NJ: Princeton University Press.

Kurasawa, S. (1986) *Social Atlas of Tokyo.* Tokyo: University of Tokyo.

MacDougall, T.E. (1980) Political opposition and big city elections in Japan, 1947–1975, in Steiner, K., Kraus, K.E. and Flanagan, S. (eds.) *Political Opposition and Local Politics in Japan.* Princeton, NJ: Princeton University Press.

McCormack, G. (2002) Breaking the Iron Triangle. *New Left Review,* **13**(1), pp. 5–23.

McGill, P. (1998) Paving Japan – the construction boondoggle. *Japan Quarterly,* **45**(4), pp. 39–48.

McKean, M. (1981) *Environmental Protest and Citizen Politics in Japan.* Berkeley, CA: University of California Press.

Mori, H. (1998) Land conversion at the urban fringe: a comparative study of Japan, Britain and the Netherlands. *Urban Studies,* **35**(9), pp. 1541–1558.

Mosk, C. (2001) *Japanese Industrial History: Technology, Urbanization, and Economic Growth.* Armonk, NY: M.E. Sharpe.

Murayama, A. (2007) Civic movement for sustainable urban regeneration: downtown Fukaya city,

Saitama prefecture, in Sorensen, A. and Funck, C. (eds.) *Living Cities in Japan: Citizens' Movements, Machizukuri and Local Environments.* London: Routledge.

Nishiyama, M. (1997) *Edo Culture: Daily Life and Diversions in Urban Japan, 1600–1868.* Honolulu: University of Hawai'i Press.

Noguchi, K. (1988) Construction of Ginza Brick Street and conditions of landowners and house owners, in Ishizuka, H. and Ishida, Y. (eds.) *Tokyo: Urban Growth and Planning 1868–1988.* Tokyo: Center for Urban Studies.

Okamoto, S. (2000) Destruction and Reconstruction of Ginza Town, in Fukui, N. and Jinnai, H. (eds.) *Destruction and Rebirth of Urban Environment.* Tokyo: Sagami Shobo Publishing.

Okata, J. and Murayama, A. (2010) Tokyo's urban growth, urban form, and sustainability, in Sorensen, A. and Okata, J. (eds.) *Megacities: Urban Form, Governance and Sustainability.* Tokyo: Springer Verlag.

Onishi, T. (1994) A capacity approach for sustainable urban development: an empirical study. *Regional Studies,* **28**(1), pp. 39–51.

Rozman, G. (1973) *Urban Networks in Ch'ing China and Tokugawa Japan.* Princeton, NJ: Princeton University Press.

Samuels, R. J. (1983) *The Politics of Regional Policy in Japan: Localities Incorporated?* Princeton, NJ: Princeton University Press.

Sassen, Saskia (2001) *The Global City: New York, London, Tokyo.* Princeton, NJ: Princeton University Press (2nd edition).

Schebath, A. (2006) Fiscal stress of Japanese local public sector in the 1990s: situation, structural reasons, solutions, in Hein, C. and Pelletier, P. (eds.) *Cities, Autonomy and Decentralization in Japan.* London: Routledge.

Seidensticker, E. (1991) *Low City, High City. Tokyo from Edo to the Earthquake: how the shogun's ancient capital became a great modern city 1867–1923.* Cambridge, MA: Harvard University Press.

Shelton, Barrie (1999) *Learning from the Japanese City:West meets East in Urban Design.* London: E & FN Spon.

Smith, H.D. (1978) Tokyo as an idea: an exploration of Japanese urban thought until 1945. *Journal of Japanese Studies,* **4**(1), pp. 45–80.

Smith, H.D. (1979) Tokyo and London: comparative conceptions of the city, in Craig, A.M. (ed.) *Japan: A Comparative View.* Princeton, NJ: Princeton University Press.

Sorensen, A. (2001) Building suburbs in Japan: continuous unplanned change on the urban fringe. *Town Planning Review,* **72**(3), pp. 247–273.

Sorensen, A. (2002) *The Making of Urban Japan: Cities and Planning from Edo to the 21st Century.* London: Routledge.

Sorensen, A. (2007) Changing governance of shared spaces: *Machizukuri* in historical, institutional perspective, in Sorensen, A. and Funck, C. (eds.) *Living Cities in Japan: Citizens' Movements, Machizukuri and Local Environments.* London: Routledge.

Sorensen, A. (2010) Urban sustainability and compact cities ideas in Japan: the diffusion, transformation and deployment of planning concepts, in Healey, P., and Upton, R. (eds.) *Crossing Borders: International Exchange and Planning Practices.* London: Routledge.

Sorensen, A., Okata, J. and Fujii, S. (2009) Urban renaissance as intensification: building regulation and the rescaling of place governance in Tokyo's high-rise Manshon boom. *Urban Studies,* **47**(3), pp. 556–584.

Sorensen, A. and Sagaris, L. (2010) From participation to the right to the city: democratic place management at the neighbourhood scale in comparative perspective. *Planning Practice and Research,* **25**(3), pp. 297–316.

Tsuru, S. (1999) *The Political Economy of the Environment: The Case of Japan.* London: The Athlone Press.

Ui, J. (ed.) (1992) *Industrial Pollution in Japan.* Tokyo: United Nations University Press.

Watanabe, S. (1999) *Citizen Participation Based Machizukuri: From the Point of View of Making Master Plans. (Shimin Sanka no Machizukuri: Masutâ Puran Zukuri No Genjô Kara).* Tokyo: Gakugei Shuppansha.

Watanabe, S. (2007) Toshi keikaku vs machizukuri (Emerging paradigm of civil society in Japan, 1950–1980) in Sorensen, A. and Funck, C. (eds.) *Living Cities in Japan: Citizens' Movements, Machizukuri and Local Environments.* London: Routledge.

Woo-Cumings, M. (ed.) (1999) *The Developmental State.* Ithaca, NY: Cornell University Press.

Woodall, B. (1996) *Japan Under Construction: Corruption, Politics and Public Works.* Berkeley, CA: University of California Press.

Yamamura, K. (1992) LDP dominance and high land price in Japan: a study in positive political economy, in Haley, J.O. and Yamamura, K. (eds.) *Land Issues in Japan: A Policy Failure?* Seattle, WA: Society for Japanese Studies, pp. 33–76.

Chapter Three

The Dragon's Head: Spatial Development of Shanghai

Susan Walcott

Booming Shanghai, the giant 'dragon's head' at the mouth of the Yangtze River,[1] has a population of over 20 million early in the twenty-first century and one that has more than doubled over the past 20 years. The city's long history as a prosperous site for business and agriculture (silk, tea and rice cultivation) has made it an enduring destination for migration and a target for exploitation. Its name is often interpreted as meaning 'on the ocean' and its importance as a port has been recognized since the eleventh century. Shanghai has transcended occupation by Western colonizers in the mid-nineteenth century and Japanese military imperialists a century later, as well as revolutionary turmoil and retarded growth during the early communist era. Drawing on its outstanding location for internal and foreign trade, and a history of entrepreneurial innovation, Shanghai now serves as China's model city and face to the world.

The Treaty of Nanjing in 1842 began Shanghai's period as the spearhead of European colonial-led and foreign concession-based industrialization (Lu, 2004). Factories in the concessions were fed by refugees from turmoil in the countryside. The founding of the Chinese Communist Party took place in Shanghai's French quarter in 1921. Shanghai's prosperity in the 1930s (the 'Paris of the East') was followed by its control by the central government during the Maoist period as a 'cash cow' providing taxes to replenish the national coffers. Until its economic liberation in 1984, Shanghai sent up to 86 per cent of its revenue to the national treasury, representing one-sixth of national income (*Ibid.*). In 1992 Deng Xiaoping, following his famous tour of southern China, expressed regret for the delay in letting Shanghai use its resources to finance its own development. The same year, the Fourteenth National Party Congress coined the term 'dragon-head' for its vision of Shanghai assuming a new (old) role as China's premier regional economic growth engine, based on its international links and strengths. This chapter profiles how Shanghai quickly regained its past prominence thereafter to become the financial and foreign investment centre of China, the nation with Asia's fastest growing economy.

Policies since the 1980s have rewarded economic success and accelerated urban development with greater local power and autonomy. Along with Beijing, Tianjin and Chongqing, Shanghai was granted special district level status in recognition of the city's ability to attract and retain global businesses, using foreign direct investment (FDI) both to boost revenues and to learn lessons for engaging with a wider and more market-oriented world. Capital investments in real estate, along with major infrastructure projects, such as bridges, tunnels and subways, encouraged the spatial inflow of labour and gave rise to significant increases in property values. Metropolitan Shanghai's political divisions are listed in table 3.1, followed by a map of their locations in figure 3.1. As this figure shows, the old, densely populated inner-city core is adjoined by the delta area of Pudong on its east and by a ring of counties anchored by satellite cities. Two particularly – and spectacularly – transformed areas, described below, are the downtown, redeveloped as upmarket retail, park and residential nodes,

Table 3.1. Shanghai districts.

Urban Districts Name	*Area* (km²)	*Population*	*Density* (persons per km²)
Pudong	570	1,770,000	3,105
Xuhui	51	884,000	17,333
Changning	37	610,000	16,486
Putuo	55	1,000,000	18,181
Zhabei	29	810,000	27,931
Hongkou	23	860,000	36,596
Yangpu	55	1,240,000	22,342
Huangpu	11	575,000	4,667
Luwan	7	328,000	43,850
Jing'an	8	351,300	46,102
Satellite Settlements Name	*Area* (km²)	*Population*	*Density* (persons per km²)
Minhang	372	750,000	2,018
Baoshan	425	1,017,600	2,394
Jiading	464	511,800	1,103
Jinshan	586	527,100	899
Songjiang	605	527,100	871
Qingpu	670	730,000	1,089
Fengxian	705	706,800	1,002
Nahui	688	856,000	1,244
Chongming Island	1,225	635,000	518

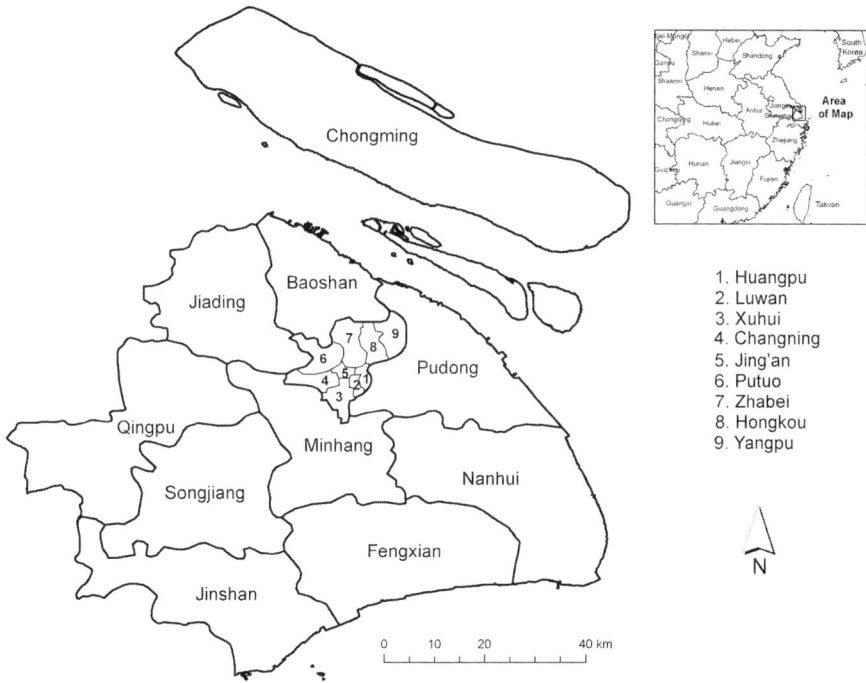

Figure 3.1. Shanghai city map.

and Pudong itself, created with enormous central government largesse as a business development centre in the early to mid-1990s.

Caohejing preceded Pudong as Shanghai's first high technology zone. It was a project of Shanghai Mayor Jiang Zemin who became President of China in 1989. A similar transition to national politics was subsequently made by former Shanghai Mayor Zhu Rongji, who served as Premier of China from 1998 to 2003. Economics led politics as a path to success and entrepreneurialism at a lower level found its reward at a higher level.[2] National politics continue to direct and validate the economic policies implemented at city level, but with a degree of flexibility, following Deng Xiaoping's famous dictum of 'crossing the river by feeling for stones'. The current stage of political decentralization permits even greater scope to China's major cities as the central government further loosens the reins of control, leading to a competition between cities for economic assets of a sort more familiar to developed countries in the West.

International funding agencies including the World Bank and the Asian Development Bank participated heavily in infrastructure construction projects in the early 1990s, but self-generated financing (local government finance or capital raised by enterprises themselves) has provided a large source of funding for urban development projects in Shanghai since then, boosted by funds derived from changes to land-use rights and property transfers (Shanghai Municipal Statistical Bureau, 2006). In the most recent stage of municipal market maturity, the Shanghai stock market also attracted private investors to support the construction and manage the revenues of large municipal projects, such as the new Pudong airport (He and Ning, 2007). Shanghai's stock exchange now rivals Shenzhen and Hong Kong as mainland Asia's busiest. Shanghai was already Asia's second busiest cargo port by 1992 with 56 million tons of cargo handled, a volume exceeded only by Singapore. Major industries include information technology, Baoshan Steel and also major automobile ventures involving General Motors and Volkswagen, the last two indicative of Shanghai's position as the largest recipient of FDI in China. As indicated in table 3.2, Shanghai's remarkable growth as a focus for foreign investment is reflected in the continuing strength of its manufacturing as well as the growth of its services sector. The increased value of its FDI and exports emphasizes Shanghai's global links, and the impressive acceleration in per capita GDP over almost four decades has underpinned the creation of a middle class and a concentration of wealth in the city.

Urban planners since the 1980s have sought to leverage Shanghai's locational endowment as an export-oriented port while readying the city to attract foreign investment and major international companies. The following sections detail the attempts by planners to alleviate the problems of very high settlement densities in inner Shanghai by promoting the growth of surrounding suburban areas while simultaneously aspiring to lift the remodelled metropolis to global city status. This is followed by an exploration of the challenges such efforts entail for infrastructure, the environment and the economy. The conclusion to the chapter offers a prognosis for the city's future, noting China's continuing struggles with corruption and its seeming willingness to endure environmental abuses for short-term economic advancement.

Shanghai's Planning History: Urban Form and Function

In their study of globalizing cities in the less-developed world, Grant and Nijman (2002, p. 320) use Accra and Mumbai to demonstrate that 'the internal spatial structure of such cities can be understood in terms of their evolving roles in the wider-world political economy'. Shanghai's spatial structure clearly proclaims the nature of its economy and its historical evolution. This is a city created by water-borne trade. The 'Many-Coloured Water Tour' that departs from the banks of the Bund (an Anglo-Indian word meaning 'embankment', referring to the waterfront strip built by the British as their colonial financial centre) graphically illustrates these linkages. Leaving the shores of the greyish Huangpu River that separates Puxi ('west of the Pu', a traditional name for Shanghai) from newly rebuilt Pudong ('east of the Pu', an area formed by centuries of alluvial deposition from the Huangpu), boats pass by the much darker effluents of Suzhou Creek which flows through the city. The third and fourth colours come at the turning point of the tour where the yellowish Yangtze, bearing silt carried down by China's longest river from its origins in Tibet, meets the blue-grey vastness of the ocean. Key elements of Shanghai's spatial structure and character also reflect its water-linked history, from the Ming Dynasty Yu Yuan garden with its zigzag bridge to a popular teahouse made famous on countless Ming-style blue and white plates, in what is now downtown, to the colonial era Bund along old Shanghai's eastern edge, across the Huangpu to Pudong's oceanside Waigaochiao export processing zone, and finally south to the island-spanning bridge to the new deep water port of Yangshan.

Shanghai's land area covers about 6,500 square kilometres, with a population in 2008 of 13,910,400 (Shanghai Municipal Statistical Bureau, 2009). Many estimates also include about 6 million unregistered migrants for a rounded total of 20 million people in the city in 2010. Shanghai's striking economic and demographic growth in recent years has given rise to 'accelerated urbanization' (Wu and Li, 2002), with new gentrified areas establishing around the old high-density, low-income city centre and outward growth incorporating outlying districts and counties (Walcott and Pannell, 2006). As recently as 2009, Nanhui County was absorbed into the burgeoning Pudong New District, exemplifying the further spread of urban development across contiguous, formerly agricultural, areas. Extension of infrastructure has integrated such peripheral areas with the city, enabling real estate construction and a more mobile population to develop less expensive land, newly available under legislation facilitating land-use changes.

As detailed by Chen (2007), economic reform, opening up China's top-down, centrally-controlled economy to domestic markets and foreign capital, had its corollary in urban planning reforms commencing in the 1980s. New mechanisms to distribute and direct the flow of population focused attention on both Shanghai's past and Western models, such as a central business district, new satellite growth centres and waterfront developments, as a basis for future settlement form. With their shared goals of economic development, local policy-makers generally initiated proposals which were brought to fruition with the assistance of central government. Various researchers termed this process one of Shanghai becoming 'a global city from below' (Zhang, 2009) through 'downward pressure, upward bubbling' (Lu and Zhigang, 2009). In a significant break from previous highly-centralized city planning frameworks, power and direction for this urban transformation came in part also from input from local

Table 3.2. Shanghai economic take-off.

Year	GDP			Utilized FDI	Export value	Per capita GDP
	Primary	Secondary	Tertiary	(US$10,000)	(US$1,000)	249
1970						
1996	71.6	1,582	1,248	751,000	13,238,000	8,447
2008	111.8	6,236	7,350	1,008,400	169,350,000	39,622

residents (Zhang, 2009). The 'two levels of government, three levels of administration' reforms formally extended a policy voice to the neighbourhood (*jiedao*) level (Wu and Li, 2002). In the post-Mao era the search continues in Chinese urban policy for a workable balance between centralized planning initiatives, market responsive measures and local inputs.

The energy and vision of Shanghai's planners focused recently on readying the city to host the World Expo, which took place between May and October 2010. The omnipresent slogan of 'Better City – Better Life' (figure 3.2) was teamed with a dancing blue figure (symbolically designed by a Taiwanese) to project friendship among modern cities across the global oceans. The idealized image of New Shanghai is one of towering office and residential blocks, interspersed with pockets of green space and linked with elevated transportation arteries. Asia's only *maglev* rail line (German-built) whisks visitors to Shanghai from Pudong's new international airport, the shape of which evokes soaring sea birds and a city taking off. The consequences of Shanghai's rapid transition to prosperity are discussed later in this chapter.

Figure 3.2. 'Better City – Better Life'.

As one urban historian has noted of Shanghai, 'moving through space is also moving through time' (Wasserstrom, 2007, p. 227). This is particularly the case when walking along Nanjing Road in the centre of the city. The walk starts at Jing'an Temple (built in the third century BCE, used as a plastics factory during the Cultural Revolution and restored in 1983), now bordered by modern glass skyscrapers containing upscale international retail outlets such as Burberry. It continues past the crisply-gardened and well-guarded city hall (a sign next to the gate identifies the handsome structure as the former Jewish Country Club) to the pedestrian mall that begins at People's Park (formerly the British racetrack). The mall extends past Shanghai's 'First Department

Figure 3.3. Across space and time.

Store', Soviet style offices, trendy global boutique outlets and hotels famous in the 1930s (started by émigré Iraqi Jews). The walk terminates at the Bund, with views across the Huangpu to ultra modern Pudong. Nanjing Road is quintessential Shanghai.

Shanghai's spatial development as a megacity is considered further below at three scales, each with a spatial form and urban functions that are dynamically evolving: the central city and its eastern extension to Pudong; the satellite towns and cities; and the extended metropolitan region.

The Central City and Pudong

Three ring roads encircle Shanghai. The inner ring road (*nei huan*) encompasses the old central core, followed by the outer ring road (*wai huan*) that was once to be the outer city limit. In the last decade the third or middle ring road (*zhong huan*) was added as a more recent attempt at containment of the expanding central business district. In the first decade of Shanghai's modern growth spurt, between 1990 and 2000, the number of residential buildings in the central city increased by 135 per cent and those used for offices and shops by almost 300 per cent (Shanghai Municipal Statistical Bureau, 2001). The quality of structures and their locations also changed dramatically due to a much needed programme to demolish substandard housing and replace it with high-rise housing, complete with modern amenities, outside the core central district.

Targeted neighbourhoods were in the six districts of Putuo, Xuhiu, Nanshi, Hongkou, Luwan and Huangpu. Unlike migrants to other major Chinese cities, those coming to Shanghai were less likely to obtain affordable dilapidated housing in the core, and either congregated in peripheral areas of rapid construction, where they were likely to find employment, or commuted from outlying agricultural areas. The Xingtiandi area provides a dramatic example of the physical transformations which accompanied these economic changes. This neighbourhood housed China's first industrial age factory workers in a hybrid form of housing called *lilong,* which combined the European terraced house with the traditional Chinese courtyard dwelling. *Lilong* areas ('li' means neighbourhood and 'long' means lane) were characteristic of the concession era in Shanghai from the mid-nineteenth century until World War II. Typically they comprised two- or three-storey houses along pedestrian alleyways. An alternative name for this form of development is *shikumen*, which refers to the stone archways at the end of the alleyways. As part of urban restoration and redevelopment in the early twenty-first century, the former residents of this area were relocated to modern apartments away from the inner city and the buildings here were reconfigured for upscale residents and tourists (see figure 3.4 and note the globally-recognized multinational company now occupying one of the traditional house sites, featuring a distinctly non-Chinese beverage as an amenity for tourists and well-to-do Chinese).

Pudong is a story in itself. The Shanghai city government's establishment in 1987 of the Pudong Development Research Consultation Group, composed of local and

Figure 3.4. Shikumen.

Western experts, was emblematic of reforms since the early 1980s in the national and regional political and economic arenas. Both Shanghai's planning department and local political leaders advanced plans early in the decade to develop Pudong along the lines already set in motion for the Pearl River Delta (PRD), opening up the Yangtze River Delta to the sort of co-ordinated development pioneered in Shenzhen. Key lessons learned from the PRD planning experience were to keep control in the hands of local planners (networked with other stakeholders) and to enforce, as far as possible, consistent standards for both structures and spatial development. The central government approved these proposals in 1990. Located on the eastern side of the Huangpu River, across from the old centre of Shanghai, Pudong's current population is about 7 million. Its core elements include the central business district of Lujiazui; the greater part of the site of the 2010 World Expo; the industrial area of Jiading (where General Motors is based); the Zhangjiang high-tech park; an export processing centre in Waigaochiao; and the eastern island of Chongming.

Pudong was developed with an openness to foreign influences, seen in part in the international competition which took place for the design of its structure plan and some of its signature architectural elements[3] (see figure 3.5). There were parallels with the sort of interaction with the West as a learning mechanism characteristic of Hong Kong (Chen, 2007). Top young talent was deliberately brought into government at this time in pursuit of new ideas on the implementation of market-oriented political and legal frameworks, while cutting-edge endogenous scientific researchers and

Figure 3.5. Pudong waterfront.

entrepreneurs were attracted to the Zhangjiang high-tech park on Pudong's eastern edge. Zhangjiang now covers 25 square kilometres and is planned to double in area in a decade. It includes several different levels of housing, a nationally competitive school system, restaurants and other amenities for China's meritocratic elite and foreign business expatriates. Government incentives offered here and elsewhere were intended to restrict foreign involvement to areas of expertise lacking in China but likely to lead to technology upgrading and to the development of sustainable employment for both highly-skilled and rural migrant workers.

Suburban Satellite Settlements

The development of Shanghai's spatial morphology is currently based on a model of ten smaller 'satellite' settlements in an 'orbit' of up to 60 kilometres around the anchoring giant (figure 3.6). The railroad running from Shanghai to Suzhou, the major outlying city to Shanghai's west, passes by Kunshan, a focus now of Taiwanese/Cross-Straits business clusters but first assigned to Shanghai's administration around 1960 as an agricultural area supplying urban residents. Towns were built here on the British 'garden city' plan with populations limited to 25,000–50,000 (Yan, 1985). Following Chinese preferences for higher density, however, and limited by *hukou* (household registration) restrictions, these areas failed to attract much population initially. Incentives for relieving population density in Shanghai now make centrally located land available for commercial 'highest and best uses' and it has become more attractive for foreign investors and migrant workers to locate in such planned peripheral urban areas (Wu, 2008). Former sharp distinctions between rural and urban areas, with the walls of workers' apartments falling away to flat agricultural fields, are now blurred, moreover, by varied commercial and mixed-use developments reflecting the new land-use laws.

Infilling of former agricultural and under-utilized land by planned 'new village' and 'new neighbourhood' complexes has provided updated and 'income-appropriate' housing for relocated urban residents, including middle-class and higher-income market niches. These instant suburban settlement areas are found in between, and on the fringes of, satellite cities. Generally located close to transportation infrastructure, these include more grassy areas and dwellings with larger floor plans than are available closer to the city (Wu and Li, 2002). Advertising for these new communities (and with the reality confirmed by visits to several) promotes the continuation of the culturally idealized, multi-generational lifestyle found in more established villages and in the former crowded inner-city neighbourhoods, with grandparents and other seniors, who are themselves still active, overlooking the activities of the youngest generation. However, development in some of the new settlement areas, and also in the older core villages and remnants of what were formerly older outlying towns, is quite uneven,

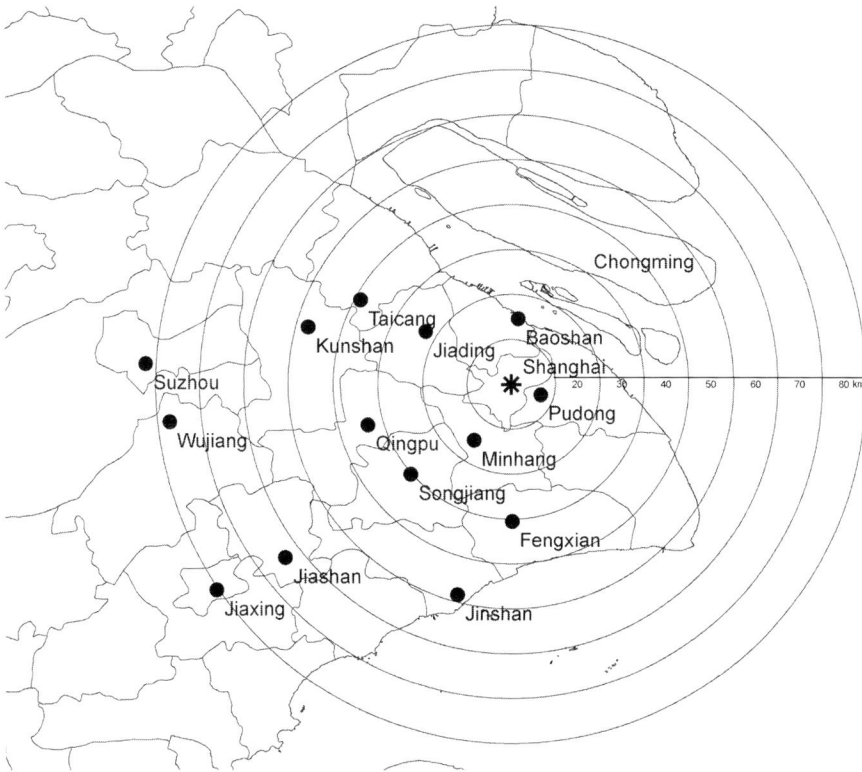

Figure 3.6. Map of Metro Shanghai.

with a lack of services, such as food retailing and entertainment, which require large markets.

As discussed below, only two satellite settlements – Baoshan, and Minhang – are currently considered to be successful projects in terms of developing an economically self-sustaining core, focused respectively around a major steel company and a university centre. Development in each involved the principle of clustering (*ji-zhong*). Several traditional villages were combined into one larger urban entity, freeing land to zone for uses such as business or housing development that would produce higher income from taxes and transfer fees. Older satellite towns, dating from before the twenty-first century, typically have an existing centre and their own established economic base, such as Fengxian's carpentry skills which became the core for a furniture industry. New economic and housing development sites in these older villages and towns are not always well integrated with the existing centre by the local municipal planning body.

Baoshan, one of the more successful satellite settlements, grew around the giant complex of Baoshan Iron and Steel, relocated from the banks of the heavily polluted Suzhou Creek in Shanghai's core. 'Bao Steel' was well integrated in its new setting with associated manufacturing. In Jinshan, however, the idea was to build a major support cluster around the automobile industry, but this was not achieved, spatially or functionally, in practice. Co-operation with Shanghai Mobile (a telephone company) eventually established a stable population and economic activities. The trick clearly lay in building on an existing flagship company with the potential to form the core of an integrated settlement. The majority of the original ten satellite cities failed to become self-sustaining after funds for satellite development began to dry up in 2004–2005. A particular disappointment (and an example of planners' architectural over-reach) was the failure of the 'global design' concept. In order to accentuate the international nature of the new satellite towns, each adopted its own theme for what ultimately was over-expensive housing. The furniture city of Fengxian, for example, has rows of empty 'Spanish' Mediterranean-style residences, while residential developments in other satellite settlements sport ill-suited Scandinavian and German motifs.

Each of the satellite settlements was designed to be linked to Shanghai by a metro line. These are now under construction. Minhang, the site of an early Economic Development Zone, has new campuses of East China Normal University and Jiaotong University, and is an indication of the likely importance into the future of university-centred economic development clusters. Universities were at first seen as a negative drain on area budgets, since the land they occupy must be provided free. However, the economic multiplier effects from purchases by students and (relatively well-compensated) academic staff, coupled with services, residential development and innovative business income possibilities, have made it worth competing to attract them. Legally, a university is one of the few entities able to be transferred from one area to another, so attracting a new campus is now hotly contested, with generous subsidies offered to relocate. Hospitals are also under increasing consideration as possible development cores.

Yangtze River Delta Metropolitan Region

Major cities in the Yangtze River Delta (YRD) include Hangzhou, Suzhou, Ningbo, Nanjing and Wuxi. By 2005 the YRD comprised 1 per cent of China's land, 6 per cent of its population, produced 18.6 per cent of its GDP, attracted 43.5 per cent of the country's FDI and enjoyed three to four times the national average GDP per capita. Out of a land area of 110,000 square kilometres, a total of 52.5 per cent could be classified as urban while out of the 87.4 million population in 2000, 88 per cent lived in urban areas.

An illustrative typology of the Yangtze River Delta's spatial organization (figure 3.7) classifies the region as 'Three Layers, Three Metropolitan Regions' with three triangles anchored by Shanghai in the east, Nanjing in the west, and Hangzhou in the south (adapted from Ning and Li, 2008). The most prosperous 'Big Eight' cities include

Suzhou, the classic garden city and now home to several successful high technology parks; Taizhou on the outskirts of Shanghai and home to FDI in the form of small and medium enterprises; and the large city of Hangzhou. The major corridors are delineated by an interlinked infrastructure of highways, ports, railroads, air transport and pipeline systems. An interesting finding of a study by Ning and Li (2008) was that distance from Shanghai correlated with greater economic development and openness. This seems to result from several factors but particularly the benefit of being located close enough for commuting but far enough away for regulatory freedom from the giant metropolis.

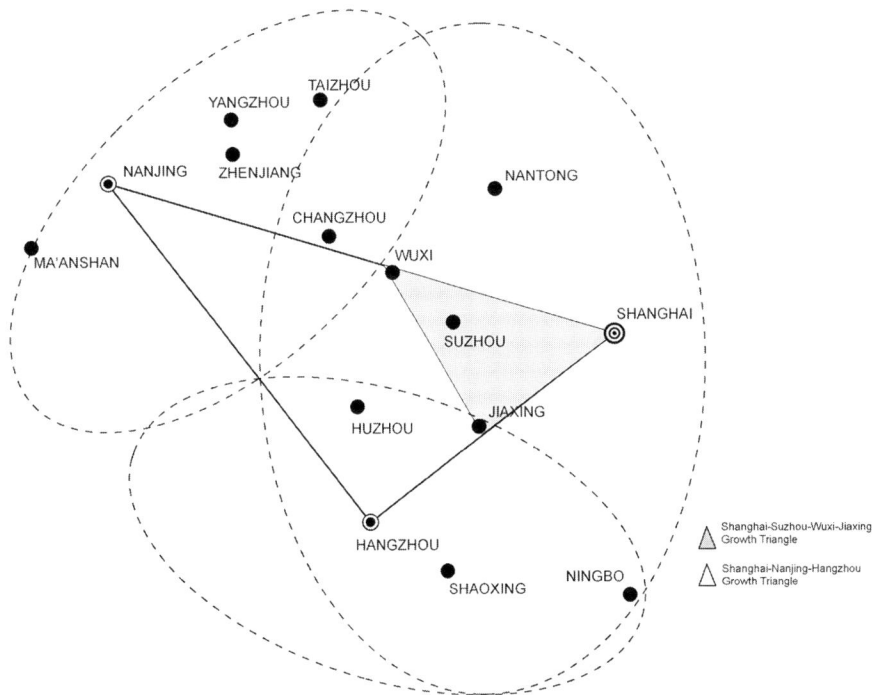

Figure 3.7. The Yangtze Delta Metropolitan Region.

The notion of an Asian extended metropolitan region draws on several strands of literature, reflecting patterns of spatially extensive and increasingly interlinked development evolving over decades. The idea of a 'Metropolitan Integrated Region' (Zhou, 1988) envisioned linked urban patterns of major city cores, primarily along China's east coast, in the first decade of accelerated urban recuperation following the Cultural Revolution. Based on Southeast Asia's exploding cities and highway network, McGee (1991) proposed the concept of *desakota*, an extended string of areas with relatively urban functions along routes linking larger cities. The notion of a global city region (Sassen, 2009; Scott, 2001) implied the elevation of major urban centres to a new level in the city hierarchy, with global cities interacting with each other, as well as with their functional near-equals within the nation and contiguous areas. Shanghai's position in this global hierarchy is further discussed in the following section.

World City Evolution

Categorization schemes are inherently problematic, reliant on their internal definitions and comparative scales for their claims to validity. One leading theorist holds Shanghai up to a set of standards, which include the number of multinational corporate regional headquarters, the type and number of service functions, and global linkages with other advanced firms and with financial networks, to measure the city's 'global city' status (Sassen, 2009). But attempts to categorize Shanghai often stumble on its unique and stubbornly resurgent local qualities as well as its post-socialist, post-colonial state of transition into a more globally-linked tool of the national state policy apparatus (Friedmann, 2005; Wu, 2009). One notable China historian traces Shanghai's 'stop and go' experience from a national trade trans-shipment point from the thirteenth to mid-nineteenth century, to an international Treaty Port, followed by an internally-focused Maoist city, and eventually characterizes its present-day manifestation as 'a reglobalizing post-socialist city' (Wasserstrom, 2007) along the lines of Budapest or Berlin.

Shanghai's current position within China is roughly equivalent to that of New York City in the United States in terms of its internal national ranking and position as the financial centre of the country (Lu, 2004; Solecki and Leichenko, 2006). Assessments of its 'world city' status vary widely, depending on the proxy measurements selected as evidence of comparative status and development level. The Loughborough Group (Beaverstock et al., 1999), for example, assigned Shanghai to the same level as Atlanta, Georgia, based on producer service headquarters (second in banking and legal services) so that, in the lowest of three categories, it just makes the grade as a world city. Other analysts propose 'world city' as an intermediate stage prior to fully assuming the more exalted title of 'global city' (Chen and Orum, 2009) and see Shanghai in that role – at least as China's best (or, after Hong Kong, next best) candidate. 'Globalizing' seems the most appropriate term at present for the Yangtze River Delta's ambitious and rapidly growing economic titan.

Shanghai's urban development has been inextricably linked with China's overall economic rise (Wu, 2009). As noted earlier, due to the prevalent national shortage of financial capital in the late 1980s, the first stage of Shanghai's rocket was fuelled by international funding bodies, such as the Asian Development Bank and the World Bank, which assisted with the construction of major urban infrastructure projects. This was followed by a second stage of more mature internally-generated financing and then by investments linked to the growth of Shanghai's booming stock market. This remains a somewhat vulnerable component of the country's development, however, as the result of an extreme reluctance to permit foreign banks to have access to the economy, and weakens Shanghai's aspirations as a global financial capital. Beaverstock *et al.* (1999) declared Shanghai not open enough to global society to merit global city designation, comparing it unfavourably to Kuala Lumpur, for example, while other, largely Chinese, critics decry Shanghai for falling too much under the sway of Western influences and losing its Chinese soul. Statistics on the nature of Shanghai's import-export components indicate that its position on the global commodity chain is a notch below that of a global city. The hi-tech export of both personal computers and electronic components, for example, is dwarfed by the import of these items, indicating that the primary role of Shanghai at present remains as a low-cost labour assembly site. This reflects China's role in comparison to the United States and Taiwan – a prosperous manufacturing entity with rapidly increasing higher level elements, but not yet in the same global value chain league.

Meeting Contemporary Challenges

Government economic and spatial policy currently focuses on creating what are termed the 'Three Stabilities' through organizing clusters based on the following categories:

1. Population by skill level (e.g. integrated high-tech parks and industrial districts);

2. Industrial location (away from environmentally sensitive areas); and

3. Agricultural land (combined into larger, more productive units suited for the introduction of new technologies and separated from urban and other economic use areas)

The intent is to remedy the classic drawbacks of mixed-use development (ironically currently seen, for other reasons, as desirable 'Smart Growth' in the West) including commuting long distances due to a spatial mismatch of jobs and skills; environmental pollution from industries located too close to residences; and inefficiencies due to small plots of agricultural land subdivided by generations of family workers. Indications are that government incentives and investments are making progress in the direction of planned land reutilization but challenges remain in Shanghai's efforts to balance the efficiencies of clustered development, described earlier, with overcoming the ills of older, excessively dense concentrations. Challenges in relation to infrastructure, environmental issues and economic development are discussed briefly below

Infrastructure: Physical and Human

Since 1991, transportation improvements have absorbed a tremendous amount of investment. Two bridges now span the Huangpu River between Shanghai and Pudong, which are also newly connected by tunnels. Many kilometres of elevated roadways and a new inner ring road help direct a greatly increased flow of traffic through the city, generated by the economic boom. Bicycles, though, are still more prevalent than in Beijing and, impressively, mass transportation improvements include several subway lines around and through the metropolitan area, linking jobs and worker residential concentrations.

Population change in the fastest growing decade of 1990–2000 revealed the spatial reconfigurations underway. While population in three core central districts dropped by a total of 692,000 during this decade, population in peripheral suburbs jumped by 2.6 million. Satellite settlement population also increased by 380,700, with the main exception being relatively distant Chongming Island (Li and Ning, 2007). Migration to major urban centres with relatively high employment possibilities, especially for low-skill, recently arrived rural workers, has helped to swell Shanghai's overall population since the easing of *hukou* restrictions. In the three decades of reform between 1978 and 2008, Shanghai's city level population increased from 5.57 million to the official census count of 13.9 million in 2008, which, as noted earlier, omits an estimated 6 million unregistered migrants (Shanghai Municipal Statistical Bureau, 2009). Impacts of this high migration are felt in many areas. Lack of legitimate *hukou* status leads to a lack of access to city services, such as education and medical care, which often remain rooted in the migrants' rural villages of origin. Although some innovative informal attempts have been made by migrants to deal with these problems, authorities have yet to institute ways to respond to the negative consequences of service denial to this population. In general, the post-1990 push by municipal and national leaders for Shanghai to excel as an economic powerhouse neglected other development priorities, such as human welfare (Tu, 2007).

Changes promulgated to land legislation in 2007–2008 allowed for compensation to resettled former inhabitants by governments exercising powers of eminent domain, in the form of either monetary compensation or housing (with the government choosing both the monetary amount or the apartment facilities and location). Local government encourages the building of apartments on abandoned former factory land, providing higher value use for often centrally-located parcels. Planning units in each sub-area have their own, occasionally competing, vision plans. Housing construction, speculation and gentrification, while inevitably subject to boom-bust cycles, have nevertheless created much individual wealth as this section of Shanghai's economy has become subject to market forces. Government-provided mortgage funds and public-private partnerships have proved successful in reducing chronic housing shortages and in providing a better residential stock for some, in healthier settings and with more living space (Wu, 1999).

Shanghai's old inner industrial areas were characterized by extremely dense housing. A major point of controversy in recent years has been the tearing down of old two- and three-storey housing blocks and the resettlement of residents to new high-rise apartments on cheaper land in outlying areas. Inside plumbing and kitchen utilities substituted for tight-knit neighbourhoods, but some recent research has found that these changes have been generally welcomed (Li and Song, 2009).

Environmental Issues

The price of rapid urban development is often paid by the environment. Since Shanghai's growth leap took place largely around the turn of the twenty-first century, with an eye on demonstrating the metropolis's ability to take its place as China's global city, much thought and technology were applied to planning urban environmental improvements and to avoiding further degradation. Measures to ameliorate the pollution that blackened the course of Suzhou Creek ranged from moving the giant Baoshan Steel plant to its new complex on the metropolitan edge and replacing it with apartment blocks – the latter largely unsuccessful as new residents substituted their own forms of pollution – to building small structures upriver for treating the water before it reached the city (Yin and Walcott, 2003). Bridge beautification included hanging flower boxes across the roadways and creating urban pocket parks underneath. New industrial estates included green areas and lakes, while the redevelopment of Pudong featured Shanghai's largest park and a French-designed strip down the middle of Pudong's main thoroughfare that incorporated a winding stream and generous landscaping.

Environmental impacts of Shanghai's development remain, however, particularly in poor air quality. Its industrial heritage and the recent surge in vehicle emissions combine to give Shanghai some of the highest levels of suspended particulate matter and sulphur dioxide in China. Regulations on the books for impact assessment, registration of polluting activities and fines for excesses are only occasionally enforced on both foreign and domestic companies. Pollution of water by industrial effluents surged in the 1990s and continues as a consequence of the continuing desire to prioritize employment and investment opportunities (Wu, 2008). Typical of a city in a rapidly developing country, manufacturing conditions directly contribute to water and air pollution (Yin and Walcott, 2003). Beginning in 2003, Shanghai received the first instalment of a three-part loan from the World Bank to address urban environmental needs such as waste water and solid waste disposal, as well as financing environmental infrastructure, government environmental management, and fuel efficiency improvements (Chen, 2007). Recent setbacks to these efforts have included the indictment of environment officials on corruption charges and the apparent stalling of much-trumpeted efforts to construct a massive eco-city in connection with the 2010 World Expo in Shanghai. Sited on outlying Chongming Island, Dongtan was funded in part by research grants from the United Kingdom and designed by UK engineering firm Arup to provide a model of sustainable design in harmony with nature. Plagued by delays, cost overruns, overambitious aims and tight deadlines, Dongtan's potential to be an eco-city with 500,000 projected occupants remains unrealized (Pearce, 2009). The project inspired now-completed major infrastructure projects, including a new bridge and tunnel, bringing Chongming Island more into Shanghai's orbit, but little else exists of Dongtan aside from some wind turbines and an organic farm. High-rise apartments, the usual response to urban fringe land availability, are under construction instead.

Shanghai's potential to serve as a model for urban environmental action in China is theoretically enhanced due to the power of the central and municipal government to influence planning and enforcement, coupled with their tacit acceptance of 'green' goals to signal their global progressive credentials and the desire to promote Shanghai on the world stage. Challenges include the legacy of extremely dense settlement intermixed with polluting industries, a hot and humid climate that retains atmospheric pollutants, and the desire to continue attracting businesses as part of the city's 'economic engine' role. Municipal authorities continue to pour resources into addressing the 'three coloured' types of environmental impacts: grey (chemical, air quality); brown (sewage); and moving toward green (plantings to reduce heat and improve the ambient environment) (Solecki and Leichenko, 2006). But there are great long-term challenges to Shanghai's environmental sustainability in accommodating a large population influx in affordable housing, connected by mass transit to employment sites, while reducing pollution and density and providing labour to fuel urban and regional growth – a set of problems shared by other large dynamic and globalizing cities.

Economic Bases

Other challenges for urban planners lie in the socio-economic sphere. Shanghai's Eighth Five Year Plan (1985–1990) prioritized development of a service sector to raise the value-added element of its growth engine and shift the economic base to a more mature level. Two-thirds of the giant, economically uncompetitive local state-owned enterprises (SOEs) have now been closed or privatized, with some 1.4 million workers being retrained in a gradual process extending over some 15 years (Tu, 2007). However, while vital service sector functions in finance, insurance and real estate (FIRE) remain at the core of Shanghai's economy, the tertiary sector's contribution has remained stuck at 50 per cent of GDP in the decade 2000–2010, up from 18 per cent at the beginning of Shanghai's modern take-off in 1978. Planners currently question whether the solution to developing Shanghai's economy further lies in working to create a more efficient growth engine, or diversifying by adding a second engine. Candidates for the latter include encouraging the growth of more research and development (R&D) concentrations, nurturing university-centred clusters of the sort discussed above.

Figure 3.8. Zhangjiang.

This sort of approach might involve, for example, the co-location of a major hospital complex in combination with a research university, specializing in biomedical research, and a medical school producing physicians, nurses and medical technicians, along with pharmaceutical manufacturers. This would integrate major life science elements along the chain of skills and would encompass economic elements from R&D through to manufacturing and service industries, as a further development of the model pioneered at Zhangjiang high-tech park, described earlier (and see figure 3.8).

With regard to Shanghai's traditional role as a port, the increasing size of ocean-going vessels, such as oil tankers and cargo ships, continues to challenge port planners. Completion of Yangshan Harbour City in 2007, at the tip of a line of islands extending off Shanghai's southeastern coast, was intended to enable the Yangtze River Delta to handle boats equal in size to those which can be berthed at Hong Kong. However, Ningbo and Yangshan are currently the only ports in the area with deep enough harbours to accommodate the largest contemporary container ships and still larger ships are likely to come into service before long.

The pressure of the current economic downturn reveals the high elasticity and fragility of diversifying into the 'creative industries', a term for economic pursuits with a high artistic and design content from painting to pottery, furniture, books and textiles, as well as performance art. Shanghai's promotion of the creative industries was largely seen as a real estate and prestige function to utilize closed factories and to add an innovative component to compete with Beijing's '798' warehouse-studio district. Rehabilitation of former factories, warehouses and industrial sites around the city has resulted in attractive artistic quarters such as M-51 and Tiantzfang in downtown pockets. Beijing features similar clusters, but they are located on the urban periphery, whereas Shanghai's now-defunct and relocated industrial areas were embedded in the urban core. Large retail showrooms for bulky items such as furniture for new apartment complexes also occupy 'greyfield' sites. Renovating formerly decaying spaces with complementary uses on the edge of residential areas proved somewhat successful in attracting visitors, but purchasers have greatly decreased with sharply curtailed personal spending in the face of economic uncertainty.

Figure 3.9. Creative industries.

Future Prospects

Shanghai attracts attention as the epicentre of 'the Chinese Dream': a *xiao kang shehui*, or middle-class society of increasing comforts, centred in the city and spreading outwards, both spatially and socially (Mars and Hornsby, 2008). Several planners have suggested that China is a great place for implementing and testing new urban planning ideas – convince the core leadership group and it is likely to happen. Shanghai's rapid rise as China's most prominent showcase of the country's path from ancient accomplishments to contemporary economic marvel demonstrates the results of a concerted national political-economic effort. As a motto broadcast during the fiftieth anniversary of the founding of the People's Republic frequently proclaimed, 'without a New China, then no New Shanghai'. The process also illustrates the opening up of China's policy-making process, with input sought from both resident communities and foreign experts, which local planners then used to forge their own response. Construction was the physical manifestation of Shanghai's – and China's – change in urban-economic policy. Pent-up needs and aspirations were addressed by new infrastructure and by huge residential developments both in outlying locations as well as through the instant gentrification of urban cores. The urban spatial outcome demonstrates the resilience of some of the city's past forms, transformed by modern uses in traditional structures, such as a Starbucks cafe in the Yu Yuan Garden, alongside the archetypal Hu Xin Ting tea house; a Ming dynasty Buddhist temple in the midst of glass skyscrapers and upscale foreign retail outlets; and a creative district and furniture showrooms in former factory areas. With its striking architecture globally recognized as a symbol of colonial domination, the recent transmutation of the Bund (figure 3.10) to be a showcase of the New China, prior to the World Expo 2010, encapsulates the dynamic changes which continue to occur to Shanghai's urban space.

The future of Shanghai's urban environmental improvement efforts depends on the quality and commitment of its leadership and on China's national commitment to

Figure 3.10. Renovated Bund, New-Old Shanghai.

such goals. The outcomes of the 2009 Copenhagen and 2010 Cancun Climate Change Conferences were, at best, only faintly encouraging and China clearly continues to put economic prosperity ahead of other goals. Since the national leadership frequently uses Shanghai as the showcase face of 'New China', however, the likelihood of its longer-term sustainability as a resilient urban centre – despite setbacks such as the ill-fated Dongtan project – remains fairly strong. China's political figures often audition in Shanghai for higher office in Beijing, so the eyes of the central government, and the large foreign presence in the city, ensure that Shanghai is likely to continue to reflect the nation's aspirations as well as its growing pains.

Notes

1. In popular English usage, the spelling of the 'long river' as 'Yangtze' under the older Wade-Giles system is preferred to the post-1949 re-Romanization as the 'Chang Jiang'.
2. As distinct from experience in the US where close-knit urban growth regimes within particular cities functioned on personal ties of long acquaintance (see Stone, 1993 and also He and Ning, 2007).
3. The latter include the Oriental Pearl Broadcasting and Television Tower (Shanghai Modern Architectural Design Co. Ltd.); the Shanghai World Financial Centre (Kohn Pedersen Fox); and the Jin Mao Tower (Skidmore Owings and Merrill) (see Marshall, 2003).

References

Beaverstock, J.V., Smith, R.G. and Taylor, P.J. (1999) A roster of world cities. *Cities*, **16**(6), pp. 445–458.

Chen, X. and Orum, A (2009) Shanghai as a new global(izing) city, in Chen, X. (ed.) *Shanghai Rising: State Power and Local Transformations in a Global Megacity*. Minneapolis, MN: University of Minnesota Press.

Chen, Y. (2007) *Shanghai Pudong: Urban Development in an Era of Global-Local Interaction*. Amsterdam: IOS Press.

Friedmann, J. (2005) *China's Urban Transition*. Minneapolis, MN: University of Minnesota Press.

Grant, R. and Nijman, J. (2002) Globalization and the corporate geography of cities in the less-developed world. *Annals of the Association of American Geographers*, **92,** pp. 320–340.

He, D. and Ning, Y. (2007) Formation of pro-growth urban regime and urban development in Shanghai since the 1990s, in Keiner, M. (ed.) *Sustainable Urban Development in China – Wishful Thinking or Reality?* Munster: Verlagshaus Monsenstein und Vannerdat OHG.

Li, J. and Ning, Y. (2007) *Population, Spatial Change and City Restructuring in Shanghai since the 1990s*. Shanghai: Center for Modern Chinese City Studies.

Li, S. and Song, Y. (2009) Redevelopment, displacement, housing conditions and residential satisfaction: a study of Shanghai. *Environment & Planning A*, **41**(5), pp. 1090–1108.

Lu, H. (2004) Shanghai rising: resurgence of China's New York City? in Chen, A., Liu, G. and Zhang, K. (eds.) *Urban Transformation in China*. Aldershot: Ashgate.

Lu, H. and Zhigang, T. (2009) Trends and determinants of China's industrial agglomeration. *Journal of Urban Economics*, **65**(2), pp.167–180.

Mars, N. and Hornsby, A. (2008) *The Chinese Dream: A Society under Construction*. Rotterdam: 010 Publishers.

Marshall, Richard (2003) *Emerging Urbanity: Global Urban Projects in the Asia Pacific Rim*. London: Spon Press.

McGee, T. (1991) The emergence of desa kota regions in Asia: expanding an hypothesis, in Ginsburg, N., Koppel, B. and McGee, T (eds.) *The Extended Metropolis: Settlement Transition in Asia*. Honolulu: University of Hawaii Press.

Ning, Y. and Li, J (2008) Urbanization progress and megalopolis restructuring in the Yangtze River Delta. *China Urban Studies*, **1**, pp.53–65.

Pearce, F. (2009) *Greenwash: The Dream of the First Eco-City was Built on a Fiction*. Available at http://www.guardian.co.uk/environment/2009/apr/23/greenwash-dongtan-ecocity. Accessed 17 January 2011.

Sassen, S. (2009) The global city perspective, in Chen, X. (ed.) *Shanghai Rising: State Power and Local Transformations in a Global Megacity*. Minneapolis: University of Minnesota Press.

Scott, A.J. (2001) Globalization and the rise of city-regions. *European Planning Studies*, **9**(7), pp. 813-826.

Shanghai Municipal Statistical Bureau (2000–2009) *Statistical Yearbook of Shanghai*. Beijing: China Statistical Publishing House.

Solecki, W. and Leichenko, R. (2006) Urbanization and the metropolitan environment: Lessons from New York and Shanghai. *Environment*, **48**(4), pp. 8–23.

Stone, C. (1993) *Regime Politics: Governing Atlanta, 1946–1988*. Lawrence, KS: University Press of Kansas.

Tu, Q. (2007) *Rethinking the Shanghai Model: Planning for China's Global City*. Shanghai: Century Publishing House.

Walcott, S. and Pannell, C. (2006) Metropolitan spatial dynamics: Shanghai. *Habitat International*, **30**(2), pp. 199–211.

Wasserstrom, J. (2007) Is global Shanghai 'good to think'? Thoughts on comparative history and post-socialist cities. *Journal of World History*, **18**(2), pp. 199–234.

Wu, D. and Li, T. (2002) The present situation and prospective development of the Shanghai urban community, in Logan, J. (ed.) *The New Chinese City: Globalization and Market Reform*. Oxford: Blackwell.

Wu, F. (2009) Globalization, the changing state, and local governance in Shanghai, in Chen, X. (ed.) *Shanghai Rising: State Power and Local Transformations in a Global Megacity*. Minneapolis, MN: University of Minnesota Press.

Wu, J. (2008) The peri-urbanisation of Shanghai: planning, growth pattern and sustainable development. *Asia Pacific Viewpoint*, **49**(2), pp. 244–253.

Wu, W. (1999) City profile: Shanghai. *Cities*, **16**(3), pp. 207–216.

Yan, Z. (1985) Shanghai: The growth and shifting emphasis of China's largest city, in Sit, V. (ed.) *Chinese Cities*. London: Oxford University Press.

Yin, Z.Y. and Walcott, S. (2003) Assessing environmental impacts of urbanization using multi-sensor data: Shanghai, 1970–2000. *Asian Geographer*, 22(1), pp. 43–60.

Zhang, T. (2009) Striving to be a global city from below: the restructuring of Shanghai's urban districts, in Chen, X. (ed.) *Shanghai Rising: State Power and Local Transformations in a Global Megacity*. Minneapolis, MN: University of Minnesota Press.

Zhou, Y.X. (1988) Definitions of urban places and statistical standards of urban population in China: problems and solutions. *Asian Geographer*, **7**(1), pp. 12–28.

Chapter Four

Beijing: Socialist Chinese Capital and New World City

Gu Chaolin and Ian G. Cook

Beijing in 1949, the year in which the new China was established, was a dilapidated old city, with a population of less than 2.1 million. By 2008, the city of Beijing had a population of more than 12 million (UN Habitat, 2010) and was at the heart of a modern metropolis of over 20 million people. Beijing is now one of the most exciting and dynamic cities on earth. The recent transformation of Beijing, from a drab, dull and austere production city to a city of hi-tech manufacturing, service provision and mass consumption, has been breathless in its speed and scope. Beijing is now the socialist capital of a modern China that is increasingly self-confident and assertive and is also becoming a new world city, engaged at a global level. The drive to build at such a rate and scale can lead to social, environmental, cultural and governmental problems, however, and planning must adapt and change in order to keep up with the new challenges of the twenty-first century.

This chapter examines the following broad themes: traditional Beijing's urban development; planning and urban development under socialism, from Mao to Deng; changes to urban internal spatial structure during the reform period since the late 1970s; Beijing's progress towards world city status; and contemporary urban challenges and prospects. Illustrations have been used extensively throughout the chapter to demonstrate Beijing's growth and transformation and how in the last two decades striking new architecture has changed the city's skyline.

Traditional Beijing's Urban Development

Beijing has 3,000 years of history as a city, and was for 800 years the capital under the

Liao, Jin, Yuan, Ming and Qing dynasties. Kublai Khan established the city of Dadu as capital of the whole of China in 1283. In 1420 the Ming Emperor Zhu Di renamed it Beijing ('Northern Capital'). There was an extension to the south in 1553. The form of the city, defined by its main north-south axis and the line of its city walls, has endured since then (see figure 4.1), surviving the 1911 Republican Revolution, the founding of the People's Republic in 1949, and the Cultural Revolution of the 1960s and 1970s. This historical form is now bounded by the Second Ring Road, within which surviving elements of old Beijing include the palace at the centre, the surrounding parks, lakes and temples, areas of old urban fabric with *hutong* alleys and courtyard houses, parts of two city gates and some very small sections of the city walls.

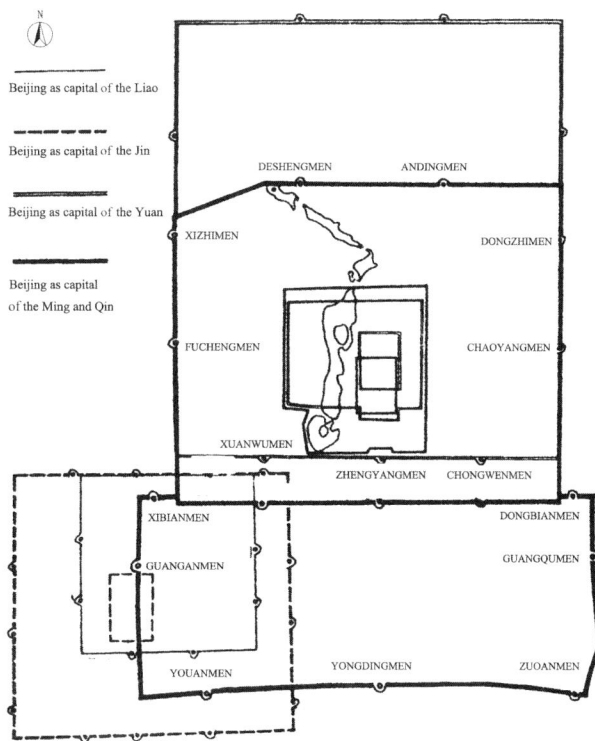

Figure 4.1. Beijing as capital of the Liao, Jin, Yuan, Ming and Qing dynasties.

At the time of Marco Polo's visit to Dadu in the late thirteenth century, the city had already become part of a process that, with hindsight, can be characterized as embryonic globalization, for goods were transhipped over vast distances along the Silk Road to and from the West, as well as north-south via the Grand Canal, completed in 1293 (Cook, 2006). The Forbidden or Palace City was constructed during this period, and was then rebuilt by the Ming (1368–1644) at the nearby location where it remains to this day. The Forbidden City formed the core of the Imperial City. It was built on a grand scale over a 14 year period, employing a million labourers and 100,000 craftsmen (Beijing Foreign Cultural Exchanges Service Center, 1997). Cosmological principles underpinned the grid layout of the Imperial City (Sit, 1996; Wheatley, 1971). High walls separated the Forbidden City from the remainder of the Imperial City, with residential quarters spread across fifty wards around the Palace. There was also an outer city to the south for the Beijing populace (Gu *et al., 2006*). The Manchu or Qing dynasty (1644–1911) retained the Forbidden City as it was but introduced ethnic segregation to Beijing. The Manchus occupied the Imperial City but the Han Chinese were largely excluded until the late nineteenth century. Beyond the Imperial City there was poverty, overcrowding and the hustle and bustle of everyday life. Within the Imperial City was a life of ease and luxury, linked to the fortunes of the Court in the Forbidden City.

Because the Ming and Qing dynasties largely closed their borders to trade and commerce with other nations, Beijing lost its early global role and functioned mainly as a trading centre for the population of the surrounding North China Plain. Western influence increased in the nineteenth century, however, following the Opium Wars and the 'Boxer' rebellion of 1900. With the dawn of the twentieth century, ideas of a substantial change in China's political and social circumstances began to take shape under the influence firstly of the Constitutional Movement and then of the revolutionary sentiments which culminated in the birth of the Republic of China under Sun Yat-sen in 1911. Modern town and country planning ideas reached China from Europe at about this time and remnants of the Western architectural influences of this period can still be seen in the churches, cathedrals and railway stations which survive as contrasting morphologies, hidden among the grand modern and post-modern structures of more recent times. Pressures for change were great during this period as the decaying Qing dynasty finally crumbled and, in the early years of the new Republic, the ancient traditional culture was shaken by further reform movements – the New Culture Movement of 1917 and the Fourth of May Movement of 1919, for example, both of which sought modernization along Western lines. This process was tempered, however, by the desire to reinforce a sense of national identity in this young, immense Republic, and a traditional-nationalist reaction emerged, aiming to 'search for a synthesis between Chinese and Western cultures that would be distinctively Chinese' (Zhu, 2009). A powerful expression of this was the New Life Movement, established by Chiang Kai-shek who succeeded Sun Yat-sen as leader of the *Guomindang*, the Republic's main political party, in 1925. One of Chiang Kai-Shek's first politically important decisions was to move the national capital back to Nanjing in 1927. Beijing was renamed Beiping (City of Heavenly Peace), as it had been at various times in its past. The loss of its status as capital, and the absence of commercial activities due to the lack of foreign concessions or a port, retarded Beijing's development and it remained largely static in size, confined within the ancient walls. Morphologically, as Gu *et al.* (2006, p. 258), citing Sit (1995), note:

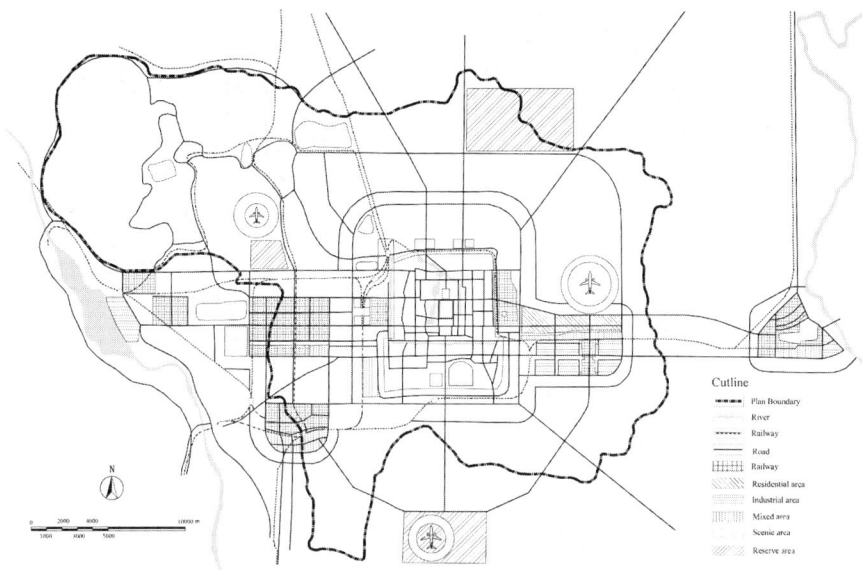

Figure 4.2. Beiping plan in1946.

Only two relatively important changes occurred between 1911 and 1949: (1) the appearance of the foreign quarters to the south-east of Tian'an Men Square that accommodated foreign embassies, banks, offices, clubs, hospitals, hotels and military garrisons; (2) the construction of Western churches, schools and hospitals in the city.

A plan for Beiping was prepared under the Nationalist *Guomindang* in 1946 (see figure 4.2). When the People's Liberation Army entered the ancient capital in 1949, however, with the Communists triumphant, Beijing embarked on a radically different path as a socialist Chinese capital. In September 1949, during the first plenary session of the National People's Political Consultative Conference, the city was chosen as the capital of the People's Republic of China and the historical name of Beijing was restored. Following the inauguration of the People's Republic of China on 1 October 1949, new structures, symbols and images of various kinds were sought to represent the new nation. Beijing's development under Mao Zedong's leadership is described in the next section.

Urban Development Planning under Mao

The basic concept for the planning of Beijing was established under the Communist Party's direction in a Master Plan developed in 1953. This drew significantly on the Moscow Plan of 1935, made under Stalin. The genesis of this plan and its consequences are set out below.

Early Plans under the Communists

There was heated debate over the early plans for the new capital of the People's Republic (Zuo, 2008; Dong, 2006). In December 1949, various planning proposals for Beijing city were discussed and two different opinions emerged. On economic and aesthetic grounds, some experts advocated an administrative centre based on the Old City (Dong, 2006, p. 4). These were mainly the Soviet experts Abramoff and

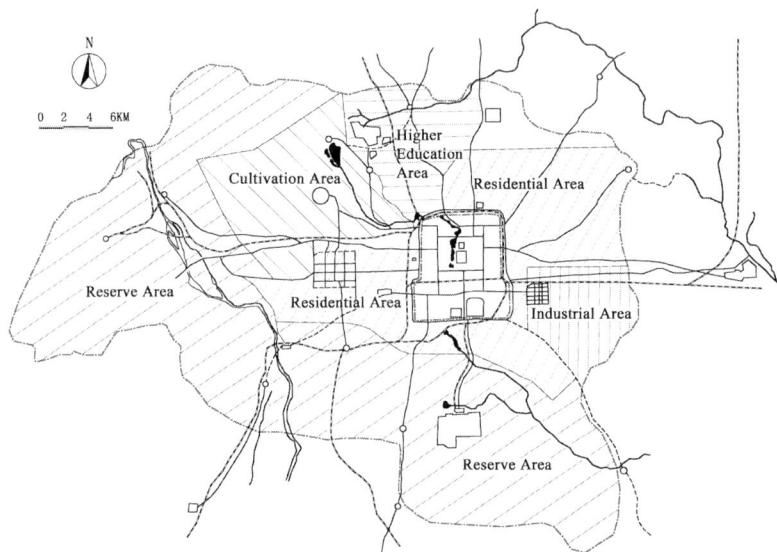

Figure 4.3. Beijing Plan: Barannukov's proposal of 1950. (*Source*: Dong, 2006, p. 6)

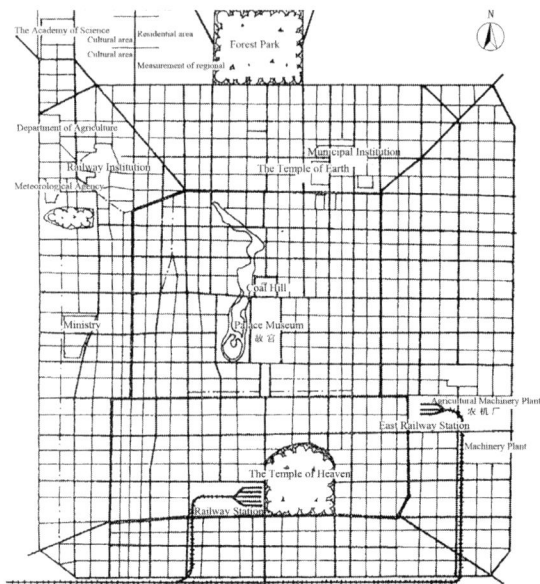

Figure 4.4. Beijing plan: Hua's proposal of 1950. (*Source*: Dong, 2006, p. 6)

Barannukov, together with a number of Chinese experts, including Hua Nankui, Zhu Zhaoxue and Zhao Dongri (see figures 4.3 and 4.4). The alternative view emphasized the protection of the Old City and the establishment of an administrative centre covering a larger area, between Yuetan and Gongzhufen (Dong, 2006, p. 4). This latter view was supported by Liang Sicheng and Chen Zhanxiang. Municipal officials displayed a clear preference for the first option.

Liang and Chen developed their 'Proposal for the Location of the Administrative Centre of the People's Government' in February 1950 (Liang, 2001). Known as the Liang-Chen Plan (figure 4.5), it proposed that the new centre be in the western

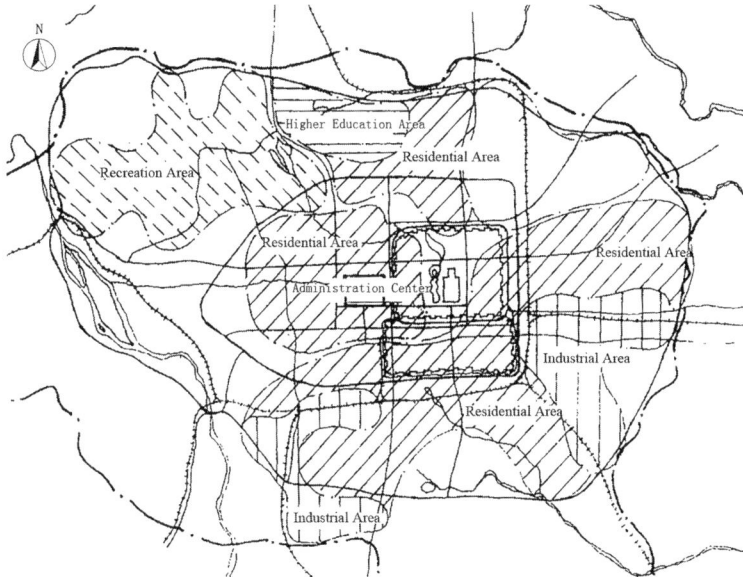

Figure 4.5. Beijing plan: Liang and Chen's proposal in 1950. (*Source*: Dong, 2006, p. 7)

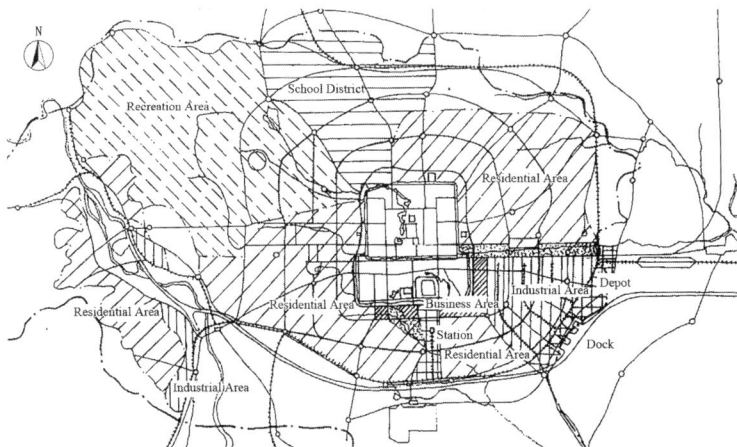

Figure 4.6. Beijing plan: Zhu and Zhao's proposal in 1950. (*Source*: Dong, 2006, p. 5)

suburbs and argued for decentralization, clear zoning, a balance between development and preservation, and conservation of the imperial city along its historic north-south axis. A counterproposal by Zhu and Zhao was made in April 1950 (figure 4.6) but construction began in Beijing's central area while these debates were under way (figure 4.7), effectively following the Russian proposals which, it was revealed later, had been accepted in a message from Mao, passed on to the Russian team via the municipal Party Secretary, Peng Zhen (Wang, 2003).

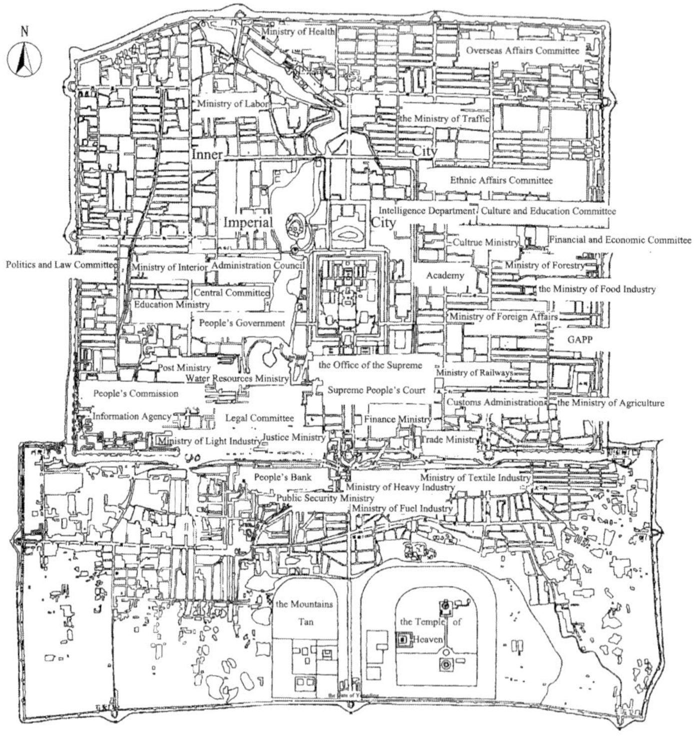

Figure 4.7. Early distribution of offices of state organs in Beijing Old City. (*Source*: Dong, 2006, p. 12)

Beijing Plan 1953

In 1952–1953 two plans, made by Hua Lanhong and Chen Zhanxiang respectively, under Liang's coordination and based on the idea of the new centre being placed inside Beijing (see figures 4.8 and 4.9), were sidelined for their lack of 'progressive' ideas. To achieve a better plan, the municipal government and the Party organized another team to work in a building called Chang-guan-lou (in a zoological garden in western Beijing) which produced a decisive master plan document entitled 'Draft Plan for Reconstruction and Expansion of Beijing' in November 1953 (see figure 4.10).

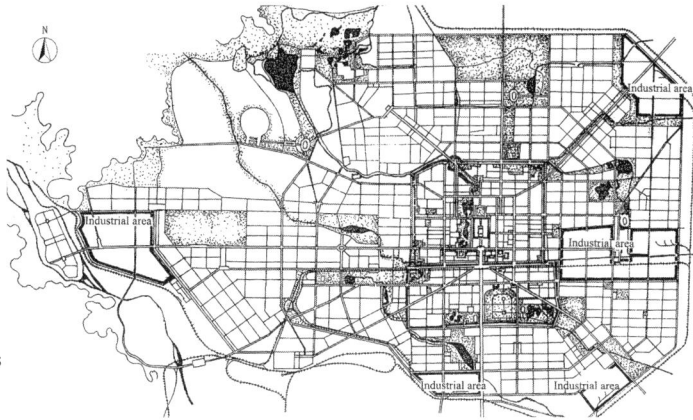

Figure 4.8. 1953 Beijing Plan (*Hua*'s proposal). (*Source*: Dong, 2006, p. 28)

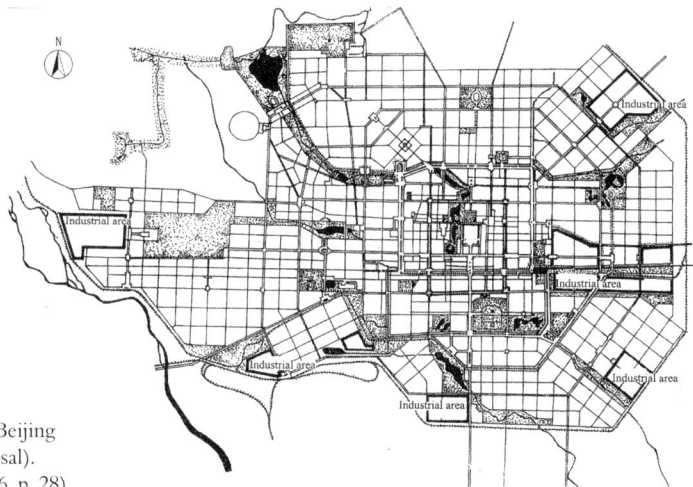

Figure 4.9. 1953 Beijing Plan (*Chen*'s proposal). (*Source*: Dong, 2006, p. 28)

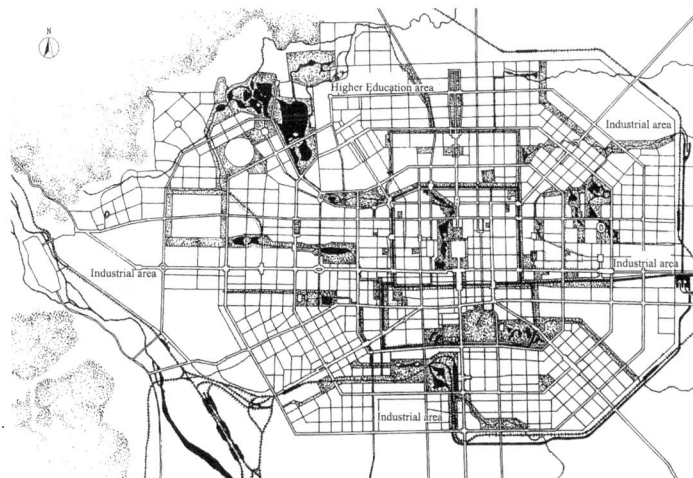

Figure 4.10. 1953 Beijing Plan (as revised in1954). (*Source*: Dong, 2006, p. 29)

In the 1953 Master Plan, functional zoning was introduced under which 'industrial zones were dispersed in suburbs, while new residential areas were located between the old city and new industrial zones' (Sit, 1995, p. 92, cited in Lu, 2011, p. 94). The zones were to be separated by greenbelts. The Master Plan privileged industry and production. It stipulated that, as well as being the political, economic and cultural centre of the nation, Beijing would be its industrial base and should strive to achieve a 'rise of efficiency in the working people's labour and production'. The Plan projected growth of Beijing's population to 5 million and the expansion of the city to a size of 600 square kilometres in 20 years, with areas for government, industry and education at the centre, southeast and northwest respectively. It required that the existing orthogonal streets be 'broadened, interlinked, and straightened' and advocated the addition of ring roads around the centre and radial avenues extending outwards in all directions. The major orthogonal avenues were to be 100 metres wide, while the radial avenues and the ring roads were to be 60 to 90 metres in width and the secondary roads 40 metres. The 1953 Master Plan suggested that Beijing should develop from its historical past, retaining important elements but eradicating the restrictions of old layouts and patterns. The preservation of heritage buildings was to be assessed on a case-by-case basis.

Known as the Chang-guan-lou Plan, the 1953 Master Plan consolidated Mao's ideas and the Russian proposals. Although further additions were made in 1954, 1957, and 1958, the 1953 Master Plan laid down the basic principles for Beijing's development for the next few decades.

Beijing Plan 1957–1958

In the 1957 Master Plan (figure 4.11) a further refinement was made through the introduction of the Soviet concept of 'micro-districts', each of 30 to 60 hectares and

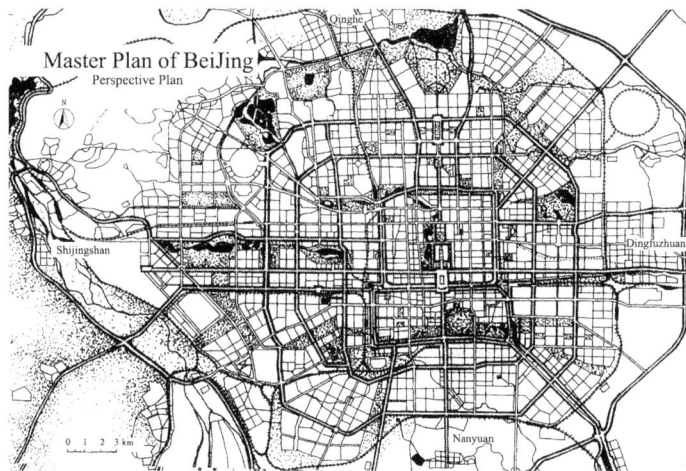

Figure 4.11. Beijing Master Plan 1957 (Draft). (*Source*: Dong, 2006, p. 30)

with 10,000–20,000 residents. An important objective of these 'micro-districts' was to limit the length of the journey to work. Even today, Beijing still has relatively short commuting distances, with the commuting ring having a radius of about 25 kilometres and most journeys to work being much shorter than this. The main reason for Beijing's shorter commutes is that government under the planned economic system has generally exercised very strong control over urban transportation facilities, employment, and the location of schools, shopping and the other daily activities of the population.

The Beijing City Master Plan of 1958 (see figures 4.12 and 4.13) further emphasized decentralized planning, based on a close relationship between work and home, and

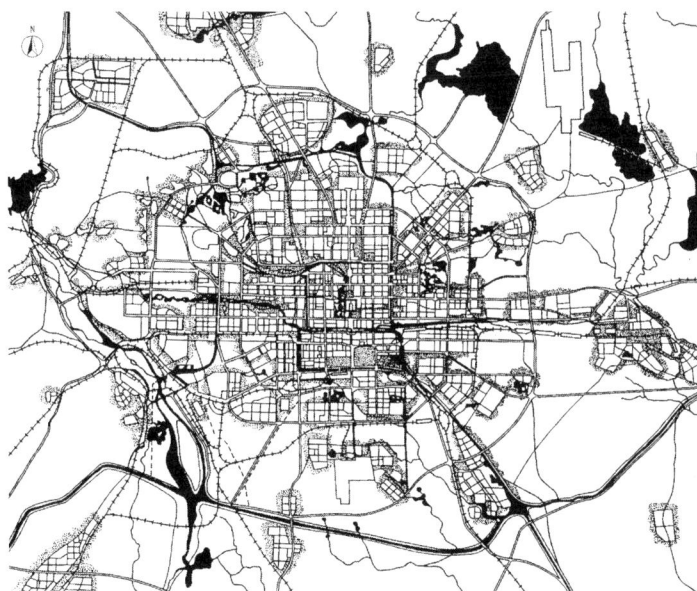

Figure 4.12. Beijing Master Plan 1958(A). (*Source*: Dong, 2006, p. 31)

Figure 4.13. Beijing Master Plan 1958(B). (*Source*: Dong, 2006, p. 31)

was somewhat critical of the principle of segregated land uses. Municipal planners struggled for many years to control urban growth, however, as work-units sought autonomy over their own areas or *danwei*. Central government oscillated between the desire to have more control over urban form and sympathy for the needs of work-units to build as they saw fit. 'At times, the Beijing Planning Bureau was dismissed by the state for being too strict and bureaucratic' (Sit, 1996, p. 96).

Under Mao Zedong, Beijing was massively transformed in both form and function. The influence of Soviet-style master planning persisted, even after the Sino-Soviet split in 1960. Major changes included the removal of the city walls that not only restricted urban development but also symbolized the previous feudal order; the layout of vast boulevards across the city, especially Chang'An, the Avenue of Heavenly Peace; and the massive expansion of Tian'an Men Square to become the focal point of the city and the largest urban square in the world. Ten huge new buildings were built (see later) to celebrate the tenth anniversary of the PRC (People's Republic of China) in 1959, including the Great Hall of the People, in which the National People's Congress is held. Beijing became a production city with a focus on heavy industries such as iron, steel and petro-chemicals (Dong, 1985). It also became one of the most austere capital cities in the world, virtually closing down by 9 o'clock in the evening, with international visitors few and far between. Until the 1990s, tourists had their own restricted currency that could only be spent in such places as the Friendship Hotel and the Friendship Store, which catered mainly for foreigners.

Urban Development Planning after the Cultural Revolution

Following the outbreak of the Cultural Revolution (1966–1976), the city's Master Plan and the Beijing Planning Bureau were both suspended in 1967 and not re-established until 1972 (Sit, 1995, p. 205). Beijing entered a period of planning anarchy and a great deal of piecemeal urban development was undertaken by individual work-units (although many outside China were completely unaware of this, viewing China still as a hierarchical monolithic state with top-down planning similar to that of the Soviet Union). The situation began to change in the early 1970s when China-US relations improved and China joined the United Nations. The pace of change increased after the death of Mao in 1976 and the assumption of the national leadership by Deng Xiaoping in 1978, following a brief power struggle with the 'Gang of Four'. Deng ushered in a period of reform that included the 'Four Modernizations' and an 'Open Door' policy, initiating a long process of economic and social transformation. Among Deng's many memorable slogans was 'to get rich is glorious', and his influence led China towards a consumption ethos. Within Beijing there were growing criticisms of the emphasis on industrial development in the early 1980s. Dong Liming argued that the city was 'poorly endowed for industrial development' (1985, p. 73) and yet was 'rich in cultural resources' (*Ibid.*, p. 75). It should therefore concentrate on expanding its role as the nation's administrative centre and should develop also as a centre of scientific research, education, cultural activities and tourism. A few industries should be permitted, but only those not consuming large amounts of water or creating pollution. This debate concerning the appropriate balance between Beijing's role as a capital and cultural centre and its role as a production centre had existed back in the 1950s (Sit, 1996) and it would continue into the twenty-first century, as concerns about pollution increased.

Beijing Plans in the 1970s and 1980s

Gan (1990) notes that a new Master Plan for Beijing was proposed by the Central Committee of the Chinese Communist Party and the State Council in 1978, in recognition of the need not only for economic development, but also for political and cultural priorities to be acknowledged (see figure 4.14). Changes to planning regulations between 1980 and 1983 were made to facilitate new forms of development and foreign investment began to flow in, funding new hotels, offices, shopping malls and exhibition centres. Large new housing districts, subway extensions and the Second and Third Ring Roads were completed, and the annual volume of construction in the city rose from 4.5 million square metres in the late 1970s to between 7 and 10 million square metres in the second half of the decade (Zhu, 2009).

Also in the 1980s the Zhongguancun area, in the north-western suburb of Haidian, began to develop as a high-tech manufacturing area. In 1988 it was formally designated

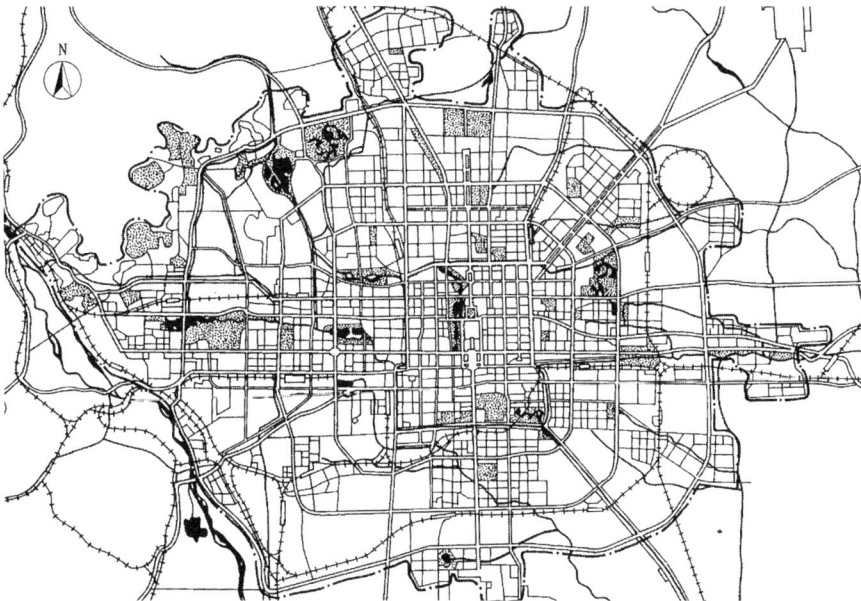

Figure 4.14. Beijing Master Plan 1982. (*Source*: Dong, 2006, p. 44)

as the Beijing High-Technology Industry Development Experimental Zone. High-tech industries have become progressively more significant in this area in recent decades and Zhongguancun, nicknamed 'China's Silicon Valley', now accommodates Microsoft and other major international companies, attracted by highly skilled Chinese labour, available from nearby prestigious universities at a far lower cost than would be the case in California or Seattle (Cook, 2006). These companies were able to locate in high-rise structures, since Zhongguancun lies outside the Second Ring Road. Under the 1982 Beijing Plan buildings within this ring road could only be up to six storeys high. In 1985, height controls were expressed more precisely: height limits in areas around the palace, the royal temples and the lakes rose gradually from 9 to 12, 18 and 45 metres as the distance of buildings increased away from the centre. In 1987, the regulations became more restrictive, such that no buildings inside the Second Ring Road could be more than 18 metres high, except those along the ring road itself, Chang'an Avenue, and another avenue further south, where maxima of 30 and 40 metres were allowed. In 1990, some twenty-five historical areas in and around the old centre were specified as zones for protection.

Urban Development under Globalization

Although the 1980s were important for the development of Beijing, it was really after 1992 that progress towards becoming a new world city accelerated. The first decade of the reform period had been characterized by a grand ideological debate over the need to balance further privatization and economic development with social stability. In the late 1980s the conservative elite became the subject of a rising social critique based on ideals of equity and democracy, voiced especially by students and culminating in the events in Tian'an Men Square in 1989. The 1990s saw both stronger state control and a further opening up of the economy to market practices and foreign investment. After Deng Xiaoping's speech during his tour of Southern China in early 1992 (Cook and Murray, 2001), the 'socialist market economy' became the established policy. Deng's 'Four Modernizations' had encouraged rural reform, and the easing of controls on the mobility of agricultural workers, together with the acceleration of economic and industrial development, led to an exponential increase in migration to the city. The 'Open Door' policy and growth in foreign investment were also of enormous significance in Beijing's development. The amount of construction per year increased to between 11 and 12 million square metres in the mid-1990s, rising later to 20 million square metres and to 30 million square metres per year after 2000. The invasion of the city centre by massive buildings, the destruction of old courtyard houses and historic urban fabric, and the breaching of height restrictions began to occur at a greater pace in the early years of the twenty-first century. The Oriental Plaza and the Financial Street, for example, which should both have had a maximum height of 45 metres under existing controls, reached 68 and 116 metres respectively, as a result of aggressive demands for profit by their developers. The development of the planning framework in this period is described below.

Beijing Plan 1991–2010

The Beijing Plan for 1991–2010 expressed the aspiration that Beijing should be an international city 'open in all aspects' by the fiftieth anniversary of the People's Republic in 1999 and that it should become a 'modernized international city of the first rank' in the period between 2010 and 2050. A new Beijing Master Plan was drawn up by the State Council for the period 1991–2010 (Zhang, 1991) (see figure 4.15). This foresaw an increase in the GDP of Beijing from 50 billion *renminbi* (RMB) (US$5.7 billion) in 1991 to 310 billion RMB or US$35 billion in 2010. This was to be achieved by, amongst other things, the following spatial policies:

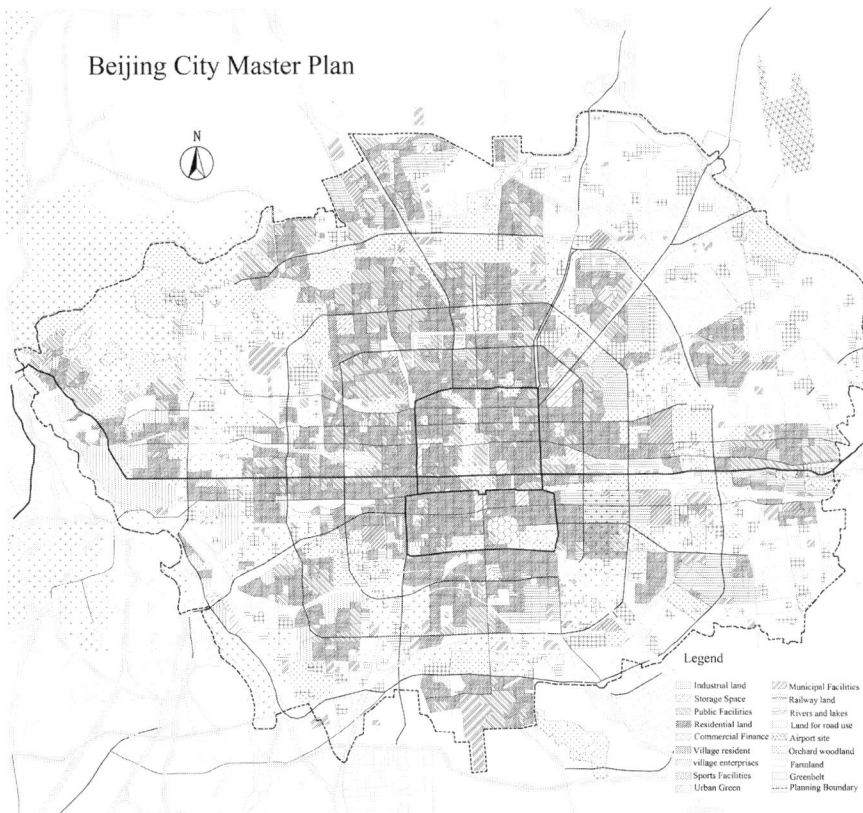

Figure 4.15. Beijing Master Plan 1991–2010.

Development of new and high-tech industries will be concentrated in Yizhuan, Zhongguancun and Shangdi economic development zones and scientific and technological parks. Secondly, major efforts will be made to develop tertiary industries to raise their position in the GDP from 38.8% to 60%. During this phase a Central Business District will be constructed in Chaoyangmenwai district. Thirdly, there will be a readjustment and distribution of secondary industries, gradually moving them out of the city proper… Measures will also be taken to prevent and control industrial pollution. Finally, the development of the agricultural sector should be accelerated to give the benefits from higher levels of technology and quality to the rural economy.[1] (Zhang 1991, p. 12)

Related features of the plan included large-scale residential building in the suburbs; satellite towns at Tongzhou and Huangcun; renovation in the old central city; improved water supply; provision of natural gas through pipelines from interior provinces; major road, telecommunications and subway developments; and a second international airport.

The Olympic Games and the Beijing Action Plan of 2002

Drawn up following China's successful bid in 2001 to host the 2008 Olympic Games (Cook, 2007b), the 2002 Action Plan sought to 'raise the level of openness in all aspects of the city of Beijing, and to display to the world a new image of the nation after reform and opening-up'. This plan was important in influencing the building of Beijing as it appears today. Its strategic conception included the theme of 'New Beijing, Great Olympics' and the organization of the Games as a national project based on 'green, science, and technology', 'humanism' principles, and a commitment to showcase a 'renowned, historical, cultural city' which, by the time of the Games in 2008, would have the 'framework of a large, modernized, international metropolis'. The general plan for the Olympic Park by the American firm, Sasaki and Associates, was based on a central Olympic area and three other zones. The 'Olympic Green' was designed as an extension of the old traditional north-south central axis, with extensive planting and water flowing through the whole area, adding life and beauty to the city's built environment. It included the three primary sporting venues, the National Stadium, the National Gymnasium and the National Swimming Centre, planned to accommodate 80,000, 1,800 and 1,700 spectators respectively. Major related infrastructure projects included four subway lines, a fast train and a second highway to the airport, the Fifth and Sixth Ring Roads, and a new terminal for the Capital Airport. The Plan also included the provision of a 'social environment' and modern cultural facilities in the city, including the National Grand Theatre, the China Central Television (CCTV) headquarters, a Capital Museum and a National Museum.

Greater Beijing Plan 2004–2020

By 2004–2005, after its dramatic opening up to the global market, Beijing, with its ambition of matching its economic level to that of the richer cities of the West, found itself also sharing some of their problems, including high pollution, a shortage of water and pressure on energy supplies. The Greater Beijing Plan 2004–2020 (see figures 4.16 and 4.17), was prepared at this time for an extended region covering 70,000 square

Figure 4.16. Greater Beijing Plan 2004–2020.

Figure 4.17. Greater Beijing Plan 2004–2020: Central Area.

kilometres and including neighbouring cities. Emphasis remained on the development of high-tech industry and a tertiary or service economy, while removing manufacturing industry to other locations away from the inner areas of Beijing. It was planned to reduce population growth and concentration within Beijing (by then already with a metropolitan population of 15 million) and to improve the city's ecological qualities.

According to the Greater Beijing Plan, the metropolitan region is to evolve into a polycentric structure, with two axes, two belts and multiple centres. The two axes run east-west and north-south, crossing at central Beijing. The two belts are the western ecological belt and the eastern development belt. Multiple centres are planned as part of the system of towns and cities, including three major new towns in the suburbs of Beijing. At Shunyi New Town, a manufacturing base is to be developed, Tongzhou New Town is to be a service centre and further high-tech development is proposed at Yizhuang. These new towns are planned to grow into cities with populations of between 700,000 and 900,000 by the year 2020.

In 2004 there were revisions to the National Constitution, bringing in a greater range of legal rights for individuals in China. Abramson (2007, p.76) notes that this has led to court decisions against local government agencies, including the Beijing Planning Bureau. Urban planners in China are now required to pay more regard to the views and opinions of various interest groups. The 'Regulating Plan of 2004–2020', introduced in 2004–2005, acknowledged these new imperatives:

The direction of development, as the Plan of 2004–5 indicates, will be increasingly about aspects of a humanistic city, including ecological and habitable qualities, rather than grand projects and heroic changes. A micro, internal and intensified urbanity at a human and walkable scale is likely to be emphasized… (This) may be remembered as the close of the first and the beginning of a new era in China's long march to modernization. (Zhu, 2009, pp. 209–210)

In 2006 the Beijing Municipal Government launched a document entitled 'Features of and evaluation of indicators for the counties and districts in Beijing'. This divides Beijing into four parts: core area, development areas, new urban area and substantial 'ecological conservation' areas. These conservation areas include the five counties of Mentougou, Pinggu, Huairou, Miyun and Yanqing and provide both an ecological barrier for Beijing and a water catchment with a combined area of more than 11,000 square kilometres. Beijing's future is now envisaged as national capital, international city, cultural city and liveable city.

Urban Internal Spatial Structure in the Reform Period

In 1990, a Land Use Reform Programme was launched and, while the State retained ownership of the land in urban areas under this programme, it became possible for urban land to be rented for speculation. An immediate consequence was the birth of construction companies and of a real estate market, as well as the initiation of important joint venture programmes with foreign companies. In the 1980s, the construction of Beijing's first large hotel and commercial complex was begun. This was the Lufthansa Centre on the Third Ring Road, conceived in co-operation with the German airline company. The foundations were also laid for the city's first steel and glass skyscrapers: the Jingguang Tower and the towers of the China World Trade Centre further to the south on Jianguomenwai, the result of a partnership between American and Japanese groups. These were followed over the next two decades by further significant developments in the central business, finance and commercial districts which have had significant effects on Beijing's urban form and internal spatial structure. The more important changes since 1990 are summarized below.

The Central Business District (CBD)

With the aim of facilitating investments by large international companies, the State Council ratified the creation in 1992 of an area called the Central Business District (CBD), situated near the Third Ring Road. This location, summed up in the slogan

Figure 4.18. China World Trade Centre Tower 3.

'the place where China meets the World', had previously been identified in 1984 as the site for the China World Trade Centre. Various complexes of buildings followed, erected by private enterprise. The Jingguang Centre was completed in 1990, the Kerry Centre in 2000 and the SOHO 'New Town' in 2001, but there was no coherent vision for this area until the 'Central Business District Plan' was approved by the Beijing Municipal Government in August 2001. The quarter currently referred to as the CBD extends over an area of about 400 hectares in Chaoyang District, delimited to the west by Dongdaqiao Road, to the east by Xidawang Road, to the south by the Tonghui River, and to the north by Chaoyang Road. It is located in a strategic position between the Cahwai Business Centre, the diplomatic quarter near Ritan Park and the residential areas of Yong'an Dongli and Xili. This area now contains the CCTV centre, to the south of Jianguomenwai (described later and see figure 4.33); the Beijing Yintai Centre, added in 2007, which is some 250 metres high; the Fortune Plaza Building 1, with a height of 260 metres; and the China World Trade Centre Tower 3 (see figure 4.18), completed in 2009, which is 330 metres high.

Finance Street

Beijing's 'Finance Street' (see figure 4.19), located at West Second Ring Road, running south from Fuxingmen Avenue, north to Fuchengmennei Street, and from west to

Figure 4.19. Beijing Finance Street.

east along Pacific Bridge Street, has a north-south extent of 1,700 metres, a width from east to west of about 600 metres, an area of 103 hectares and a total floor area of more than 300 million square metres. The core area of Finance Street has office buildings, apartments, a five-star hotel and an international conference centre as well as a variety of commercial, dining, entertainment, leisure, educational and health facilities.

Commercial Areas

Wangfujing Business Street (see figure 4.20) is one of the best known shopping streets in all of China, visited daily by an average of half a million people. It has been the site of markets since the middle of the Ming dynasty. The central area was expanded in 1999–2000 and the enormous Oriental Plaza was inaugurated, a complex of ten buildings by the Hong Kong firm P&T Architects and Engineers Limited.

Figure 4.20. Wangfujing Business Street.

Wangfujing Business Street is to the east of the Forbidden City, while to the west is Xidan Avenue where the redevelopment of an area of about 80 hectares commenced in 1999 and is still underway. Sixteen large stores are concentrated here in addition to the inevitable hotels and restaurants, giving a total floor area of 1.5 million square metres which also receives some half a million visitors daily. An underground shopping area is planned here to connect directly with a subway station. Above ground is the Xidan Culture Plaza with the Bank of China's main headquarters, as well as the capital's largest multi-storey bookstore.

The Historic Centre

The Beijing central area is the site where the ancient capital of Dadu was founded in 1283. The Ming period city walls were demolished in the 1950s, as mentioned earlier, but this area still retains its historic urban layout (see figure 4.21). Within the central area, the Greater Forbidden City (figure 4.22) remains with its dense network of

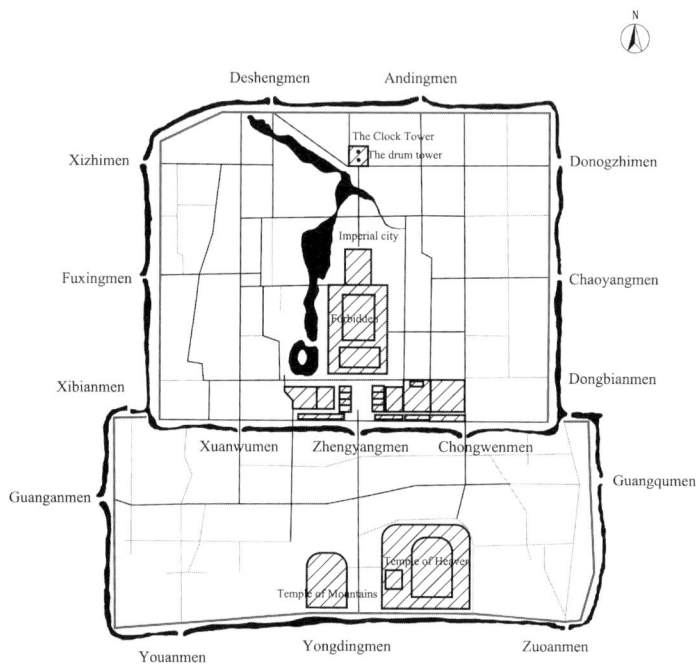

Figure 4.21. The Beijing central area and the ancient capital.

Figure 4.22. The Greater Forbidden City.

perpendicular streets and its ordered chequerboard of rectangular blocks. At the heart is Tian'an Men Square, still the focal point of the two main orthogonal axes oriented north-south and east-west. The line of the old walls is now followed by the Second Ring Road. In the Regulating Plan 1992–2010, referred to earlier, which also addressed conservation of the historic centre, the majority of new investments, mostly private, were directed to the city outskirts beyond this ring road and away from the central area.

Shichahai Bar Streets. In the Shichahai quarter of the Xicheng District, near the artificial lakes Houhai, Xihai, and Qianhai, the first 'No Name Bar' opened in 2001 and this was followed by the development of a series of 'bar streets' (see figure 4.23) which now dominate the character of an area which is one of the oldest remaining parts of Beijing.

Figure 4.23. Shichahai Bar Street.

Qianmen Street (see figure 4.24) is a pedestrian zone with 400 stores, along with antique shops, cafes and restaurants. There are also premises of some fifty major art galleries, including the Pompidou Centre, designed by Jean Nouvel, with an area of 10,000 square metres, and also the Maeght Foundation, which held its first exhibition in March 2008 with works by Miro. Qianmen Street, through its urban transformation and renewal, has become a 'brand new old place'. Both sides of the shopping street are of three-storey traditional Chinese buildings, with balconies above the first floor, lattice railings and red columns. These antique buildings also contain a shopping mall and the outlets of well-known international brands.

Figure 4.24. Qianmen Street.

Nanluoguxiang Street (see figure 4.25) is one of the most ancient avenues of Beijing. It runs north-south for 768 metres in the Dongcheng district. Here there is a requirement to respect traditional style and for the fronts of constructed buildings to observe the historic layout which dates back to the Yuan period.

Figure 4.25. Nanluoguxiang Street.

New Beijing Central Axis. As noted earlier, the historic north-south central axis (figure 4.26) has been extended further north through the centre of the Olympic Green with the intention of both emphasizing and renewing the symbolic spirit of the city's traditional axis.

There is also a strengthened east-west central axis which extends Chang'an Avenue along Fuxingmenwai and Jianguomenwai. Today this has become a broad six-lane boulevard, lined with architecturally striking, prestigious hotels, major company headquarters and up-market shopping centres.

Figure 4.26. The extended north-south central axis.

Developments in the Suburbs

Beijing is much more than just its central urban area. The city extends over a territory of some 16,800 square kilometres, subdivided into sixteen districts and two counties. Some of the more significant recent developments in this wider area are as follows.

Zhongguancun Science Park. As noted earlier, the Beijing High-Technology Industrial Development Experimental Zone was set up in 1988 at Zhongguancun in Haidian district. At the beginning of 1996 the new library of the Academy of Sciences at Zhongguancun was designed in association with the Fourth Ring Road. This library was the first building of a new campus, which was intended to modernize the scientific research quarter. Five development zones were identified under the overall name of Zhongguancun Science Park, comprising various sites across the city linked by the Fourth Ring Road. Since 2000, a further plan has sought to interconnect Beijing's major universities and research institutes. The evidence suggests that the creation of the Zhongguancun Science Park has turned out to be a successful strategy – some 60 per cent of Beijing's GDP now derives from over 7,000 companies concentrated here, linked mainly to information processing, pharmaceutical and medical research. There are also sixty-eight universities and 230 independent research institutions, which fund and support 36 per cent of all China's researchers today.

Yuanmingyuan Relics Park (The Old Summer Palace). Yuanmingyuan is located in the western suburbs of Beijing. It was originally the Royal Regency during the Qing Dynasty, comprising three parks – Yuanming Park, Changchun Park and Yee-Chun Park – with a total area of 350 hectares. The southern area of Yuanmingyuan was the imperial court, the official seat of the emperor. The rest was given over to parks and scenic features. In 1860, British and French troops sacked the Summer Palace and further damage was done during the Boxer Rebellion. In recent years, however, the park environment has been restored, all residents have been moved out and Yuanmingyuan is becoming a beautiful 'relics' park.

The Restructuring and Relocation of Shijingshan Steel Industrial Area. As described earlier, Beijing changed during the period of socialist construction from being a consumer-oriented city to a productive city, and the Shijingshan Iron and Steel plant was developed as a major Chinese manufacturing base for iron and steel. Its output grew from 260,000 tons of steel in 1949 to 1.79 million tons in 1978 and 8.24 million tons in 1994, making it the largest in the country. Since the beginning of the 1990s, however, many manufacturing enterprises have been moved out of Beijing and, in order to host the 2008 Olympic Games successfully, China's State Council approved a 'relocation, structural adjustment and environmental governance programme' for the Shijingshan Iron and Steel Plant in February 2005. Under this programme, the plant has moved to Tangshan City in Hebei Province and the Shijingshan District has become the home for a number of theme parks and entertainment facilities, including the Shijingshan Amusement Park, the Long Yang seawater swimming pool, a '4D' Cinema and an International Sculpture Park.

Yizhuang New Town: Beijing Economic and Technological Development Area (BDA). The Beijing Economic and Technological Development Area (BDA) (see figure 4.27) was established at Yizhuang in south-east Beijing in 1992. In 2007 it was incorporated by the Beijing Municipal Government as the core of an 'e-Town New City', planned to be a key new site for further economic growth in Beijing's eastern belt as part of the Yizhuang New Town development (see earlier). The development of a number of industrial clusters has been vigorously promoted at BDA. These include an ICT industrial cluster (with Nokia), an electronic industrial cluster, a medical equipment industrial cluster, a biopharmaceutical industrial cluster (with Bayer) and an auto-industrial cluster (with Mercedes-Benz-DaimlerChrysler). More than 2,000 enterprises from thirty countries and regions all over the world have established themselves at BDA with total investments exceeding US$15 billion, over seventy per cent of which are investments by foreign enterprises.

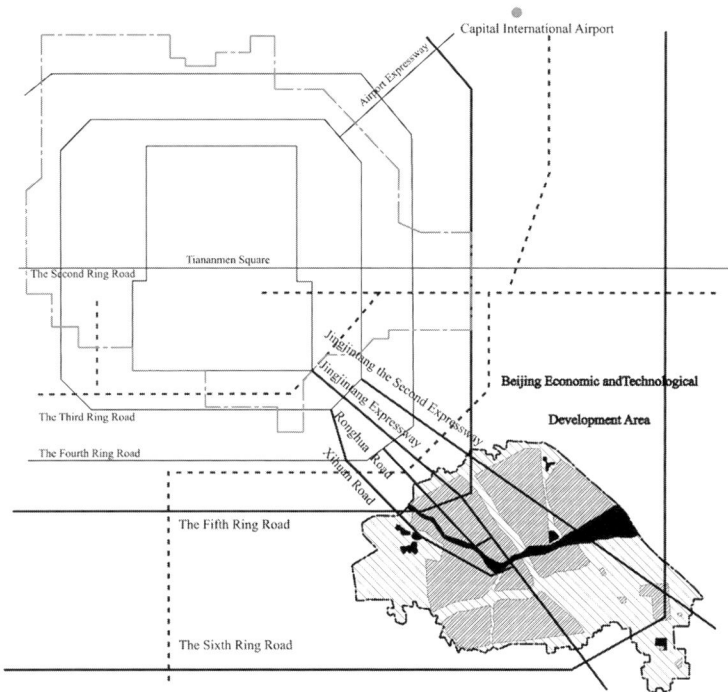

Figure 4.27. Location of Beijing Economic and Technological Development Area.

The 798 Factory: Cultural and Creative Industrial Cluster. The 798 industrial complex was originally a gift in 1956 from the government of East Germany. In 2002 a number of artists, intellectuals, writers and musicians discovered the charm of this large, abandoned factory area to the east of the city. They installed themselves in some disused production sheds to isolate themselves from the noise and confusion of the metropolis and the 798 Factory was born (see figure 4.28). Artists of all sorts began occupying the available spaces, the area was enlivened by the opening of the first bars and restaurants, and the former factory became a hotbed of ideas and culture. A modern Museum of Film was established in 2006 and the Belgian Ullens Centre for Contemporary Art was

Figure 4.28. The 798 Factory.

set up in 2007 with over 6000 square metres of land for development. This new 'Artist Village' has quickly become known internationally as an *avant-garde* cultural centre.

Fangzhuang. Some rich and successful citizens of Beijing, as well as central government ministries and their affiliated companies, began acquiring property here in the 1980s and early 1990s, so that Fangzhuang became the capital's first modern residential area or 'rich man's zone'. After several years of development, the surrounding environment of the Fangzhuang community is already quite mature and, with convenient transportation, relatively low-cost property management fees and heating costs, it has become a very large and successful residential area.

Wangjing New Town. Located in the Chaoyang district between Central Beijing and the Capital Airport, Wangjing New Town was planned in 1992. High residential tower blocks began appearing in the countryside, replacing austere, disused factories dating from the 1950s and the first tenants moved into Wangjing in 1996. It has now become Beijing's main Korean colony, with 80 per cent of its tenants, some 60,000 people, being South Koreans, mostly of lower middle class. There was substantial expansion after the Wangjing Science and Technology Park was launched in 1999. Sony-Ericsson, one of the larger international telephone companies, is based here and it is also an area for businesses directed by entrepreneurial Chinese who have returned from abroad.

Urban Transport Planning

At the end of 2007, there were 330 kilometres of urban expressway in the Beijing central area. Beijing Capital International Airport is China's largest airport, with an annual passenger throughput of 60 million passengers, 1.8 million tons of cargo and mail and some 500,000 planes taking off and landing annually. It was important to guarantee the quality of transportation for the 2008 Olympic Games and Beijing therefore accelerated infrastructure investment to build up the integrated carrying capacity and emergency response ability of its city transport system. As a result, the overall quality of transport facilities is now high and the system achieves good levels of service in relation to objectives of safety, speed, convenience, economy and environmental protection. New facilities provided in the run-up to the Olympics included Terminal Three of the Beijing Capital International Airport, the Beijing South Railway Station and Metro Line 10.

Terminal Three of the Beijing Capital Airport. The commission for the new Terminal Three was awarded to the architectural firm of Norman Foster. The design, with its curvilinear profile, was chosen for its appealing resemblance to an enormous red dragon. The terminal uses passive heating systems, the window openings face southwest, and the climate control system minimizes carbon-dioxide emissions.

Beijing South Railway Station. Beijing South Station is the terminus for the departure and arrival of most Beijing-Tianjin inter-city passenger trains. This large traffic hub integrates multiple transport modes: railway, metro, suburban railway, bus, taxi and private car. Peak-hour passenger arrivals at Beijing South Station are expected to reach 30,800 by the end of 2030.

Subway Expansion. There were only 42 kilometres of subway line in operation in 1999. As preparations for the Olympic Games proceeded, substantial extensions were made to the network. In 2003, Line 13 was opened, together with extensions in an eastern direction, toward the Tongzhou District, to Lines 1 and 2. It is intended that the planned network will be completed by 2015, by which time there will be nineteen operating lines with a total length of 561.5 kilometres.

The recently built metro line from Dongzhimen to Terminals Two and Three of the Beijing Capital International Airport has a main line 28 kilometres long and only four stations. It is the first driverless metro line in China and can reach a speed of 115 kilometres an hour, making the trip from Dongzhimen to the airport as short as 16 minutes.

Changing Urban Landscapes, Architecture and Social Spaces: Towards a World City

Before the Reform Period, as noted above, the domination of production units in the *danwei* resulted in Chinese cities being full of 'functionally and visually homogeneous landscapes' (Gaubatz, 1995, p. 31) extending over vast areas. As Gaubatz notes, the Reform Period saw a reduction in the planning powers of the work-units and an increase in the strength of municipal planning that was to lead to a growing separation of work and residential areas. However, although the influence of work-units decreased, the power of the market system increased in the 1990s under the 'mixed system of plans and markets' (Chan, 1994, p. 98), leading to a number of new challenges for planners.

For example, Liu *et al.* (2002) noted that new urban growth in the outer area in the years 1982–1992 was concentric, at an average distance of 7.5 kilometres from the city centre. By 1992–1997, however, new growth was mainly to the north, and to a lesser extent to the south, at an average distance of 10.8 kilometres. Such 'sprawl' to the north contradicted the intent of directing growth towards the southern and eastern parts of the city (*Ibid.*, p. 272). Other problems included the fact that this suburban growth was leading to a loss of high-quality arable lands, loss of already limited open space and increased traffic congestion (Jiang *et al.*, 2007, p. 475). In part, such changes were the result of the development of a land market that:

drives the spatial separation of land use. In other words, office and commercial development have an economic advantage in locations close to the city centre, whereas industrial development is pushed farther away towards the suburbs. Residential development is most likely to take place in between. (Ding, 2004, p. 1904)

Such separation of form and function has led to Beijing suffering from many of the problems of other great cities in Asia, with the city being assessed as the most polluted city on earth in 2005, causing great concern in advance of the 2008 Olympics (Cook, 2007*b*). It has also encouraged the development of a substantial number of tall, architecturally-distinctive commercial buildings in the inner parts of the city which increasingly establish the image of Beijing.

There are different Beijing landmarks that typify different periods of its development. These landmark buildings (see table 4.1) and spaces reflect the spirit that animated the city at the time of their construction, recording economic, technological and cultural characteristics as well as the pattern of architectural and urban evolution. The next section of this chapter traces the significant stages in Beijing's architectural development since the 1950s and describes some of its more important buildings.

A Modern Tradition before 1976

In the 1950s, under the Communist Party, an architecture of socialist realism or 'Socialist Content with National Form' was promoted in Beijing. Under socialist modernism, architecture was to serve large, public, collective functions. The celebration project for the tenth anniversary of the People's Republic in 1959 under

Table 4.1. Beijing's grand buildings.

	Grand Buildings
1950s	The Great Hall of the People, Museum of Chinese History and Revolution, the Chinese People's Revolutionary Military Museum, the Cultural Palace of Nationalities, Ethnic Hotel, the Diaoyutai State Guesthouse, Overseas Chinese Building, Beijing Railway Station, National Agricultural Exhibition Centre and the Beijing Workers Stadium
1980s	National Library, the China International Exhibition Centre, the Central Colour TV Centre, the Capital International Airport Terminal 2, Beijing International Hotel, Grand View Garden, the Great Wall Hotel, China Theatre, the Chinese People's Anti-Japanese War Memorial Hall and the Beijing Subway Dongsishitiao Station
1990s	Central Radio and TV Tower, the National Olympic Sports Centre and the Asian Games Village, Beijing New World Centre, the Beijing Botanical Gardens Exhibition Greenhouse, New Library of Tsinghua University, Office Building of Foreign Language Teaching and Research Press, Beijing Henderson Centre, New DongAn Market, International Financial Building, the Capital New Library
2000s	The National Grand Theatre, Bird's Nest, Water Cube, the new CCTV building, T3 terminal, Beijing South Railway Station, Capital International Airport, Beijing South Railway Station, Capital Museum, Beijing TV Centre, National Library (2), Beijing New Poly Plaza

Mao was a major architectural event. In addition to the expansion of Tian'an Men Square, it required the completion of 'Ten Grand Buildings' by 1 October 1959, the PRC National Day. The Ten Grand Buildings included the Friendship Hotel by Zhang Bo (1956), the Palace of Nationalities and the Beijing Railway Station (by Zhang Bo and Yang Tingbao, respectively, 1959). There were also two Grand Buildings to the east and west of Tian'an Men Square, the Museum of Revolution and History (Zhang Kaiji, 1959) and the Great Hall of the People (Zhao Dongri and Zhang Bo, Beijing, 1959).

In the sixties, as a consequence of the radical campaigns of the Cultural Revolution, popular icons such as large portraits of Mao, revolutionary slogans and sculptures of red flags, red torches and other symbols of the 'revolutionary masses' in a socialist-realist genre were applied to or around buildings, as in the Memorial Hall of Chairman Mao to the south of Tian'an Men Square, completed in 1977 (Zhu, 2009).

'Neo-National Style' and 'Modern Vernacular' in the 1980s

The 1960s were a fairly quiet period in construction. After the Cultural Revolution, however, there was an eclectic period of architecture which drew on examples from straight revivalism to the creative use of traditional Chinese elements with post-modern influences (such as the Western Railway Station). Another style was modern vernacular. Instead of employing traditional Chinese roofs, as in the National Style, a regional, vernacular language (including pitched roofs, traditional window patterns and textured walls in vernacular houses) was employed in designs which were also consciously modern or abstract (see, for example, Denton Corker Marshall's Australian Embassy in Beijing, 1982–1992) (Zhu, 2009).

Neo-Classical or Late Modern in the 1990s

The 1990s saw more neo-classical or late modern architecture with representative examples being Wu Liangrong's Juer Hutong Houses (1992), Guan Zhaoye's New Library at Tsinghua University (1991) and Liu Li's Yanhuang Art Gallery (1991).

The International Style in the 2000s

The first decade of the twenty-first century saw an increase in the scale of new developments in Beijing. Through design competitions, foreign architects with established portfolios in the new and radical modernism of the 1990s won many commissions for projects which are now under construction or have recently been completed. These include the new headquarters for China Central Television (CCTV) by Rem Koolhaas (with ECADI), the National Olympic Stadium by Herzog and de Meuron, with China Architecture Design and Research Group (CAG), and the National Grand Theatre by Paul Andreu. Some modernist structures have also been designed by Chinese architects, including the China Academy of Urban Planning and Research by Cui Kai (2003).

What is emerging in Beijing today is a distinctive landscape as the Chinese state has invited these international architects to create the landmarks of an 'open, modern, international city'. This landscape provides a spectacle intended to showcase Beijing and China in the global media. These buildings are symbols and material components of the real socio-economic and political transformations underway as China has shifted from being a closed state administrative centre to becoming a hub city in the Asian region (Zhu, 2009). Beijing's key role in these new processes of globalization gained further impetus from the urban redevelopment driven by the 2008 Olympic Games, the results of which, in architecture and urban design, were spectacular. Figure 4.29 provides a view of the public realm of the Olympic Park, referred to earlier, while the following two figures show the striking architecture of the Olympic Stadium and the Aquatics Centre.

The Bird's Nest. The National Olympic Stadium (see figure 4.30) is the realization of the idea, simple but closely bound to traditional Chinese culture, of a 'bird's nest' or, alternatively, as described by artist Ai Weiwei,[2] who collaborated with the Swiss designers Herzog and de Meuron, of traditional local porcelain bowls with their networks of tiny cracks. The functional sporting venue is thus transformed into a poetic yet powerful and internationally recognized architectural symbol of Beijing.

Figure 4.29. Beijing Olympic Park.

Figure 4.30. The Bird's Nest.

The Water Cube. The National Aquatics Centre is affectionately dubbed the 'Water Cube' (see figure 4.31). Designed by the Australian firm PTW, the Water Cube is a large enclosure with walls that look like water bubbles, an effect obtained by using a complex spatial structure of steel tubing, over which is extended an innovative, translucent, inflatable material. As the sunlight filters through, spectators and athletes feel that they are inside a magical underwater world and, by night, the Cube is transformed into a luminescent aquarium. North of the Water Cube is 'Digital Beijing', designed as a new aesthetic image for the Olympic Games by Zhu Pei. This looks like an enlarged part of an integrated circuit board or microchip, with water pouring down from the top like a waterfall, and gradually changing into a star shower. It has now become the headquarters for the Municipal Office for Information Systems, as well as an exhibition centre for creators of digital products.

Figure 4.31. The Water Cube.

National Grand Theatre. Paul Andreu's National Grand Theatre (figure 4.32) has a radical and controversial design (see Broudehoux, 2004) with a huge curved dome of titanium plates and glass panels above an underwater tunnel. The performance complex is conceived as a citadel of theatres enclosed by an elliptical, semi-transparent cupola in glass and grey titanium, surrounded by water. The entrance to the north on Chang'an Avenue is connected to a subway station and to a huge parking garage for 2,500 vehicles. The interior contains an opera house, a concert hall and a theatre.

The CCTV Tower. The headquarters for China Central Television, this project by Rem Koolhaas, begun in 2002 and completed in 2010, has three main elements – the CCTV Tower, the Television Cultural Centre (TVCC) and the Media Park. The striking CCTV Tower itself in fact comprises two towers, which lean towards each other and eventually join dramatically in a cantilevered form (see figure 4.33).

Figure 4.32. National Grand Theatre.

Figure 4.33. The CCTV
Tower.

Urban Challenges and Prospects

This chapter has outlined the development of planning within the amazing city
of Beijing, a city that faces the demands of being the capital of the most populous

nation on earth, of being run within a socialist system, albeit one in which the market economy plays a major role, and of being increasingly presented as a new world city that is recognized across the globe. As the earlier parts of this chapter have shown, there has been a dramatic change and transition in Beijing in a fairly short period of time from being in the radical vanguard of Marxism-Leninism after 1949 to the socialist market economy since the 1990s. The combination of market economy and strong centralist state has delivered an apparent social stability together with a staggering economic growth rate. A different urban society has emerged. Yet underlying problems, notably income disparity, environmental pollution and the neglect of social welfare, are accumulating.

The Beijing Olympic Games of 2008 were generally very successful and demonstrated the high level of internationalization to which the city aspires, although some outstanding issues still exist, including debt, the reuse of the Bird's Nest and other facilities, and continuing jobs for Olympic volunteers (Cook and Miles, 2010). The Olympics helped improve transport infrastructure, communication facilities, housing, stadiums, the old city, environmental governance and ecological protection, as well as stimulating consumption. All of this helped to expand China's 'soft power' overseas, as well as contributing to the city's overall development. However, Beijing's planning processes have not always been successful in guiding and directing the incredible pace of change in recent years. This penultimate section examines some of the main challenges that remain for the next decades of the twenty-first century.

Social Polarization and Social Injustice

Beijing is a city of increasing contrasts in income, with marked inequalities between different groups within the population. Under Mao, China introduced the *hukou* registration system that entitled officially sanctioned residents to gain access to housing, health, food rations and other benefits. In the Maoist period migration was tightly controlled by this system but, with the introduction of the market economy, rural migrants have flooded into China's cities, including Beijing. Known as the *luidong renkou* (floating population) they are 'in' the city but not 'of' the city (Tang and Parish, 2000, p. 31). Gu *et al.* (2006, p. 275) showed that there were over a million of these people in Beijing by 1989, followed by a dip in 1990 and then a steady expansion up to 3.85 million by 2004. These are generally people without *hukou* registration, employed, for example, in the construction industry, as stallholders selling a range of clothing and foods, or as taxi drivers – all jobs which are highly demanding, and from which the existing Beijing population has tended to move on.

Through their visibility to tourists and outside visitors, these 'marginal' groups disturb the modernization project and question Beijing's alleged modernity, symbolically regaining their right to the city. (Broudehoux, 2004, p. 136)

These are people who would be regarded as comprising the informal sector and living in shanty towns in other large cities of Asia. But these migrants do not live in shanty towns. Rather, they are likely to live in crowded conditions with people from their own town or province within urban enclaves in Beijing, where there are over a hundred migrant villages (Gu *et al.*, 2006). They face a regular threat of eviction and demolition if their dwelling-place or street location is required for new urban development or if the authorities seek to regulate the informal system by providing new covered markets as occurred, for example, at the Silk Markets in Jianguomenwai. With the recession in 2009, many went back to their home villages and towns but they are likely to return when the economy recovers in the next few years. When they do, they will face issues of lack of access to schooling for their children, plus lack of 'access to low-cost health services and housing, and equal employment opportunities in state-owned enterprises or foreign and joint-venture companies' (Gu *et al.*, 2006, p. 288).

Singly, and with others, Gu Chaolin has conducted a range of studies into the spatial concentration and segregation of these migrant groups, showing clearly that social polarization is a key feature of Beijing's current situation (Chan *et al.*, 2000; Gu, 1998, 1999 and 2001; Gu and Kesteloot, 1997, 2001, 2002; Gu and Liu, 2001; Gu and Shen, 2003; Gu *et al.*, 2005). The challenge that the floating population poses is not just one for Beijing, but for the PRC government more generally.

Environmental Injustice

Planning also has to contend with the unequal impacts of environmental problems. Over the years Beijing has had many successes in the struggle to overcome such environmental negatives as water shortages, desertification and air and water pollution (Cook, 2007*a*, 2008; Murray and Cook, 2002, 2004). Massive afforestation and 'greening' programmes, increased use of liquid petroleum gas, restrictions on trucks entering the central area during daytime, regulations to limit the use of poor quality coal and the like have increased the number of 'blue sky days' that the city has annually. An analysis by the United Nations Environmental Programme (UNEP) complimented the Beijing authorities for the improvements that were made to raise the environmental bar in the lead up to the Olympics. Restrictions on vehicle use proved so effective at this time that these were extended after the Olympics (Cook and Miles, 2010). Nonetheless, as noted earlier, Beijing was identified by the European Space Agency in October 2005 not just as a polluted city, but as the most polluted city on earth (Cook, 2007*a*) and air pollution remains a significant problem. 'Beijing throat' is a regular threat to the visitor and resident, and coal remains the main energy source in Beijing as it is in the rest of China. Vehicle emissions also remain a major contributor to air pollution, and to a high level of respiratory disease, but the growth in vehicle numbers continues to outstrip forecasts. Water quality is also an issue – even in the capital of an increasingly wealthy China, a country with the third largest economy in the world.

Wealthier residents can afford to relocate further away from pollution sources or can apply pressure on the authorities to ameliorate the worst impacts of pollution; poorer people are much more likely to have to put up with such conditions with detrimental effects on their health (Cook, 2007a, 2008). There is an ongoing debate over such issues in Beijing, and in China's megacities in general, as seen, for example, in the *China Daily* during 2010. Some argue that urbanization permits improved environmental conditions as a result of the incorporation of better environmental technologies, such as smarter air conditioning units, solar power and rainfall collection techniques in modern high-rise buildings. This was an important emphasis of Shanghai's 2010 World Expo. Until recently, one of the current authors, Cook, was moderately optimistic that Chinese and foreign ingenuity would combine to develop effective solutions to environmental constraints in Beijing, but this confidence was shaken by the severe dust storms of March 2010. These were a reminder that Beijing's regional location remains a significant factor in shaping current and future environmental conditions, notwithstanding the massive tree planting programmes of recent decades.

The Commodification of Culture

After 20 years of urban renewal and transformation, Beijing has improved parts of its cultural environment, but it is obvious that, overall, there is a continuing deterioration of the Old City. Beijing Old City has been over-developed in commercially intensive ways, with damage to more than half its historic buildings and a drastic reduction in the number of *hutong*s. The Olympics were the latest and grandest example of how culture is being commodified in Beijing, as it has been in other world cities. This process began some years ago, paradoxically with Mao becoming a tourist commodity through the marketing of Mao caps, watches, cigarette lighters and street sales of the Little Red Book (Cook, 2008). The *hutong*s have also become an attraction, with the '*hutong* tour' by pedicab a popular part of the tourist experience. Gu Huimin and Chris Ryan (2009) have studied the impact of tourism on the largest of the *hutong* protection zones, Shichahai, based on interviews with residents and local tourism business people. Scores on Likert-type scales identified traffic congestion as the main problem with tourism, and locals tended to agree with the statement that 'local residents mainly suffer from living in a tourist area'. Many recognized that the area was cleaner than before, yet 48 per cent 'strongly agreed' with the statement that 'I feel tourism is growing too fast for the *hutong* to cope with' (Gu and Ryan, 2009, p. 318). Longer-term residents, in particular, were likely to be more negative towards tourism developments. *Hutong*s are laid out along narrow lanes, so it is small wonder that traffic congestion is a problem for residents, but congestion is also increasingly a problem for other tourist sites in and around Beijing, as at the Great Wall at Badaling, for example. Beijing is a growing cultural attraction for people from within and outside China, and increasingly planning will have to confront the congestion and other issues that cultural tourism raises.

Questions of Governance

Prior to the Olympics, concerns were raised in the West about the thorny question of human rights. The memory of the events in Tian'an Men Square in 1989 has proved hard for the PRC to shake off, and such issues as the use of the death penalty, Tibet and treatment of the Buddhist sect Falun Gong have drawn criticism from Western human rights activists (Cook, 2007*b*). Prior to the 2008 Olympics, the Olympic Torch was attacked by Free Tibet supporters and this led to counter-protests by PRC supporters in Hong Kong and China itself (Cook and Miles, 2010). Journalists were also concerned in 2008 that they would not be able to report freely from Beijing because of restrictions placed upon them. In the event, there were few problems at the Olympic Games themselves, and the PRC government continues to defend itself against critics of its record by suggesting that human rights are less than ideal in some of those countries which are most vociferous in their criticism of China. It is the case that the rights of the individual are gradually expanding in China while the use of the death penalty is reducing. China is changing, and so is its governance.

There can be tensions, however, between Beijing's local (district) authorities and the municipal or central government, as Huang (2004) notes. The local authorities are particularly driven by economic imperatives, seeking to maximize local income and investment, and this can go against broader citywide strategies that are concerned with some of the issues of environmental quality and liveability noted above. Further, population forecasts are difficult to make precisely at the citywide level, allowing local authorities with control of their local area's forecasts to develop housing programmes that may be at odds with wider needs or demands. Because the ultimate responsibility lies with the State Council, such contradictions may be resolved by higher level intervention, but central government finds it difficult in Beijing, as elsewhere in China, to keep up with the rapid pace of local change. There have been some moves in recent times in the direction of a planning system which places more emphasis on a more humanistic, liveable city and it is to be hoped that these continue. But in order for this to occur, local people, *hukou* and non-*hukou* alike, will need to be more fully involved in planning for their own areas and in helping the planners to decide between alternative models of planning development. The seeds of such a system have already been planted in the growing number of non-government organizations (NGOs) and community groups found in China generally and in Beijing in particular (Cook, 2008). As in other countries, NGOs fill a key space that government is unable or unwilling to occupy, and can act as pressure groups to try to ensure that local or municipal authorities deal more sensitively with issues faced by local people. These seeds will require careful nurturing, however, in order to deal with the issues of social inequality, environmental threats, periodic economic downturns and political change that will arise in the decades to come.

Conclusion

Beijing, the ancient capital of China, is a city with a long history of civilization. Contemporary Beijing has grown into what it is now on the foundations of the old city of the Ming and Qing Dynasties. After hundreds of years of development and change, Beijing has now emerged as a megacity, which continues to grow very rapidly with each passing day. At the time of the 2008 Olympic Games, Beijing displayed both the cultural heritage of the ancient capital and the elegant charisma of the modern metropolis. It is becoming both a Chinese socialist capital and a new world city – a new world city with a long history. There is no doubt that Beijing will continue to develop at a rapid pace and, as this chapter has shown, it faces some substantial challenges and problems in becoming a sustainable world city. These include social, spatial, environmental, transport and housing problems. Backed by the power and wealth of the People's Republic of China, the prognosis for the near future at least is promising. In the longer term the wider-scale regional, national and global context of change for megacities will increasingly determine the ways in which Beijing deals with complex questions of climate change, urban food resources, inequalities, liveability and governance in the twenty-first century. In this respect it will face similar long-term challenges to the other cities in this book, and it is likely that international co-operation between these cities and their national governments will be required to ensure survival and prosperity in potentially difficult, uncertain and risky situations as the twenty-first century unfolds.

Notes

1. This last point may seem strange to outside observers, particularly from the West, but in the People's Republic agricultural areas are often included within city boundaries.
2. Ai Weiwei (born in 1957), who had previously founded the Star Group (1979–1980), the first 'non-aligned' movement of Chinese art after modernization, returned to his homeland, after spending more than 10 years in the United States.

References

Abramson, D.B. (2007) The dialectics of urban planning in China, in Wu, F. (ed.) *China's Emerging Cities: The Making of New Urbanism*. London: Routledge.

Beijing Foreign Cultural Exchanges Service Center (1997) *Beijing*. Beijing: Information Office of Beijing Municipal Government and Beijing Tourism Administration.

Broudehoux, A.-M. (2004) *The Making and Selling of Post-Mao Beijing*. London: Routledge.

Chan, K.W. (1994) *Cities with Invisible Walls: Reinterpreting Urbanization in Post-1949 China*. Hong Kong: Oxford University Press.

Chan, R., Gu, C. and Breitung, W. (2000) Immigration, neue Armut und Segregation in Peking. *Geographica Helvetica*, Haft 1, pp. 13–22.

Cook, I.G. (2006) Beijing as an 'internationalized metropolis', in Wu, F. (ed.) *Globalisation and China's Cities*. London: Routledge.

Cook, I.G. (2007a) Environment, health and sustainability in twenty-first century China, in Chen Y. and Sanders, R. (eds.) *China's Post-Reform Economy: Achieving Harmony, Sustaining Growth*. London: Routledge.

Cook, I.G. (2007b) Beijing 2008, in Gold, J.R. and Gold, M.M. (eds.) *Olympic Cities: City Agendas, Planning and the World's Games, 1896–2012*. London: Routledge.

Cook, I.G. (2008) 21st century issues and challenges in Chinese urbanization, in Wagner, L.N. (ed.) *Urbanization: 21st Century Issues and Challenges*. New York: Nova Science Press.

Cook, I.G. and Miles, S. (2010) Beijing 2008, in Gold, J.R. and Gold, M.M. (eds.) *Olympic Cities: City Agendas, Planning and the World's Games, 1896–2016*. London: Routledge.

Cook, I.G. and Murray, G. (2001) *China's Third Revolution: Tensions in the Transition to Post-Communism*. Richmond: Curzon.

Ding, C. (2004) Urban spatial development in the land policy reform era: evidence from Beijing. *Urban Studies*, **41**(10), pp. 1889–1907.

Dong, G. (2006) *Historical Capital Beijing: Changes and Evolution over 50 Years*. Nanjing: Dongnan University Press.

Dong, Liming (1985) Beijing: the development of a socialist capital, in Sit, V.F.S. (ed.) *Chinese Cities: The Growth of the Metropolis Since 1949*. Oxford: Oxford University Press.

Gan, G.-H. (1990) Perspective of urban land use in Beijing. *Geojournal*, **20**(4), pp. 359–364.

Gaubatz, P.R. (1995) Urban transformation in post-Mao China: impacts of the reform era on China's urban form, in Davis, D.S., Kraus, R., Naughton, B. and Perry, E.J. (eds.) *Urban Spaces in Contemporary China: the Potential for Autonomy and Community in Post-Mao China*. New York: Woodrow Wilson Center Press.

Gu, C. (1998) Beijing's socio-spatial structure in transition, in Breuste, J., Feldmann, H. and Ohlmann, O. (eds.) *Urban Ecology*. Berlin: Springer Verlag.

Gu, C. (1999) Social polarization and segregation phenomenon in Beijing, in Guillermo, A. (ed.) *Problems of Megacities: Social Inequalities, Environmental Risk and Urban Governance*. Mexico City: Universidad Nacional Autonoma de Mexico.

Gu, C. (2001) Social polarization and segregation in Beijing. *Chinese Geographical Science*, **11**(1), pp.17–26.

Gu, C. and Kesteloot, C. (1997) Peasant immigrants and their concentration areas in Beijing. *Belgische Vereniging Voor Aardrijksundige Studies*, **1**, pp. 107–119.

Gu, C. and Kesteloot, C. (2001) Beijing's social-spatial structure in transition. *TRIALOG*, **68**(1), pp. 17–24.

Gu, C. and Kesteloot, C. (2002) Beijing's socio-spatial structure in transition, in Schnell I. and Ostendorf, W. (eds.) *Studies in Segregation and Desegregation*. Aldershot: Ashgate.

Gu, C. and Liu, H. (2001) Social polarization and segregation in Beijing, in Logan, J.R. (ed.) *The New Chinese City: Globalization and Market Reform*. Oxford: Blackwell.

Gu, C. and Shen, J. (2003) Transformation of urban socio-spatial structure in socialist market economies: the case of Beijing. *Habitat International*, **27**(1), pp. 107–122.

Gu, C., Chan, R.C.K., Liu, J. and Kesteloot, C. (2006) Beijing's socio-spatial restructuring: immigration and social transformation in the epoch of national economic reformation. *Progress in Planning*, **66**, pp. 249–310.

Gu, C., Wang, F. and Liu, G. (2005) The structure of space in Beijing in 1998: a socialist city in transition. *Urban Geography*, **26**(2), pp. 167–192.

Gu, Huimin and Ryan, C. (2009) Place attachment, identity and community impacts of tourism: the case of a Beijing hutong, in Ryan, C. and Gu, H. (eds.) *Tourism in China: Destination, Cultures and Communities*. London: Routledge.

Huang, Y. (2004) Urban spatial patterns and infrastructure in Beijing. *Land Lines*, **16** (4). Available at www.lincolninst.edu/pubs/969_Urban-Spatial-Patterns-and-Infrastructure-in-Beijing. Accessed 10 June 2010.

Jiang, F., Liu, S., Yuan, H. and Zhang, Q. (2007) Measuring urban sprawl in Beijing with geo-spatial indices. *Journal of Geographical Sciences* **17**(4), pp. 469–478.

Liang , S.(2001) *Liang's Works*, Vol. 5. Beijing: China Construction Press.

Liu, S., Prieler, S. and Li, X. (2002) Spatial patterns of urban land use growth in Beijing. *Journal of Geographical Sciences*, **12**(3), pp. 266–274.

Lu, D. (2011) *Remaking Chinese Urban Form: Modernity, Scarcity and Space, 1949–2005*. London: Routledge.

Murray, G. and Cook, I.G. (2002) *Green China: Seeking Ecological Alternatives*. London: Routledge Curzon.

Murray, G. and Cook, I.G. (2004) *The Greening of China*. Beijing: China Intercontinental Press.

Sit (1995) *Beijing: The Nature and Planning of a Chinese Capital City*. Chichester: John Wiley.

Sit (1996) Soviet influence on urban planning in Beijing, 1949–1991. *Town Planning Review*, **67**(4), pp. 457–484.

Tang, W. and Parish, W.L. (2000) *Chinese Urban Life under Reform: the Changing Social Contract*. Cambridge: Cambridge University Press.

UN Habitat (2010) *The State of Asian Cities 2010/11*. Fukuoka: UN Human Settlements Programme.

Wang, Jun, (2003) *Beijing City* (Cheng Ji). Beijing: SDX Joint Publishing.

Wheatley, P. (1971) *The Pivot of the Four Quarters: A Preliminary Enquiry into the Origins and Character of the Ancient Chinese City*. Edinburgh: Edinburgh University Press.

Zhang, J. (1991) Beijing's 20 Year Master Plan. *China Monitor*, pp.12–14.

Zhu, J. (2009) *Architecture of Modern China: A Historical Critique*. London: Routledge.

Zuo, Chuan,(2008) Location selection of National Administration Centre in Beijing. *Journal of Urban and Regional Planning*, **1**(3), pp. 34–53 (in Chinese).

Chapter Five

Taipei's Metropolitan Development: Dynamics of Cross-Strait Political Economy, Globalization and National Identity

Liling Huang and Reginald Yin-Wang Kwok

Taipei is the capital of Taiwan, the primary city of the urban hierarchy and the nodal point of national economy and politics. Taiwan's most important national issues and policies unfold and are contested in Taipei, and are thus reflected in its development. Three sets of economic-political events have been of particular significance in providing the context for Taipei's development – Cross-Strait political economy, globalization and national identity. This chapter reviews and analyzes how these sets of events, independently and in combination, have conditioned the evolution of Taipei's development.

Cross-Strait interaction with China (see figure 5.1) is one of the primary economic and political issues in Taiwan, with major impacts on Taipei. Many scholarly works have been published on the recent development of economic links across the Taiwan Strait (for example, Chen, 1996; Leng, 2002; Lin and Lin, 2001; Naughton, 1997), and political relations (for example, Breslin, 2004; Chao, 2004; Dent, 2005; and Gold, 1993). Shifting economic and political trends pose a constant predicament for the development of Taiwan, which has to navigate its way through these contradictory trajectories. At each turn of events, Taipei's developmental policy has to be adjusted and its spatial organization remodelled.

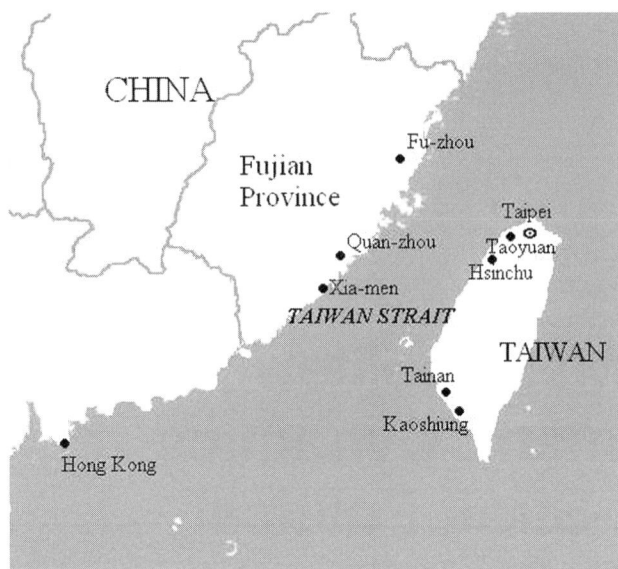

Figure 5.1. Taipei, Taiwan and the Taiwan Strait. (*Source*: Authors)

Ever since Taiwan was ranked as one of the four 'Asian Dragons', its development has been immersed in the global economy (Kwok, 2005*b*). Selected large cities, with appropriate locations, institutional and national support, and infrastructure are the spatial targets for global economic flows. Taipei, the prime city in Taiwan, is an obvious locational choice. As the global economy expands and changes, Taiwan has had to make periodic adjustments to occupy new production niches in order to survive and thrive. Much of this economic and institutional restructuring is manifested in Taipei. Taipei's metropolitan development, therefore, has to facilitate not only domestic, but also global demands.

National identity is a key component of nation building and development. Taiwan has had several ruling regimes, and each has imposed a specific identity on local citizens. Identity politics, a constant in Taiwan's history and now a major domestic political issue, has imprinted its mark on developmental policies, urban development and urban space. Urban landmarks are the identity symbols by which the state projects its historical, cultural and constitutional legitimacy (Kim, 1997; Kwok, 2003). The state regularly makes use of urban form to transmit its ideological messages. This chapter demonstrates the evolution of Taiwan's national identity by reference to urban landmarks and state-built urban forms which serve as icons of state power (Vale, 1992).

Taiwan's history has witnessed several drastic regime changes. Each change fundamentally altered state-society relationships, ideology and cultural norms. Each regime period brought a radically different set of governance, economic system and developmental goals. Since the founding of Taiwan, the island has experienced five economic-political regimes: Dynastic Incorporation, Japanese Colonization,

Kuomintang Rule, Global Competition, and Cross-Strait Re-opening. Taiwan's development can be understood only through these historical periods. Consequently, they are discussed in some detail below to provide the contextual framework for analyzing Taipei's evolution.

Historical Evolution of Urban Development

Dynastic Incorporation: From Neglect to Development (1683–1894)

Taiwan's early development can be traced back to more than 6,000 years ago, when Austronesian people lived on the island. However, today the majority of Taiwan's residents are the descendants of immigrants from Mainland China who have arrived since the mid seventeenth century. Not long after the Manchu took over China and established the Qing Dynasty (1644–1911), Zheng Chenggung, a Ming general committed to restoring the Han's rule over China, led loyalists from the collapsed Ming Dynasty, and retreated to Taiwan to establish a military base. Zheng Chenggung and his troops settled in the southern part of Taiwan, but his campaign eventually failed.

The Qing court initially ignored Taiwan, as it considered the island insignificant economically and politically, but it finally took over Taiwan in 1683 and attached it as a territory to Fujian province. Due to the geographic remoteness of Taiwan and the consequent difficulties in controlling its local communities, the Qing court adopted a passive approach to the government of the island (Lin and Keating, 2000). The real governance of the island was left to trade guilds, in association with social and religious networks. Territory ties and blood relationships played key roles in organizing the local affairs of the immigrant society. Apart from their centrality to commercial activities, guilds had the autonomous power to organize urban affairs and militia defence (Ho, 1989). Through the ruling gentry class, which controlled the guilds, state and society managed to keep a 'separated but collaborative relationship' (Hsu, Shuo-bin, 2005).

Compared with southern Taiwan, the north was developed relatively late. Early Chinese migrants did not settle down in the Taipei (literally, northern Taiwan) basin until the early eighteenth century. Based on a trading economy with China's coastal cities across the strait, Manjia, located in the western part of present Taipei, became the first major settlement to be developed along the shore of the Tamsui River, the most important commercial river in the Taipei basin (see figure 5.2). With the renowned Longshan Temple and its surrounding popular temple streets as the urban centre, Manjia thrived through a hundred years of prosperity until the mid nineteenth century. In 1853, feuds between local guilds over commercial interests led to large-scale collective violence, and a group of guilds retreated north along the riverbank

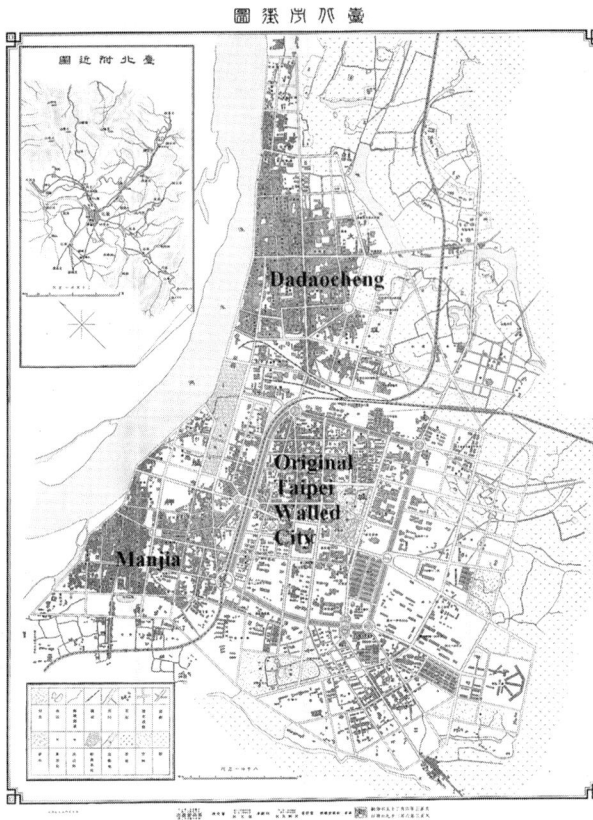

Figure 5.2. Manjia, Dadaocheng and Taipei, 1914. (*Source*: Shin-kao-tan Book Store, 1914)

from Manjia to Dadaocheng (see figure 5.2). Shortly thereafter, a new trading port was developed to counter the status of Manjia (Lin and Keating, 2000). Manjia and Dadaocheng became competitive rival ports for Cross-Strait trade, and together developed commerce as the economic base of early Taipei.

In 1862, the Qing Dynasty, under pressure from Western colonial powers, was forced to open the seaport of Tamsui, along with all other port cities along the river, including Dadaocheng and Manjia. The trade guilds in Manjia were antagonistic and refused to co-operate with foreign merchants, leading to the port's inevitable decline. The merchants in Dadaocheng, however, grasped the opportunity, accepting foreign residents and marketing tea products through foreign companies to the Western world. This open approach helped Dadaocheng to surpass Manjia and it soon became the most important settlement in the Taipei basin (Huang, Fu-san, 1995).

The rise of Western colonialism alarmed the Qing court and the threat of Western forces prompted China to adopt a more aggressive foreign policy. Taiwan, especially its northern region, was fortified as the gateway to China's coastal area. In 1885, the Qing government raised Taiwan's administrative status to that of a province and Taipei walled city was built as the administrative centre of northern Taiwan. To avoid the two antagonistic existing settlements, the Taipei walled city was built on swampland between Dadaocheng and Manjia (see figure 5.2). As with other Chinese administrative cities, the construction of the Taipei walled city closely followed traditional concepts of geomancy and political order based on Confucian ideology. The city had a well defined north-south central axis, with clearly demarcated functions and locations for main buildings on both sides, including administration halls, official temples, an Examination Hall and a Confucian School as the indispensible urban spatial landmarks (Hechutushe, 2004).

Liu Ming-chuan became Magistrate in 1884. Rather than continuing the passive policy of previous times, it was his aim to build Taiwan as the most advanced province in China by modernization (Ho, 1989). He launched massive institutional reforms and infrastructure construction. For example, he completed the first railway in Taiwan, with Dadaocheng as the hub of a network which reached Keelung in the north and Hsinchu in the south, thus linking the entire north-west coastal region. A telegraph system was set up across the strait to facilitate communication with China for the tea trade. He also established a Mechanics Bureau and modern weapons were produced locally for the defence of Taiwan. Under Liu's administration, Taipei's urban development made great strides.

Thus, Taipei's urban origins were based on Cross-Strait trade, and the triad of the two rival trade ports of Manjia and Dadaocheng, together with the administrative centre of Taipei, converged to become an international port city, based on commerce. Liu Ming-chuan's progressive and ambitious transportation plan made the trading port the centre of regional infrastructure, and Taipei became a potential hub for industrial development. However, only 15 years after Liu Ming-chuan's arrival, the modernization programme which he established was terminated as Taiwan came under Japanese control.

Japanese Colonization: Cultural Subjugation and Resource Extraction (1895–1945)

In 1895, with the loss of the Sino-Japan War, the Qing Dynasty signed the Shimonoseki Treaty, which ceded Taiwan and the Penghu islands to Japan. This concession shook the people of Taiwan. An independent government was organized by local gentry and government officers to stop the handover of Taiwan to Japan (Lin and Keating, 2000) but, after a short campaign, Japan claimed its colony. The Japanese government chose the newly built Taipei walled city as the site for the capital, away from the southern part of Taiwan, which was still an area of strong resistance. Taipei, with its seaport facilities, would facilitate the export of the island's resources to mainland Japan (Wei and Gao, 2004).

By the end of the nineteenth century, Manjia, Dadaocheng and the walled city formed three key centres in the Taipei basin, with a total local population of approximately 50,000 people. Taipei quickly rose to become the most important city in Taiwan and, with its unusual combination of traditional Chinese and colonial Japanese influences (Wen, 1986), the foundations were established for its rise as a significant East Asian international commercial hub.

Japan introduced modern technology and Western ideas, and ushered Taiwan into the modern era. Infrastructure development and early industrialization were centred on the capital. The island's infrastructure system, with Taipei as the main node, was gradually developed as Japanese rule over Taiwan society became firmly established. In 1909, an island railway was completed, connecting Taipei to the other major cities on the west coast. This railway extended Taipei's influence to the southern tip of the island, further affirming Taipei's status as the gateway city to mainland Japan. Urban construction in Taipei was required to comply with the same rules applying to all other cities in Taiwan which included: (1) improving 'unhealthy' and 'unsanitary' environments to prevent epidemics; (2) the building of Shinto shrines; (3) setting up educational institutes, mainly primary schools, to assist with cultural assimilation; (4) building military facilities to control local insurgencies; (5) setting up prisons; (6) ensuring the provision of administrative buildings and residences for government workers and officers; and (7) constructing infrastructure for industrialization and commerce (Huang, Lan-shiang, 1995). The main aim was to subjugate the Taiwanese to Japanese culture and nationalism, with a secondary goal of industrialization. These rules were more rigorously implemented in Taipei, as the colonial capital.

The Japanese development of Taipei had important symbolic elements. Within the original walled city area, the traditional Chinese official buildings were all demolished. In their place, a Presidential Hall and Presidential Residence were built on a grand scale. These followed contemporary Japanese style, with cutting edge technology and extravagant European motifs and decorations. The Mazu Temple in the city centre, housing the most popular goddess for Chinese immigrants, was torn down to make way for the Taipei Park, serving the leisure demands of Japanese officers. Later, in 1915, a grand museum, designed in the European Renaissance style, was completed on the exact site of the original temple to celebrate the achievements of the former Japanese Governor-General, Kodama Gentaro, and his head of Civilian Affairs, Goto Shinpei, in developing Taiwan. In 1935, Taipei Park and its neighbouring streets were specially decorated to host the great Taiwan Exposition, celebrating the fortieth anniversary of Japanese colonization. This received 2.75 million visitors (Cheng, 2004). In these and other ways the colonial power, through the manipulation of urban landmarks, supplanted the local citizens' Chinese identity with the symbols of Japanese nationalism. Taipei's cityscape became the backdrop to Japanese colonial rule.

In 1905, the Japanese government conducted a modern urban planning exercise to revamp the city. New areas were developed, mostly for Japanese residences, close to government agencies. A modern grid pattern of roads and a sewer system were developed in these Japanese residential areas, although the Taiwanese settlements in Manjia and Dadaocheng received little government investment in infrastructure. This was a clear attempt at residential segregation between the colonizers and the locals (Huang, Lan-shiang, 1995). The old city walls were also torn down, to be replaced by boulevards in the modern European mode. The most significant boulevard linked the Governor's Hall in the city centre to the Taiwan Shinto Shrine, located on a mountain at the northern end of the city, symbolizing the combined military and political power of the Japanese Emperor. In 1897, its surrounding site was developed as the Grand Park (In, 2006). This segregated planned development made clear the racial and power divide. Taipei became an experimental site for urban planning as a tool for development along social and ethnic lines.

Opening up the walled city enabled the growing areas of Dadaocheng, Manjia and the capital to merge gradually, while maintaining the racial segregation between Japanese and Taiwanese. In 1932, the Taipei Expansion Plan was announced. This sought to extend urban growth within a total planning area covering over 800 square kilometres and extending 20 kilometres east from the city centre. This plan was projected to accommodate a population of 600,000 people. Major parks and roads

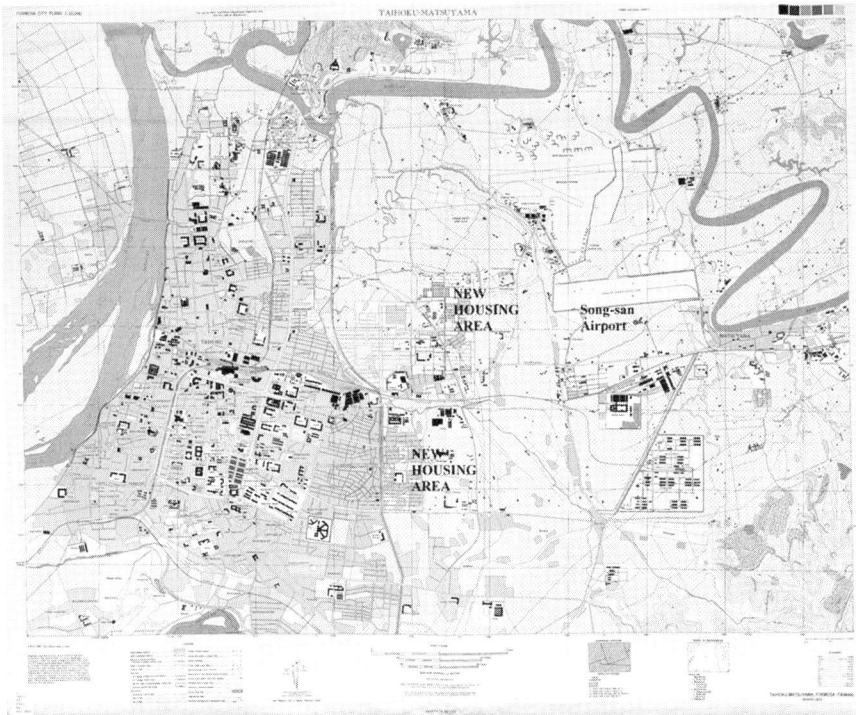

Figure 5.3. Taipei City, 1945. (*Source*: US Army Map Service, 1945)

were designated as the backbone for urban functions (Zhang, 1991). Urban parks were planned to provide beautiful landscapes as well as open recreational space, once again mostly for the Japanese residential areas. Under the 1905 and then the 1932 plans, Taipei grew significantly, reinforcing its position as the largest and most important city in Taiwan (see figure 5.3), with an urban form best described as 'colonial modernity' and a style of planning which suppressed the nationalist sentiments of the local society (Hsia, 2000). The 50 years of colonial rule transformed Taiwan into an outer national territory of Japan. The cultural conversion project was disseminated from the new colonial capital, Taipei, which was also the chief port for exports to Japan. Both roles enhanced the city's cultural, political, trade and commerce functions and underpinned its rapid urban expansion.

The introduction of modern technology and Western ideas made Taipei the demonstration centre for new innovations and adaptation and it became established under the Japanese as the undisputed primate city in the urban hierarchy and as a global city in the making.

Kuomintang Rule: From Authoritarianism (1945–1975) to Conciliation (1975–1988)

After World War II, Taiwan was restored to China, at that time governed by the Nationalist Party or Kuomintang (KMT). Initially, the Taiwanese people eagerly welcomed this new chapter in their history, but the corruption of the KMT and its inability to control inflation soon turned optimism to disappointment, and discontent grew in local society. On 28 February 1947 ('228') police arrested an illegal tobacco vendor at Dadaocheng and the incident flared into island-wide riots (see figures 5.4*a* and 5.4*b*). Within a couple of months the '228' movement was suppressed, political dissent was forbidden (Rigger, 1999) and many local people, including local leaders and members of elites, were imprisoned or executed. In 1949, at the peak of the civil war between the KMT and the Communists, the KMT imposed martial law and a curfew in Taiwan. This gave the state absolute power to limit freedom of speech and to suppress all demands for political reform.

In 1949, the KMT lost the civil war to the Chinese Communist Party, and its remnants retreated to Taiwan under the leadership of Chiang Kai-shek. More than a million mainland refugees, military and civilian, migrated to the island. In the hope of returning to China shortly, the KMT designated Taipei as the 'provisional' capital of Taiwan. In the 1950s, under the shadow of imminent war, the military claimed 80 to 90 per cent of national government expenditure, so that government investment in social and economic development was grossly under-funded. US aid became the critical source of funding for economic development and the maintenance of social stability from 1950 until the mid 1960s (Ho, 1978). From the 1960s to the 1980s, by adopting

Figure 5.4a. The terrible inspection. (*Source*: Huang, 1947. Woodcut print of '228' incident)

Figure 5.4b. Crowd protesting '228' incident: angry demonstration at the Taipei branch, Bureau of Monopoly. (*Source*: Tsu, 2007)

an export-oriented industrialization model, Taiwan successfully developed industries in food, garments, footwear, toys and electronics for the international market. Under strong state leadership, an 'economic miracle' had occurred by the end of the 1980s, with Taiwan recognized as one of the four Asian Tigers (Wang, 2004; Castells, 1992).

The KMT, after losing the civil war and retreating to Taiwan, set up a dual society on the island – KMT mainlanders and local citizens. The KMT not only controlled politics and resources, but also imposed authoritarian rule over the local citizens. Taipei's development was determined with little local participation, and planning reflected national rather than local interests (Huang, 2005). During the Cold War, Taiwan became a capitalist bulwark to communism in Asia. The KMT secured US protection and international legitimacy, with Taiwan given United Nations membership as representing the whole of China. From the 1950s to the early 1970s, the widespread imprisonment and execution of local dissidents kept domestic politics stable. The 1970s, however, were a decade of political turmoil (Rigger, 1999). Two international events – the loss of the UN seat to Mainland China in 1971, and the oil shock in 1973, seriously weakened the KMT administration. New domestic political challenges emerged, with the growing 'Dangwai' movement mobilizing the opposition to the KMT and pressing for political reform. A strong wave of cultural awareness also arose simultaneously in literature, dance and music, representing a search for a new collective Taiwanese identity to challenge nationalist China (Hsiao, 2008).

In 1972, Chiang Kai-shek appointed his son, Chiang Ching-kuo, as the premier. Facing a legitimacy crisis, the KMT initiated a series of policies to ward off local resentment and to recover party prestige. By redistributing resources to local capitalists, collaborating with local factions, and co-opting young elites into government agencies, the KMT managed to retain control. Between 1974 and 1979, Chiang Ching-kuo launched the 'Ten Great Constructions', major infrastructure projects, such as the Taoyuan International Airport, Zhongsan Highway and investments in state-owned companies in the heavy industries of shipbuilding, petroleum-refining and steel production. After Chiang Kai-shek's death in 1975, he was succeeded as President by Yen Chia-kan and then, in 1978, by Chiang Ching-kuo who, from 1980 to 1985, continued the national building programme through the 'Twelve Constructions'. Apart from consolidating infrastructure for economic development, these also included cultural centres, meeting a demand from Taiwan's growing consumer society and, more importantly, strengthening nation building.

Taipei's economic growth from the 1960s to the 1980s was accompanied by unprecedented urban growth as the capital attracted a large number of rural-urban migrants. The urban area quadrupled in size by incorporating a large part of the surrounding Taipei County area and the population doubled from 1.2 million people at the end of the 1960s to over 2.6 million by 1987 (Taipei City Government, 1988) (see figure 5.5). A number of satellite cities – Sanchung, Xingzhuang, Younghe, and Zhonghe – were planned for Taipei County to absorb the spillover of industrial activities and to provide housing for the new migrants. Meanwhile, within Taipei's industrial areas, illegal factories were hidden in residential neighbourhoods. By 1991, Taipei County had a population of over 3.1 million. By the end of the 1980s, the Taipei Metropolitan area was clearly part of the global capitalist system and the most prosperous part of Taiwan. Its intense industrialization had also led to severe environmental pollution, overcrowding and an under-provision of infrastructure – characteristics found in other rapidly growing cities of the developing world (NTUBP, 1991).

Taiwan's economy in the post-war period was deeply dependent on the Western countries and on Japan. With the UN embargo on China after its involvement in the Korean War, Taiwan was cut off from all connections with Mainland China. Paradoxically, however, in terms of urban form China had a strong influence. In the nineteen-fifties, the KMT administration adapted most of the public buildings left

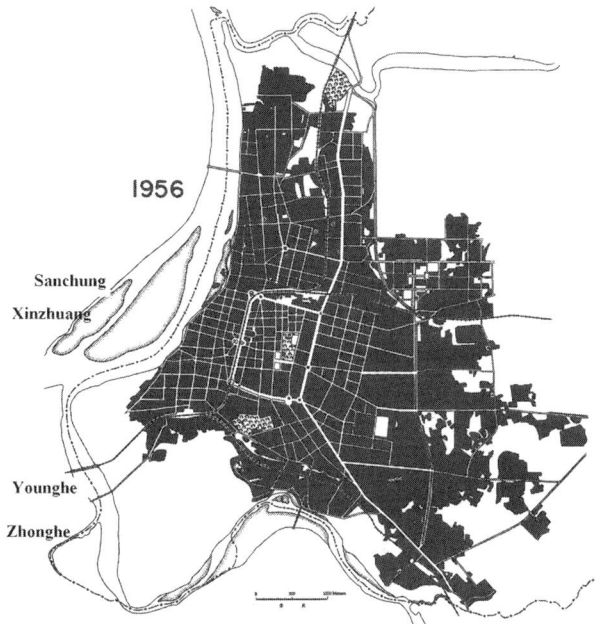

Figure 5.5. Taipei City, 1956. (*Source*: Chen, 1957)

by the Japanese colonizers, partly because of financial constraints and partly because of the lack of any long-term plan to stay in Taiwan. By the mid 1960s, the hope of returning to China had faded and new public buildings had to be constructed under the pressure of rapid urban growth. For new buildings for the national government, a Chinese Classic (Palace) Style was consistently adopted, as the KMT considered itself as the rightful heir to Chinese history and culture. At the same time, across the strait, Communist China was embroiled in the Cultural Revolution, which refuted all Chinese traditions (Fu, 1993). The National Palace Museum in Taipei was built in 1965 to rival the National Museum in Beijing. Chiang Kai-shek's favourite architect, Huang Bao-yu, who specialized in the traditional palace style, designed this project.

Another example of symbolic architecture was the Grand Hotel, which functioned as a showcase for the KMT, and was the principal hotel for visiting international politicians. Originally built in 1963, and expanded in 1973 to its current fourteen storeys, the Grand Hotel was erected to replace the Taiwan Shinto Shrine, signalling the removal of elements of the former colonial regime (see figure 5.6). It became the new symbol of Taiwan in international politics – an extravagant compound designed with red columns, golden roofs, colourful carvings and paintings on the façades to impress foreign guests. This architectural icon informed the visitors that they were now in the domain of true Chinese tradition and legitimate political succession.

In the 1970s, the Chinese Classic Style gradually evolved into the Chinese Modern Style, reflecting the rising influence of industrial modernization. The most

Figure 5.6. The Grand Hotel. (*Source*: Authors)

outstanding landmark of this period was the Dr. Sun Yat-sen Memorial Hall, built to commemorate the 100th birthday of the founding father of the Republic of China. This building, designed for exhibitions and cultural events, was completed in 1972 in the newly-developed area of east Taipei. It was designed by Wang Dahung, who broke with previous motifs by simplifying building decoration and proposing a modern roof design, while keeping the traditional Chinese structural form. Chiang Kai-shek intervened, however, and insisted on a large palace roof of traditional character (Hsu, 2007). Once again, symbolism dictated style.

The national government decided to build a memorial hall for Chiang Kai-shek after his death in 1975, and the building was completed in 1980. According to the designer, Yang Zuo-cheng, the design of the Chiang Kai-shek Memorial Hall was based on the motifs of the Heaven Temple in Beijing, the Sun Yet-sen Mausoleum in Nanjing and the Lincoln Memorial Hall in Washington DC (Liu, 1976). The memorial hall combined the traditional mandate of the emperor with national revolutionary and civil war leadership (see figure 5.7). Later, in 1987, two annexe buildings, the National Theatre and Concert Hall, were completed to complement the memorial hall. Both buildings were modelled on the palaces in the Forbidden City in Beijing and were further evidence of the KMT's ideological mimicry of Chinese orthodoxy (Kwok, 2009).

Figure 5.7. Chiang Kai-shek Memorial Hall. (*Source*: Authors)

For the KMT to maintain its dominant position, it had to stamp its cultural mark on the island. The KMT had two missions – to replace the Communist Chinese government as the legitimate heir to the rule of all China, and to infuse Taiwanese citizens with the traditional Chinese identity of their provincial ancestors from southern Fujian. The state pursued a public building programme which celebrated and spread the Mainland Chinese identity. Chiang Kai-shek was the gatekeeper of the historical Chinese tradition, which allowed him and the mainland migrants to assert cultural superiority. He pursued this conviction strongly through state architecture. By reproducing traditional Chinese icons on institutional buildings, where artistic and performance activities attracted people from all levels of society, the intention was to expose the public to traditional mainland culture. The local citizens, in other words, were subjected to another round of identity indoctrination.

When Chiang Ching-kuo succeeded to the Presidency, he recognized that prosperity would spread greater resources to the local communities. He also acknowledged that the Dangwai movement was creating an increasing demand for political reform. Political compromises and conciliatory measures had to be accepted to appease the rising demand for democracy from the indigenous majority. As the seat of power of the KMT, Taipei was the centre where political transition took place. Taipei now entered an era where its political position and identity became a site for contest between the power groups of party, state and local factions, represented by the KMT, and the rising forces of local, intellectual and grassroots power, represented by the Democratic Progressive Party (DPP), formed in 1986.

Global Competition: Democracy and Independence (1988–2008)

By the mid 1980s, the appreciation in value of the Taiwan New Dollar, tightening environmental regulations and increasing labour costs caused Taiwan to re-evaluate the export-oriented industrialization model, which had been effective to that point. With more Third World nations entering the global production system, international competition intensified. Consequently, industrial restructuring had to take place to improve Taiwan's position in the global economy. Improvements to production technology were necessary, resulting in the establishment of Hsinchu Science Park, as part of a Research and Development (R&D) corridor for technological research to the west of Taipei. But a more profound shift was towards a new transnational economy model in which Taiwan played the role of foreign investor. Taiwan had been receiving foreign direct investment for some decades but, by the mid 1980s, Taiwanese industrial firms had accumulated sufficient capital to begin to invest abroad themselves. Most of Taiwan's industrial firms were of small and medium size, using labour-intensive technology. They now had the necessary investment capital to move to South-East Asia and to coastal China where they could take advantage of low land and labour costs (Chou, 2005; Hsu, Jinn-yuh, 2005). When China opened itself up for foreign investment and trade in the late 1970s, it was a very convenient nearby location for Taiwanese production and a market for goods (Kwok, 2005*a*). The Cross-Strait economy flourished, with a steady stream of industries relocating to coastal China.

The large-scale move of Taiwanese investment to Mainland China gave rise to strong economic ties across the Strait. Cross-Strait political development, by contrast, was deliberately prevented. The fast growth of Taiwan in economic globalization coincided with the birth of a new independent national identity, and led to opposite trajectories of Cross-Strait economic networking and political tension. After Chiang Ching-kuo died in 1988, Lee Teng-hui succeeded him as President. Born in Taiwan, he received his early education during the Japanese colonial period and later gained a PhD in the United States. Lee had a legendary background of supporting the underground Taiwan independence movement in his early years and, although a long-term party member of the KMT, he had also maintained a close relationship with the recently founded DPP. During his term as President, from the late 1980s to the mid 1990s, Lee promulgated a series of policies to sever Cross-Strait economic connections and to bolster national independence. He claimed that the continuous exodus of Taiwanese industrial plants to China would hollow out the economy and threaten national security. In 1994, to counter the trend of industrial moves to China, he proposed a policy of 'Moving Southward' (*Nanjin*), in which the national government provided incentives for Taiwan enterprises to invest in South-East Asia, with the KMT party-related companies leading the way. In 1996, aiming to keep industry in Taiwan, Lee further espoused a policy called 'No Haste, Be Patient' (*Jieji yongren*), which banned hi-tech industries, infrastructure developments and large projects worth more than US$50 million from going to China. In 1999, he promoted a 'two-state discourse' (*Liangguo lun*), claiming that relationships and negotiations between Taiwan and China should be on a state-to-state level, as Taiwan was a *de facto* state. This announcement seriously damaged Cross-Strait economic links (Chou, 2005).

In 2000, Chen Sui-bian, a former Taipei City Mayor and a DPP leader, was democratically elected as President of Taiwan and introduced a new period of further change to Taiwanese identity. DPP started as a voice for local citizens. It aspired to correct social inequity, to promote local interests and to restore local identity. The advocacy of a localized Taiwanese national cultural model sought to distance Taiwan from China. Fearing that Cross-Strait industrialization would strengthen Taiwan's dependence on China, thus increasing Chinese dominance and diminishing the island's political flexibility and independence, the DPP government deliberately and consistently resisted political, economic and cultural integration with China. Chen's policy brought to a head the dichotomy in Taiwan's economic and political development – its economic interest in co-operation with China and its desire to maintain its political and cultural independence (Ching, 2005; Chou, 2005).

Chen's presidency coincided with a critical period of intensifying global competition in Asia. With less developed economies, such as Vietnam and China, entering into the global market, the growing competition within the global economy increased the pressure on enterprises to relocate to lower cost locations. Despite successive prohibition policies by Presidents Lee and Chen, industrial out-migration to China could not be prevented. Cross-Strait economic collaboration was not only going to stay, it was the new Taiwan development model which sustained Taiwan's economy in an increasingly competitive global market. Out of necessity, industrial entrepreneurs, finding foreign investment loopholes, continued to expand the Cross-Strait production network. To reach China, enterprises found ways to bypass restrictive regulations, usually by funnelling investment through Hong Kong or a third country. Despite Chen's overt ideological stand, the 'No Haste, Be Patient' policy was ineffective. By 2000, the continuous and growing pressure from business and industrial sectors forced Chen to launch a limited Cross-Strait policy – the 'small three links' (*Xiao Santung*) policy which allowed Cross-Strait business, transportation and mail activities, but limited them to links between the offshore islands of Taiwan and nearby coastal port cities in China. By 2004, the notebook computer industry, a product in which Taiwan had been the top exporter in the world, had moved entirely to China. According to government statistics, from 1991 to 2004, the total amount of Cross-Strait investment approved by the Ministry of Economic Affairs amounted to US$41 billion (Huang, 2006).

During the DPP's rule from 2000 to 2008, a domestic policy was enacted to decentralize some power from Taipei, the primate city in Taiwan's urban system. The national government and the Taipei City government were ruled by opposing parties. Consequently, the state, controlled by DPP, and the capital city, ruled by the KMT, would not collaborate to explore a new urban development path for Taipei (Chou, 2005). Under the KMT regime, development had been concentrated in the north and the south had been neglected. Chen decided to reverse this trend by channelling development to Kaohsiung as the 'capital in the south'. A series of policies sent national investment, infrastructure construction and international mega-events to the south. The ideological mobilization in election campaigns was often sharply divisive on social and ethnic issues and deepened the conflict over national identity. A north-south political split emerged.

The Lee-Chen regimes, spanning from the late 1980s to 2008, witnessed a major economic restructuring of the information technology (IT) industries – the dominant production sector. By the end of 1980s, Hsinchu had developed its R&D capacity and had well-developed technological links with California's Silicon Valley (Saxenian and Hsu, 2001; Hsu, Jinn-yuh, 2005). From the early 1990s, many hi-tech manufacturing firms began relocating to China, mostly clustering around Shanghai in the Yangtze River Delta. A new global production region, comprising Silicon Valley, Taipei and the Yangtze River Delta, emerged. The Cross-Strait connection was based not only on production resource advantages, but also on common language, cultural and social networks. Thus Taipei and Southern China formed a cross-border economic region (Wang, 1997) and, by 2006, it was estimated that more than 500,000 Taiwan citizens had moved to Shanghai and its neighbouring areas (*Dajiyuan News*, 2006).

Domestically, Taipei and Hsinchu formed an industrial region or 'growth corridor', based on production clustering complemented by intensive social and learning networks (Saxenian and Hsu, 2001). A new spatial form developed as the Taipei Metropolitan area expanded into a mega-region, covering the hinterland of Taoyuan International Airport and Hsinchu County (Huang, 2008). The completion of the Taiwan Speed Rail in 2007, which reduced the travel time between Taipei and Hsinchu to half an hour, further integrated the growth corridor. The Speed Rail also reduced travel time between Taipei and Kaohsiung to one and a half hours, extending Taipei's economic influence to southern Taiwan.

In the central district of Taipei, the Xinyi Planning District was aggressively developed in the late 1990s. The state envisaged Taipei as a global financial centre serving East Asia. This major project, supported by a rare collaboration between the national and city governments, was intended to build an international business centre in Xinyi in the heart of the capital. The plan transformed an old military camp and housing area to a central business district. To the national government, Xinyi was to turn Taipei into an Asia-Pacific financial operations centre. To the city government, the Xinyi Planning District would become Taipei's Manhattan (Jou, 2005). This district was to be the home of the headquarters of international and domestic corporations in Taiwan and was intended to divert Taipei to a different developmental path, away from Cross-Strait business. However, Hong Kong and Singapore, which had geographic advantages and well-trained human resources, were already well-entrenched as global financial centres and the megaproject failed to live up to its intentions. Instead, Xinyi ended up housing the Taipei City Hall, several flagship department stores and an agglomeration of entertainment activities and 'high-end' real estate developments, including the Taipei 101 Building, the tallest building in the world at the time of its opening in 2004 with 101 floors (see figure 5.8).

Urban landmarks in Taipei were inevitably included in the ideological struggle. The Chiang Kai-shek Memorial Hall Park area was completed in the late 1980s and

Figure 5.8. Taipei 101.
(*Source*: Authors)

its forecourt became a popular site for organized or spontaneous political and social gatherings, parades, sit-ins and night vigils. In May 2007, the national government declared its intention to rename it 'The National Taiwan Democracy Memorial Hall' and the forecourt 'Freedom Square', as symbols of 'breaking the authoritarian rule of KMT' (Chen, 2007). The national government also planned to remove the statue of Chiang Kai-shek and to demolish the surrounding walls. These proposals caused long public disputes, with critics commenting that the plan was 'a gesture of symbolic mobilization' in the post-authoritarian era (Nan, 2007). Identity politics between the DPP and KMT, the national state and the capital city continue to be played out in the public arena. Taipei is the key national identity battleground, with public buildings as the targets.

At the broader scale, the KMT's role in the government of Taipei has been in direct confrontation with the DPP's role at the national level. This has led the DPP to divert resources away from the capital in order to rein in the KMT's economic and political power. In order to keep Taiwan competitive in the global economy, however, Chen's government continued to support technological research and development at Hsinchu. This strategy strengthened Taipei as a centre of innovation, while increasing the need to spread production abroad, as Taiwan had lost its manufacturing advantages (Kwok, 2005a). Despite the lack of support in other respects from the central government, Taipei continued to grow as a metropolitan region with strengthening ties with China's southern coastal region. In reality, Taipei continues to dominate the Taiwan economy and Taiwan has no option but to rely on the Cross-Strait economy.

Cross-Strait Re-Opening: Political-Economic Peace? (2008–Present)

After 8 years of DPP rule, Ma Ying-Jeou of the KMT was elected President of Taiwan in March 2008. Ma's election campaign platform was based on saving the Taiwan economy, which had become sluggish in the preceding years. The pragmatic new KMT government argued that Taiwan's economic crisis, caused by global economic recession and intensified global competition, had to be tackled. Strengthening the Cross-Strait economic link was seen as the key to reviving the domestic market, increasing capital supply and stimulating consumer demand. After taking the presidency, Ma worked immediately to revive the Cross-Strait economy, opening up Taiwan to new Chinese investment and tourism. Cross-Strait interaction expanded and the pace of economic development quickly accelerated.

Ma pursued Cross-Strait policy by developing an 'Economic Co-operation Framework Agreement' (ECFA) with China. Since May 2008, top officials of the Straits Exchange Foundation (SEF), from Taiwan, and the Association for Relations Across the Taiwan Straits (ARATS), from China, have met regularly to progress Cross-

Strait negotiations (ECFA, 2010; Zhang, 2010). On 29 June 2010, representatives from Taiwan and China signed the ECFA in Chongqing, China, marking a new historical period in Cross-Strait relationships.

Domestically, the KMT national government announced a 'New Twelve Constructions' plan in April 2009 to inject NT$858.5 billion (approximately US$26 billion) over the following 4 years into a series of national construction projects, including a new science park in Taichung, the Honhai (Foxconn) Technology Park in Kaohsiung, and upgrading of the island-wide transportation network (Executive Yuan, 2009a). Among the mega construction projects announced, the top priority went to the expansion of Taoyuan International Airport and its merger with neighbouring areas to form Taoyuan Aviation City (or 'Taoyuan Aerotropolis' in official documents). The Aviation City covered over 6,000 hectares and included a free trade zone, an exhibition zone, new towns, a value-added agricultural area and recreational belts (Taoyuan County Government, 2009). This planned area will form part of a metropolitan conurbation linking Taipei City, Taipei County and Taoyuan County, extending to Hsinchu (see figure 5.9).

The local governments included in this plan had earlier established a 'Northern Taiwan Regional Development Commission' in 2005 to co-ordinate industrial growth, tourism and transportation networks (Northern Taiwan Development Commission, 2005). Later, the Construction and Planning Bureau of the Ministry of the Interior produced the Greater Taipei Plan to co-ordinate the development of the metropolitan region. The Greater Taipei Plan was not a statutory plan, but a development guide for private and public development, providing flexibility for

Figure 5.9. Hsinchu-Taoyuan-Taipei Metropolitan Region. (*Source*: Authors)

accommodating global enterprises. Key mega construction projects were centrally controlled in order to avoid local competition for political and financial support from the central government. Collaboration with local governments was considered to be effective, however, in certain regional projects, such as river conservation and disaster relief. The Greater Taipei Plan was a major political initiative by the KMT.

However, regional collaboration can be administratively challenging. In January 2011, the adjacent Taipei County, previously under the jurisdiction of Taipei City, became a separate municipality, renamed as 'New Taipei City'. This upgrading has put the new municipality directly under the jurisdiction of the central government. The flagship projects, increased budget and power for the new municipality, which are likely to follow, will give it greater local autonomy. The consequent duality of Taipei City and 'New Taipei City' are likely to change the dynamics of the regional plan (New Taipei City, 2010).

The plan for Taoyuan Aviation City was intended to build up the potential of Taoyuan Airport as an international air hub and to raise its competitive position in relation to the other major airports of East Asia. Unlike the earlier proposed development of Xinyi Planning District, which aimed to build Taipei as an international financial centre in Asia, as an alternative to Taipei's Cross-Strait economic integration, Taoyuan Aviation City is intended to reposition Taipei as a transportation node with strong connections to China. Parallel to the development of the Taoyuan Aviation City Plan, a new 'Agreement on Air Transportation across the Strait' was ratified by the Legislative Yuan in December 2008. All Taiwan airports thereafter were to be opened for more than a thousand Cross-Strait flights per month (Executive Yuan, 2009*b*), enabling Taiwan to serve as a trans-shipment and distribution point between China and the world. Thus, Taiwan's economic position would be elevated as part of China's global trade. This agreement brought a new political and economic vision to Taiwan, with China acknowledged as a key factor in Taiwan's global strategy.

Also in 2008, group tourism from China and Chinese investment in Taiwan were opened up. In 2008, the national government deregulated the 'small three links', and allowed ordinary Taiwanese citizens and certain categories of Chinese to travel freely across the straits. In less than half a year, approximately 600,000 trips took place (Executive Yuan, 2009*b*).

The deepening global economic crisis impelled the DPP to reassess its position on Cross-Strait issues. Commissioners of some DPP-controlled southern counties came under increased pressure from their constituencies to solicit Chinese capital to shore up their declining local economies. To respond to these demands, they visited China to promote agricultural products and tourism for their counties. Officially, the DPP still held a firm position of pursuing Taiwan's independent sovereignty and opposing the ECFA. In practice, however, local DPP leaders employed strategies to develop the China market and Chinese tourism. This tension between DPP ideology and economic reality remains to be resolved. A well-respected public intellectual and pro-DPP scholar, Huang Wu-shung, recently put forward an 'Agreement for Fifty Years of Cross-Strait Peace'. His proposal was to resolve conflict incrementally, and he suggested that the Cross-Strait relationship should start with a peace agreement based on the status quo. At the end of 50 years, the question of Cross-Strait integration could then be deliberated and decided (Huang, 2009; Tong, 2009). The DPP, now an opposition party, continues to protest against KMT policies and to warn that opening-up to China will bring increasing unemployment and social inequality. In reality, however, with the sluggish state of the economy, opening up Taiwan to Chinese capital and tourists appears to be generally popular and it will be hard to reverse these developments in future.

With the arrival of Chinese capital and tourists, the meaning of Taipei's landmarks has come under further reconsideration. In July 2009, Taipei 101 was purchased by the Ting Hsin International Group, an enterprise owned by a Taiwanese migrant to China, Master Kong, now established as the top maker of instant noodles in China. Many economic observers saw this ownership change as a sign of an impending invasion of Taipei's commercial real estate by Chinese capital (Ien, 2009). There is some considerable irony in Taipei 101 becoming a Chinese real estate investment, given its original intent of symbolizing Taiwan's financial independence on the global stage.

Cross-Strait tourism has boomed, with group tourist numbers from China reaching over 300,000 within the first year of the new open policy. The Grand Hotel in Taipei has become a popular choice of accommodation for Chinese tourists, but it is no longer a sumptuous representative of orthodox traditional Chinese culture. Instead, its theatre now presents shows from Las Vegas. Other locations related to Chiang-kai Shek's legacy, such as the CKS Memorial Hall, and the CKS Residence and Summer Residence in Taipei, as well as his Mausoleum in Taoyuan, have been turned into popular Chinese tourist destinations (Jiang, 2010). The iconic significance of these locations has diminished as they have become business opportunities to meet the burgeoning Chinese demand.

Thus, in recent times, there have been some significant changes to long-standing attitudes, symbols and political ideologies as Taiwan seeks to exploit its Chinese connections by strengthening Taipei's position, at the heart of the emerging Hsinchu-Taoyuan-Taipei global urban region, as an East Asian transportation and exchange hub for China's global trade.

Sustainable Development

The shift in the economy of the Taipei region towards high-tech industries and producer services was accompanied by a growing awareness, on the part of both citizens and government, that the old mode of development at the cost of the environment and the quality of life needed to be rethought. Thus, as part of Taipei's development as a global city, the sustainable development concept was embraced as a key element on the agenda of urban and regional governance (Huang, 2005). For decades, the 'growth first' model of urban development in the Taipei Metropolitan Region had left the city with a significant deficit in green infrastructure. In order to meet growing citizen complaints, an urban park programme was launched and, starting in the late 1980s, riverside parks and urban parks were established, including the No. 7 Park in Daan District, Longsan Park in Wanhua (Manjia) within Taipei City, the No. 4 Park in Zhonghe and the Zhenai Park in Yongher, within Taipei County. This urban park programme has greatly enhanced the urban amenity of Taipei and its surroundings.

The change in lifestyle caused by rising income led to a huge increase in household waste. Taiwan started a nation-wide garbage recycling programme in 1998, which partially alleviated Taipei's waste management problem. Taipei's policy of garbage collecting and recycling impressed the international media as an example of policy innovation and inter-governmental agency cooperation. A journalist from the *Washington Post* noted that

the garbage trucks collect garbage five nights per week and you'll recognize the truck … because it plays music – a tinny version of the Beethoven classic 'For Elise'… Waiting for the garbage truck is one of Taiwan's liveliest communal rites. Many evenings I watched food vendors from the night markets, buckets of eggshells in hand, chat up convenience store clerks alongside Filipina nannies who traded kitchen appliances as if they were at a Sunday morning swap meet. Freelance recyclers keen to make a few dollars showed up to collect cardboard and newspapers, which they would sell back to the city. An alderman with a whistle kept traffic at bay…'. (Ross, 2007)

Entering the new millennium, a more radical policy was initiated to connect individual households to environmental protection. In July 2000, under the strong support of Mayor Ma Ying-jou, the Taipei City government started to charge a 'per-bag' trash collection fee to curb the production of household waste. Citizens bought specially designed bags for different categories of waste. This policy raised citizen awareness of the economic advantages of reducing the amount of daily waste. The city government claimed that this policy was immediately effective, achieving a 25 per cent reduction in garbage volume in the first 18 months (Taipei City Government, 2010). Almost 10 years later, Taipei County adopted the same policy in 2009, and reported that, in some townships, the garbage volume was dramatically reduced by over 50 per cent (*United Evening News*, 2009).

The need to clean up rivers was another important issue. Taipei's rivers, including the Tamsui, Dahan and Hsintien in the west, and the Keelong in the north, traditionally received untreated household and industrial discharges. Cleaning the rivers, in earlier decades, was costly to both the Taipei City and County governments, but tangible improvements were limited. In addition to the construction of a sewer system, which was fast-tracked between 2007 and 2010, Taipei County introduced an innovative approach to cleaning polluted water using natural processes in constructed wetlands. Fifteen new wetlands, covering 300 hectares, were developed along the Dahan River. This project successfully upgraded the water quality, increased the biodiversity of the region and restored the habitat for birds and insects. Moreover, these wetlands were connected to riverside bike routes and to the mass rapid transit system, gradually forming new leisure zones for the Taipei Metropolitan Region. These environmental improvements were achieved through collaboration between the city and county governments (despite them being under the political control of different parties) and local non-government organizations (NGOs), such as the Wilderness Society.

As 'green ideology' has been adopted widely by global cities, experience sharing among major cities has become more common in recent years. Chinese cities across the strait are also part of this trend. Government officers from Nanjing and Shanghai have visited the wetland projects in Taipei County and have initiated further co-operation and exchanges to compare experiences. These exchanges led to the inclusion of a Taiwan Pavilion and a Taipei Pavilion in the 2010 Shanghai World Expo. This was the first time that Taiwan participated in a World Expo since it left the United Nations in 1971 (CRN, 2009). The same architect who had designed Taipei 101, C.Y. Lee, designed the Taiwan Pavilion. In the Taipei Pavilion, there was an 'Urban Best Practices' exhibit, where Taipei City displayed its environmental sustainability projects under the 'three-Rs' policy (reuse, recycle and reduce). By marketing itself as a green city, the Taipei City government aims to attract more Chinese and international tourists to Taipei (CTS, 2009).

The recent Cross-Strait environmental co-operation on technical and non-political issues between China and Taiwan has further strengthened links while bypassing more disputed questions of sovereignty. And it can be argued, of course, that the deepening environmental crisis is a major issue of common concern on both sides of the strait. The Taipei Metropolitan Region has accumulated rich experiences based on strong synergies between government, communities and NGOs, while coastal cities in China represent a model of strong local government leadership and entrepreneurship. Going beyond the current interaction at the level of mutual visits and promotion, governmental collaboration on environmental issues offers great potential for a new and stronger Taiwan-China relationship.

Synthesis and Future Prospects

During the era of Dynastic Incorporation, when the Chinese Qing government incorporated Taiwan as part of Fujian province, Taipei emerged as a port city for Cross-Strait trade in the early 1700s. Throughout the subsequent periods – Japanese Colonization, Kuomintang Rule, Global Competition and the Cross-Strait Re-opening – Taipei has maintained its roles as the international commerce, political and administrative centre of Taiwan. From the onset, these urban functions and activities have been fixed as the intrinsic characteristics of the city's evolution. Over two centuries of development, these roles have remained unchanged but they have mutated, often drastically, according to national trends and international and domestic political circumstances. Taipei today, as a global city deeply enmeshed in the global economy, retains these roles at a time when new developments in the Cross-Strait relationship are having profound effects.

With the exception of the early part of the period of KMT rule, the Cross-Strait economic link has been a constant and crucial urban function. This link started in trade, but changed later into one of transnational industrial production. During the early phase of Kuomintang Rule, between 1945 and 1979, when China was closed to the outside world, Cross-Strait interactions ceased. This was a critical point when Taiwan redirected its trade away from China and steered its production and industries towards the global economy. By joining the global economy, Taiwan prospered. As incomes and prices increased and production costs rose, Taiwan soon lost its low-cost labour and manufacturing advantages. With the intensifying competitive global market, Taiwan's industries, in order to recover these production advantages, had to cross the strait. Since China adopted the 'Open-Door' policy, the Cross-Strait economy has grown and changed in the ways described in this chapter. Taipei, the transportation and communication centre, continues to be the key transition point, and its future growth will depend on how globalization further shapes the Cross-Strait relationship. At present, the global financial crisis, which has tarnished Taiwan's economic miracle, has made further integration with China imperative and, partly as a consequence, Taiwan is now at a critical stage for a decisive political and economic re-orientation. The expanded Taipei Metropolitan Region along the Hsinchu-Taoyuan-Taipei Corridor will be at the heart of this new Cross-Strait and global spatial dynamic.

National identity is a perpetual issue that haunts Taiwan. Identity politics and contests are often played out in the capital. The divide between mainland immigrants and local inhabitants has led to identity swings and major points of contention expressed in urban planning and design. Taipei's urban space has been the stage set for these disputes and, as national identity questions are unlikely to be resolved in the near future, while social compromise is difficult to reach, Taipei's landmarks and structures will continue to be targets for political image manipulation. Most recently, with the arrival of group tourists from China, urban landmarks, which used to carry messages of political contestation, have now become icons of consumption. How Taiwanese society, mainland migrants and local citizens alike, handle the 'China factor' will be a challenge for the future urban and regional governance of Taipei.

Globalization provides cities with unprecedented economic opportunities and gives rise to unprecedented environmental risks. Sustainable development presents new challenges to urban and regional governance. Pollution, energy shortages and the greenhouse effect have potential for urban disaster that has to be prevented. The coastal cities in China and the Taipei Metropolitan Region have begun to explore these issues collaboratively in pursuit of viable solutions. Policy innovations in curbing air pollution and in improving water quality through wetlands have been experimented with in Taipei with considerable success. City collaboration for green development appears to offer a new prospect for the urban societies on both sides of the strait to achieve greater mutual understanding and collaboration. Sustainable development may be the arena that leads to a smoother path for the Taiwan-China relationship, and a new direction for Taipei's future development.

References

Breslin, Shaun (2004) Greater China and the political economy of regionalization. *East Asia*, **21**(1), pp. 7–23.

Castells, Manuel (1992) Four Asian tigers with a dragon head: A comparative analysis of the state, economy, and society in the Asian Pacific rim, in Appelbaum, R.P. and Henderson, J. (eds.) *States and Development in the Asian Pacific Rim*. Newbury Park, CA: Sage.

Chao, Chien-Min (2004) National security vs. economic interests: reassessing Taiwan's mainland policy under Chen Shui-bian. *Journal of Contemporary China*, **13**(41), pp. 687–704.

Chen, E. Y. (1996) Greater China: an emerging technological giant? in Fung-yee Ng, Linda and Chyau Tuan (eds.) *Three Chinese Economies: China, Hong Kong and Taiwan: Challenges and Opportunities*. Hong Kong: The Chinese University Press.

Chen, Lo-wei (2007) The disputes over designating the C.K.S Memorial Hall as heritage. *China Times*, 13 March, p. C2.

Chen, Zhen-shiang (1957) *The Taipei City History*. Taipei: Nan-tien Book Store.

Cheng, Chia-huei (2004) *The First Exposition in Taipei*. Taipei: Yuan-Liou.

Ching, Chia-ho (2005) The development of economic structure: producer services and growth constraints, in Kwok, Reginald Yin-Wang (ed.) *Globalizing Taipei*. London: Routledge.

Chou, Tsu-lung (2005) The transformation of spatial structure: from a monocentric to a polycentric city, in Kwok, Reginald Yin-Wang (ed.) *Globalizing Taipei*. London: Routledge.

CRN (China Review News) (2009) *Participating in the Shanghai Expo*. 20 October. Available at http://www.chinareviewnews.com/doc/1011/0/9/7/101109782.html?coluid=7&kindid=0&docid=101109782. Accessed 20 February 2010.

CTS (Chinese Television System) (2009) The Taipei Pavilion Opened – Marketing Green Taipei. Report by Yang Zeng-hai, 20 October. Available at http://news.cts.com.tw/udn/international/200910/200910200332881.html. Accessed 20 February 2010.

Dajiyuan News (2006) While many Taiwan people live in Shanghai, going to local school is the trend. 26 March. Available at http://www.epochtimes.com/b5/6/3/26/n1266997.htm. Accessed 30 January 2010.

Dent, Christopher M. (2005) Taiwan and the new regional political economy of East Asia. *The China Quarterly*, **182**, June, pp. 385–406.

ECFA (2010) Introducing Economic Framework Corporation Agreement. Available athttp://www.ecfa.org.tw/. Accessed 30 January 2010.

Executive Yuan (2009*a*) Premier Liu: Government Will Directly Invest NT$858.5 billion to Stimulate Economy. News Release, 13 April. Available at http://www.ey.gov.tw/ct.asp?xItem=52919&ctNode=2877&mp=11. Accessed 20 January 2010.

Executive Yuan (2009*b*) Deregulation on the People's Travel Cross Straits. News Release, 2 October 2009. Available at http://www.ey.gov.tw/ct.asp?xItem=58436&ctNode=3062&mp=99. Accessed 20 January 2010.

Fu, Tsao-ching (1993) *The New Taiwan Architectures of Chinese Classic Style*. Taipei: Nan-tien Publisher.

Gold, Thomas (1993) Taiwan's Quest for Identity in the Shadow of China, in Tsang, Steve (ed.) *In the Shadow of China: Political Developments in Taiwan since 1949*. London: Hurst.

Hechutushe (2004) *Old Map Taipei Walk: The Taipei Old City of Qing Dynasty in 1895*. Taipei: Hechutushe Publisher.

Ho, I-hong (1989) *The Urban Development of Taiwan in Early Times*. Taipei: Taiwan History Archive.

Ho, Samuel P.S. (1978) *Economic Development of Taiwan 1860–1970*. New Haven, CT: Yale University Press.

Hsia, Chu-Joe (2000) Building colonial modernity: rewriting histories of architecture and urbanism in colonial Taiwan. *Taiwan: A Radical Quarterly of Social Studies*, **40**, pp. 47–82.

Hsiao, A-Ching (2008) *Return to Reality: Political and Cultural Change in 1970s Taiwan and the Postwar Generation*. Taipei: Institute of Sociology, Academic Sinica.

Hsu, Jinn-yuh (2005) The evolution of economic base: from industrial city, post-industrial city to interface city, in Kwok, Reginald Yin-Wang (ed.) *Globalizing Taipei*. London: Routledge.

Hsu, Ming-sung (ed.) (2007) *How the Sun Yat-Sen Memorial Hall Was Built. The Agony and Compromise of Architect Wang Da-hung*. Taipei: The Sun Yat-Sen Memorial Hall.

Hsu, Shuo-bin (2005) *The Visible and Invisible Taipei: The Transformation of Space and Power in Taipei during the Later Qing Dynasty and the Japanese Period of Time*. Taipei: Rive Gauche Publishing House.

Huang, Fu-san (ed.) (1995) *The One Hundred Years of History of Taipei*. Taipei: The Archive of Taipei City.

Huang, Lan-shiang (1995) The study of urban planning of Taipei at the beginning of the colonial period. *Taiwan: A Radical Quarterly of Social Studies*, **18**, pp. 189–213.

Huang, Liling (2005) Urban policies and spatial development: the emergence of participatory planning, in Kwok, Reginald Yin-Wang (ed.) *Globalizing Taipei*. London: Routledge.

Huang, Liling (2006) A world without strangers? Taiwan's new households in the nexus of China and Southeast Asia relations. *International Development Planning Review*, **28**(4), pp. 448–473.

Huang, Liling (2008) Taipei: post industrial globalisation, in Jones, Gavin and Douglass, Mike (eds.) *Mega-Urban Regions in Pacific Asia: Urban Dynamics in a Global Era*. Singapore: NUS Press.

Huang, Wu-Shung (2009) *Looking for Peace: the Proposal of 50 Years Peace Across Straits*. Available at http://www.wretch.cc/blog/fortwpeace/9459222. Accessed 30 January 2010.

Huang, Zhuong-tsang (1947) The terrible inspection. *Shanghai Wenhui Bao (Wenhui Daily)*, No. 231, 28 April. (Reproduction of woodcut print of 228 Incident).

Ien,Chuong-Zhen (2009) Ding-Xing bought the building of Taipei 101. *Business Week*, No. 463. Available at http://mag.chinatimes.cpm/mag-cnt.aspx?artid=1145. Accessed 25 January 2010.

In, Bao-ning (2006) *Sexuality, Nationalism and Post-colonialism: Whose Zhong-San North Road?* Taipei: Rive Gauche Publishing House.

Jiang, Xun (2010) Popularity of Chang Kai-sek spread over China. *Yazhou Zhoukan (International ChineseNewsweek)*, **24**(10), 14 March, p. 45.

Jou, Sue-ching (2005) Domestic politics in urban image creation: Xinyi as the Manhattan of Taipei, in Kwok, Reginald Yin-Wang (ed.) *Globalizing Taipei*. London: Routledge.

Kim, Won Bae (1997) Culture, history, and the city in East Asia, in Kim, Won Bae, Douglass, Mike, Choe, Sang-Chuel and Ho, Kong Chong (eds.) *Culture and the City in East Asia*. Oxford: Clarendon Press.

Kwok, Chao-lee (2009) Conservation of city architecture and public sphere in postwar Taiwan. *Journal of Architecture*, Taipei, **67**, March, pp. 81–96.

Kwok, Reginald Yin-Wang (2003) A paradigm for Asian Pacific urban form: cultural contention of globalization and vernacularization, in Park, Jin-Ho (ed.) *Creating Livable Communities in Asia Pacific: Value, Relevance and Connectivity*. Honolulu: Proceedings of the Fifth International Symposium on Asian Pacific Architecture.

Kwok, Reginald Yin-Wang (ed.) (2005*a*) *Globalizing Taipei*. London: Routledge.

Kwok, Reginald Yin-Wang (2005*b*) Introduction: Asian dragons, South China growth triangle, developmental governance and globalizing Taipei, in Kwok, Reginald Yin-Wang (ed.) *Globalizing Taipei*. London: Routledge (with Hsu, Jinn-Yuh).

Leng, Tse-Kang (2002) Securing economic relations across the Taiwan straits: new challenges and opportunities. *Journal of Contemporary China*, **11**(31), pp. 261–279.

Lin, April C.J. and Keating, J.F. (2000) *Island in the Stream: A Quick Study of Taiwan's Complex History*. Taipei: SMC Publishing Inc.

Lin, Wuu-Long and Lin, Pansy (2001) Emergence of the greater China circle economies: co-operation versus competition. *Journal of Contemporary China*, **10**(29), pp. 695–710.

Liu, Li-mei (1976) The design competition of the C.K.S Memorial Hall. *The Journal of Architects*, August, pp. 2–4.

Nan, Fang-shuo (2007) The sign politics of changing names and the Party constitution. *China Times*, 22 March, p. A4.

Naughton, Barry (1997) (ed.) *The China Circle: Economic and Electronics in the PRC, Taiwan and Hong Kong*. Washington DC: Brookings Institution Press.

New Taipei City (2010) *Welcome to New Taipei City: flagship projects and other plans*. Available at http://www.tpc.gov.tw/web66/_file/2890/upload/new_tp_city/ntp-index.html. Accessed 25 June 2010.

Northern Taiwan Development Commission (NTDC) (2005) *The Founding History of NTDC*. Available at http://www.ntdc.org.tw/01_1.htm. Accessed 25 June 2010.

NTUBP (National Taiwan University, Institute of Building of Planning) (1991) *The Comprehensive Development Plan of Taipei County*. Available at http://gisapsrv01.cpami.gov.tw/cpis/Cprpts/taipei_county/. Accessed 25 January 2010.

Rigger, Shelley (1999) *Politics in Taiwan: Voting for Democracy*. New York: Routledge.

Ross, Julia (2007) What I picked up about trash in Taipei. *The Washington Post*, 2 December. Available at http://www.washingtonpost.com/wp-dyn/content/article/2007/11/29/AR2007112901887.html. Accessed 25 January 2010.

Saxenian. A. and Hsu, Jinn-yuh (2001) The Silicon Valley-Hsinchu connection: technical communities and industrial upgrading. *Industrial and Corporate Change*, **10**(4), pp. 893–920.

Shin-kao-tan Book Store (1914) *The Map of Taipei Streets*. Taipei: Shin-kao-tan Book Store.

Taipei City Government (1988) *The 20th Year of Anniversary of Taipei Jurisdiction Expansion*. Taipei: Bureau of Information, Taipei City Government.

Taipei City Government (2010) *The Results of the Policy of 'Per-Bag' Trash Collection Fee*. Available at http://www.epb.taipei.gov.tw/olddep/Eia_Report/report/90rpt/pdf/5-1-1.pdf. Accessed 20 February 2010.

Taoyuan County Government (2009) *Introducing the Aerotropolis*. Available at http://aerotropolis.tycg.gov.tw/page_base.php?id=2&lang=0. Accessed 20 January 2010.

Tong, Qingfeng (2009) 50 Years of Cross-Strait peace, and strive for a million signatures. *Yazhou Zhoukan (International Chinese Newsweek)*, **23**(8), pp. 12–14.

Tsu, Jing-tao (2007) *The 228 Incident Documents*. Taipei: The Cross-Strait Publisher.

United Evening News (2009) The practice of per-bag trash collection fee reduced the garbage by 54% in Senken township. Report by Wang Chang-ding, 21 March . Available at http://udn.com/NEWS/DOMESTIC/DOM2/4802262.shtml. Accessed 20 February 2010.

US Army Map Service (1945) *The Taihoku Map*. Available at http://www.lib.utexas.edu/maps/ams/formosa_city_plans/.

Vale, Lawrence J. (1992) *Architecture, Power and National Identity*. New Haven, CT: Yale University Press.

Wang, Jenn-hwan (1997) Governance of a cross-border economic region: Taiwan and Southern China. *Taiwan: A Radical Quarterly of Social Studies*, **27**, pp. 1–36.

Wang, Jenn-hwan (2004) World city formation: geopolitics and local political process: Taipei's ambiguous development. *International Journal of Urban and Regional Research*, **28**(2), pp. 1–17.

Wei, Wen-de and Gao, Chuan-chi (2004) (eds.) *The 120th Anniversary of the Construction of Taipei City: Exhibition of Old Maps, Images, Documents and Artifacts*. Taipei: Bureau of Cultural Affairs, Taipei City Government.

Wen, Zhen-hua (1986) The Urbanization of Taipei in the Early 20th Century. PhD Dissertation, Department of History, Taiwan Normal University.

Zhang, Dianwen (2010) Economic great leap on two shores: rise of Western Strait Economic Zone. *Yazhou Zhoukan (International Chinese Newsweek)*, **24**(10), pp. 20–25.

Zhang, In-hua (1991) The urban system of Taiwan, in Tsai, Uong-mei and Zhang, In-hua (eds.) *The Urban Society of Taiwan*. Taipei: Jui-Luo.

Chapter Six

Seoul as a World City: The Challenge of Balanced Development

Seong-Kyu Ha

Following the devastation wrought by the Korean War in the early 1950s, the city of Seoul became a magnet for people, particularly from rural areas, searching for work and for a better life. It is now an important city in the global urban hierarchy – a 'world city' – with 10 million inhabitants. However, excessive concentration of population in Seoul and in the extended Seoul Metropolitan Region (hereafter SMR) has created many problems, including housing shortages, inflated land prices and traffic congestion. Disparities between the capital region and other parts of South Korea have grown over time, despite periodic attempts by government to direct development away from Seoul.

There are those in South Korea who advocate a policy of more 'balanced regional development' – that is, a more even distribution of economic activity and wealth between Seoul and the rest of the country. In 2003, in response to the perception of excessive concentration in the SMR, relocation of the administrative capital of South Korea to Chungcheong Province was proposed by the (late) President Roh Moo-Hyun. It was suggested that Seoul and a new separate capital could play similar complementary roles to those of New York City and Washington DC in the United States. The case of Seoul, however, illustrates some problems with this idea of separating city functions. World cities typically experience continued pressure to concentrate economic activities in the heart of their metropolitan areas. In Seoul, there is a clear tension between these centripetal forces of world city growth and the arguments for population decentralization and growth containment put forward by supporters of balanced regional development. Can this tension be resolved? This chapter examines this question as part of a broader review of Seoul's development organized around three principal themes – Seoul's planning history and urban form; Seoul's status as a world city; and contemporary challenges facing Seoul, including the attempts of its urban policy-makers to balance continued growth with the management of environmental risks.

History of Seoul and its Urban Planning

Seoul's Urban History

Hanyang, on the site of modern day Seoul, became the capital of Korea under Yi Seong-gye at the beginning of the Joseon Dynasty[1] in 1394 and retained that position until the fall of the dynasty in 1910. In the final years of the nineteenth century Korea opened itself to foreigners and began to modernize its economy and military, as well as its legal and education systems. However, in 1910 the country was occupied by the Japanese and from then until 1945 it remained under Japanese colonial rule. The Japanese adopted the name of Gyeongseong for the capital.

Following World War II and Korea's liberation, the city took its present name of Seoul on 15 August 1945. The country split into North Korea and South Korea in 1948, with Seoul becoming the capital of the Republic of Korea (South Korea). In 1950, the Korean War broke out and caused tremendous damage to Seoul. At least 191,000 buildings, 55,000 houses, and 1,000 factories in the capital were destroyed. The war also gave rise to a flood of refugees from the North,[2] swelling the city's population to more than one and a half million by 1955.

Following the Korean War, Seoul became the focus of an immense reconstruction and modernization effort, driven by government-led programmes to establish heavy and export-oriented industries. In the 1960s and 1970s South Korea was ruled by a military-led government under Park Chung-hee. Rapid economic growth achieved during these decades raised the living standards of residents considerably. Described as the 'miracle of the Han River', this period saw South Korea quickly achieve the type of industrial development that had taken most advanced countries over a century. Per capita GDP rose from US$81 in 1960 to US$1,580 by 1980. The proportion of urban residents in South Korea increased with industrial development, passing 50 per cent in the late 1970s, and eventually reaching 80 per cent by 2000. The majority of migrants from rural areas headed for Seoul or other larger cities, such as Busan and Ulsan in the south-east, so that the share of the total population in larger cities also increased progressively. The 2005 census showed population losses in most rural counties, except for those close to the larger cities.

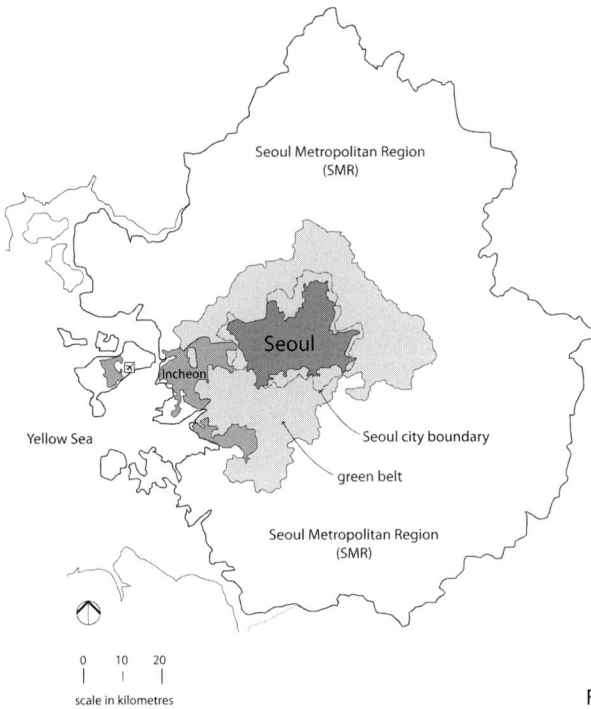

Figure 6.1. Seoul and the SMR.

Rural to urban migration was the major component of rapid urbanization in its earlier stages, but its contribution has fallen since the 1980s and urban development has been driven by the outward growth of the larger cities into surrounding areas. Seoul, in particular, has spread to encompass Incheon, a separate city of some 2 million people, and neighbouring areas of Gyeonggi Province. The extended SMR (see figure 6.1) has a total area of 11,745 square kilometres and had a population in 2007 of 24.5 million – about half of the national total. As well as being one of the largest cities in the world, the SMR is also amongst the most crowded, with a metropolitan population density only slightly less than that of Tokyo and a net density in its inner city which is higher than that of comparable areas of the Japanese capital (Kim *et al.*, 2002).

A distinctive feature of South Korea's population change in the second half of the twentieth century has been the rapidity of its 'demographic transition', with fertility rates declining steeply from the 1960s and reaching bare population replacement level as early as the mid-1980s. The national population growth rate is expected to be zero in 2021. Figure 6.2 shows the projected decline of South Korea's population to 2035. Seoul's population growth rate actually reached zero in 1990. There was a brief recovery in the 1990s, with population growing from 1995 to 1999, but thereafter the city's growth rate fell again. The forecast population of the city of Seoul in 2015 is less than its present 10 million. The percentage of people over 65 years old in Seoul

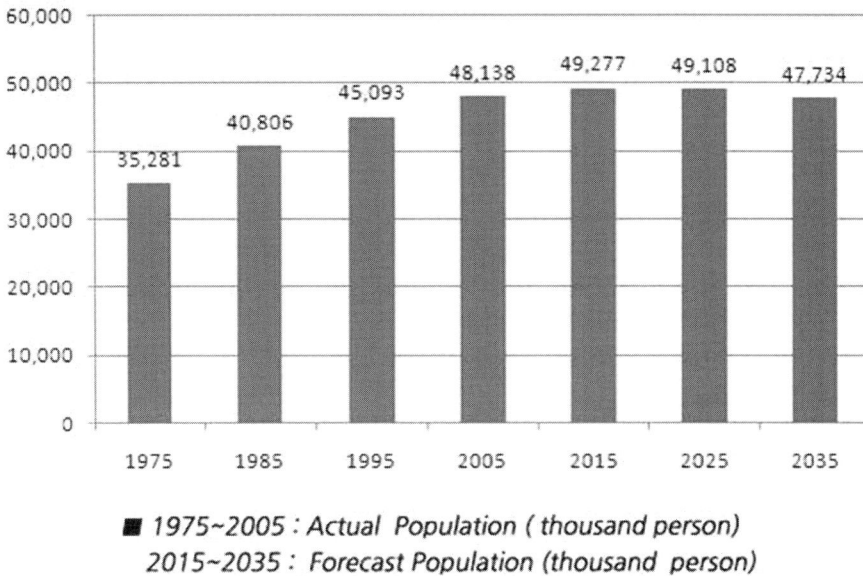

Figure 6.2. Population of South Korea 1975–2035. (*Source*: Statistics Korea, http://kostat.go.kr)

is increasing and had reached 7.2 per cent by 2000. More than 14 per cent of the population is forecast to be older than 65 by 2019.

Planning and Urban Form

The first modern urban planning legislation in Korea was the 'City Planning Act of Joseon', introduced by the Japanese colonial government in 1934. It served as the only statutory urban planning framework for nearly three decades until the South Korean government introduced its own planning legislation in 1962. Following liberation in 1945, city planning was unable to keep up with the pace of rapid social upheaval and, in the early 1950s, after the Korean War, the main emphasis was on reconstructing the many devastated areas of Seoul. It was not until the 1960s that significant spatial plans for Seoul emerged, as part of the strong drive for national development under the military regime of Park Jung-hee. The national government embarked on a series of 5-year plans, intended to put the capital city in the vanguard of national economic growth. Seoul was placed under the direct control of the prime minister at this time, in accord with the 'Special Act for the Administration of Seoul City' of 1962. A City Master Plan was adopted in 1966, with the aim of turning Seoul into a modern city with a population of 5 million by 1985. This population target was reached by 1970! The rapid growth of Seoul in the 1960s and 1970s led to the development of substantial informal settlements, housing an estimated 320,000 households by 1976 (Ha, 2004, p. 128).

New residential development within Seoul during the 1960s and 1970s mainly took the form of planned suburbs just beyond the traditional centre. The beginnings of outward sprawl were soon evident, however, and an attempt was made to control this by establishing a green belt with a total area of 1,566 square kilometres around the edge of the city. During the 1970s much of the city's growth, involving an additional 3 million or so people, was accommodated within this green belt which was located at an average distance of about 15 kilometres from the city centre. New laws were adopted in the late 1970s and early 1980s to encourage heavy and polluting industries away from the city and subsequent plans proposed the relocation of substantial amounts of employment to newly designated areas south of the Han River.

Pressure for new housing development and redevelopment at higher densities within the green belt grew during the 1980s, turning Seoul from a fairly low-rise city into a high-density city of high-rise apartments (Kim, 2004). Further large-scale urban renewal projects were also undertaken in the 1990s and early 2000s (see figure 6.3).

Figure 6.3. High-rise apartments in the Boramae urban renewal district.

Densities in some of Seoul's urban renewal areas are as high as 400 persons per hectare. The 1980s also saw suburban development jump the green belt and high rates of development were experienced in the SMR within a commuting distance of 25 kilometres or so from the city (Kim and Jung, 2001). By the second half of the decade there was a severe housing shortage in Seoul and the central government designated five new towns to boost supply. Target populations for these towns were between 160,000 and 400,000, with most housing in the form of high-rise, high-density apartments. Within 7 years these towns were providing housing for more than a million residents, many of whom commuted to Seoul and added to its traffic congestion. A second round of new towns, some 30 to 40 kilometres from Seoul, was announced in 2001. Proposed densities are lower here and there is more obvious attention to the environmental, social and economic dimensions of sustainability in the planning of these towns. Most will not be served by rail lines in the foreseeable future, however, making them dependent on cars and buses for any journeys which are not local.

Seoul's green belt has been regarded as a qualified success by some observers (Yokohari *et al.*, 2000; Choe, 2004). It has not been effective in stopping the outward growth of the city, but its steep slopes have been largely protected from development and it now provides a substantial area of green, open space for both the residents of the city of Seoul and of the new towns on its outer side. However, the green belt intensified the pressure for very high-density development in the inner areas of Seoul. It also encouraged the metropolitan region to spread further out, with development leapfrogging the green belt to satellite towns and to speculative developments around these towns. Densities are also high outside the green belt, especially in the earlier new towns. Seoul, therefore, has a somewhat unusual form with a strong high-density centre, and also high-density, high-rise satellite towns some considerable distance away from the city along major highway corridors. Public transport provision is not uniformly good. There is an excellent subway network, which serves inner areas and much of the commuting from the older new towns is now by public transport, but, as noted above, most of the more recent new towns are likely to be auto-dependent for the foreseeable future.

Balanced Development at the National Scale

The rapid growth of South Korea since the 1960s has been accompanied by regional economic disparities that some argue have been detrimental to the nation's overall development. Table 6.1 gives some measures of the SMR's concentration of economic activities. In addition to growing inequality between the SMR and other regions (Mawson, 1997), the concentration of growth in the SMR has led to serious problems of housing shortage, traffic congestion, high land prices, a low level of urban amenity and environmental pollution. In response, the South Korean government has developed a number of policies over time to try to moderate Seoul's dominance by steering people and industries away from the SMR.

During the period 1960–1971, primarily for reasons of national security, the government attempted to restrict the growth of Seoul by relocating government offices to other major cities, by developing 'garden city' growth poles, and by establishing new

Table 6.1 Indicators of Concentration in the SMR (2005).

Variables		Nation (A)	SMR (B)	Seoul (C)	Share of A B (%)	C (%)
Size (km²)		99,646	11,745	605	11.8	0.6
Population (000s)		48,782	23,465	10,167	48.1	20.8
Density (persons/km²)			490	2,000	16,805	
Industry	Number of workers (000s)	22,856	11,135	4,890	48.7	21.4
	Gross regional product (billion won)	817,812	386,990	185,091	47.3	22.6
Manufacturing	Number of establishments (000s)	340	177	70	52.1	20.6
	Number of workers (000s)	3,451	1,694	474	49.1	13.7
Services	Number of establishments (000s)	2,287	1,033	526	45.2	23.0
	Number of workers (000s)	9,092	4,713	2,700	51.8	29.7
Universities	Number of universities	173	68	38	39.3	22.0
	Number of students (000s)	1,264	496	320	39.2	25.3
Banks	Deposits (10 billion won)	561,946	381,040	279,344	67.1	49.7
	Loan (10 billion won)	613,922	409,655	262,826	66.7	42.8
Vehicles	Number of vehicles (000s)	15,397	7,115	2,809	46.2	18.2

Sources: National Statistical Office (2007) *Social Indicators in Korea*. Seoul: NSO. Korea Statistical Information Service, http://www.kosis.kr/search/totalSearch2.jsp.

industrial cities away from Seoul (Ahn and Ohn, 2001, p. 56). The first comprehensive National Physical Development Plan, covering the period 1972 to 1981, made provision for the establishment of large-scale industrial complexes, and for the expansion of power, water, roads, harbours, communications and other infrastructure in regional areas. The National Land Use Management Law (1972) provided powers to control land use and to regulate the distribution of industrial estates nationwide, and the Industrial Estate Promotion Law (1973) created incentives for the development of industrial estates in strategic locations outside the SMR. In order to promote dispersal, some existing tax laws were also revised to penalize industries in urban centres and a 'citizen tax' was also imposed on residents of the SMR.

The Second National Physical Development Plan of 1981 contained further measures to redress regional imbalance and to promote growth centres. The Capital Region Management Law was enacted in 1982 to guide development within the SMR. This was followed in 1984 by the first Capital Region Plan, under which industrial plant and large-scale office buildings seeking to relocate from the city of Seoul were prohibited in all parts of the SMR, except in a 'Development Promotion Zone'[3] (Park, 1985). In 1994 there was some deregulation of land-use policy and some of the stricter regulations, on private development and on smaller firms, were relaxed or phased out.[4] A new development charge was also levied on large-scale office and commercial developments in Seoul, with the revenue to be used to provide incentives to locate elsewhere.

However, despite policies over several decades attempting to achieve more balanced regional development, it has not proved possible to narrow the gap between the metropolis and the other regions in income and living standards. With the exception of Busan and other parts of the south-eastern coastal area which have been prime beneficiaries of government industrialization policy since the 1960s, the rest of the country generally lags well behind the SMR (Kim and Cha, 1998; Ahn and Ohn, 2001, p. 67).

Some critics argue that government policies for balanced regional development have failed because of a bias towards physical measures, such as the development of industrial parks and the construction of infrastructure, rather than towards fiscal instruments and economic policies (Lee and Kim, 1995; Park and Kim, 2001). Perhaps the most ambitious physical development proposal to redress regional imbalance was the planned relocation of the administrative capital, announced by President Roh Moo-hyun in 2003, referred to at the beginning of this chapter. This would have involved relocating 245 government agencies and state-run organizations to central Chungcheong province (Embassy of the Republic of Korea, 2003) (see figure 6.4).[5]

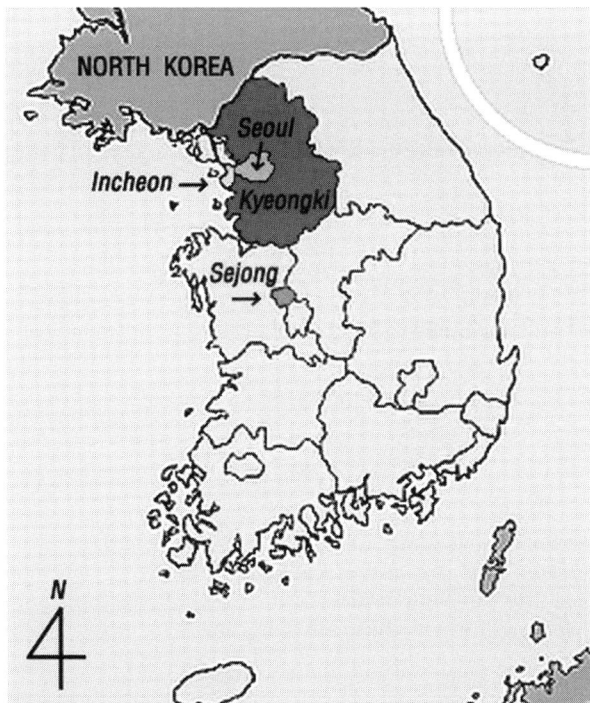

Figure 6.4. Seoul Metropolitan Region and the proposed alternative site for the administrative capital at Sejong.

Not surprisingly, this proposal was strongly contested, with the mayor of Seoul prominent amongst its opponents (Joongang Ilbo, 2003). Supporters endorsed the idea that relocating central government offices would relieve pressure on the SMR and stimulate the regional economy (Ko, 2003; Nam, 2003). Critics questioned whether these benefits would really be achieved and argued that moving the capital to Chungcheong province would simply lead to the further outward expansion of the Seoul Metropolitan Region along a corridor of some 150 kilometres between Seoul and the new capital. The proposal stalled after it was blocked by the Constitutional Court in October 2004 but it was subsequently revived in modified form, first as a proposal to relocate some government ministries away from Seoul and subsequently as a planned hub for economic and scientific development. It remains alive as a controversial topic for political groups with varying regional constituencies. Overall, however, the prospects of challenging the SMR's dominance seem slim in the foreseeable future.

The World City Concept

In 2006 the Seoul Metropolitan Government[6] set out its general vision for Seoul's future city planning for the period to 2020. Figure 6.5 summarizes the intent of the government at this time to repair the damage done to the urban environment during the rapid growth of the previous decades and to embrace a new style of planning which places more emphasis on qualitative aspects of urban development.

Figure 6.5 indicates that one of Seoul's explicit goals is to be a 'World City leading the northeast Asian economy'. The concept of a 'world city' is generally traced back to

Seoul, a world city in harmony with nature, human beings, history and technology

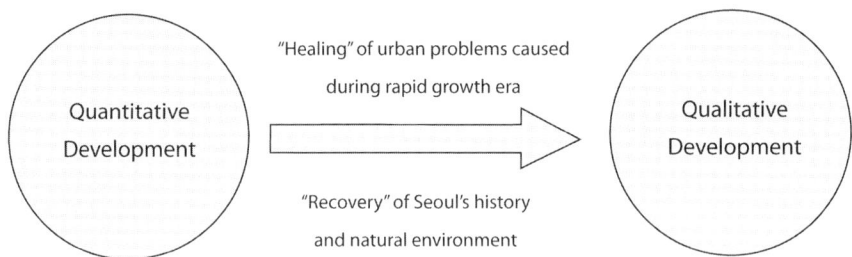

"World City" leading the Northeast Asian Economy
"Culture City" presenting a unique character
"Eco City" reviving nature
 "Welfare City" enriching everyday life

Figure 6.5. Vision and goals for Seoul's city planning. (*Source*: Seoul Metropolitan Government, 2006)

John Friedmann who, in 1986, set out his 'world city hypothesis' as a number of loosely joined statements about the links between global economic forces and urbanization processes (Friedmann, 1986). The other classic work on the 'global city' (essentially an interchangeable term with 'world city') remains Saskia Sassen's book *The Global City*, with its detailed studies of New York, London and Tokyo as the prime examples of this recent urban phenomenon (Sassen, 1991). The global or world city results from a new logic of agglomeration which concentrates certain financial and producer service functions in a relatively small number of sites around the world with advanced telecommunications facilities. As global corporations spread economic production through cross-border sub-contracting and other transnational business alliances, they need a specialized infrastructure to control their globally-dispersed operations (Sassen, 1995, p. 63). World cities provide this infrastructure. Because financial and producer service firms benefit from proximity to one another, they cluster together in major cities. People working in high-end financial and producer services jobs are attracted to big city amenities and lifestyle, so they also promote clustering (Sassen, 1995, p. 68; Leyshon and Thrift, 1997).

World cities form a new level in the international urban hierarchy, with rank in this hierarchy determined by a city's ability to attract global investments and services (Friedmann, 1995, p. 26). Competition for places in this hierarchy is fierce and reputable ranking surveys are eagerly awaited and closely scrutinized. As a further element in the description of world cities, Sassen observed that they tend to form a new urban hierarchy under the influence of a transnational capitalist class and, as a consequence, become more like each other and less like other areas in their own country. According to Sassen (1991, p. 8), 'there is a systemic discontinuity between national growth and forms of global city growth'. Global cities are not just competing with one another but form a system of their own.

It is possible to discern a 'systemic discontinuity' between the SMR and most of the rest of South Korea. However, in other respects, the applicability of the concept of the world city, as developed by Friedmann and Sassen, to East Asian cities has been challenged by some authors. In particular, in an influential paper written in 2000, Richard Child Hill and Jun Woo Kim presented evidence to suggest that there are at least two types of world city – 'market-centered and bourgeois' Western capitalist cities, such as New York, and 'state-centered and bureaucratic' Asian world cities, such as Tokyo and Seoul.[7] Whether this distinction is as valid now as it was 10 years ago is debatable, but certainly Seoul's development for much of the latter part of the twentieth century was led predominantly by government ministries under authoritarian regimes in the style associated with a 'developmental state' (Hill and Kim, 2000, p. 2168).

Douglass (2000) made a further significant contribution to the world cities' literature by examining how globalization was producing a new physical geography of Asian world cities as 'mega-urban regions' extending for a hundred kilometres or more from traditional metropolitan centres. These mega-urban regions grew faster than national population growth rates and continued to increase their shares of national population (McGee, 1991; McGee and Robinson, 1995). According to Douglass, 'specifically, in the higher-income Asian NIEs [Newly Industrializing Economies], governments are intentionally promoting a remaking of key metropolitan regions into world cities' (2000, p. 2322).

Douglass identified eight types of function commonly associated with Asian world cities. These were finance (banks, stock, real estate, insurance); transnational corporate headquarters functions (commodity production and distribution); global service (education, high-tech producer services); transport (world hub airports, very fast trains, super container ports); information (creation, processing, screening, dissemination); political-ideological management of state-economy-society relations; culture (production-commodification- dissemination of cultural icons, practices, events); and spectacular world events (Olympics, World Expo, conventions, music concerts) (*Ibid.*, p. 2323). It seems clear that much of Seoul's recent development has been driven by the explicit desire of national and metropolitan policy-makers to increase its standing as a world city. The next section considers Seoul's world city status against some of the dimensions set out above and describes some of the more significant initiatives which have been pursued to enhance this status.

Seoul as a World City

Seoul now ranks sixth among the world's cities in the number of transnational companies listed by *Fortune 500* that have headquarters in the city. Eleven global *Fortune 500* companies were located in Seoul in 2009 (table 6.2). In the Worldwide Centers of

Table 6.2. Cities with the most Global 500 companies (2009).

Rank	City	Country	Number of Global 500 companies	Global 500 revenues ($ millions)
1	Tokyo	Japan	51	2,237,560
2	Paris	France	27	1,399,172
3	Beijing	China	26	1,361,407
4	New York	United States	18	869,150
5	London	United Kingdom	15	994,772
6	*Seoul*	*South Korea*	*11*	*519,351*
7	Madrid	Spain	9	434,393
= 8	Toronto	Canada	7	195,510
= 8	Zürich	Switzerland	7	242,595
= 8	Osaka	Japan	7	291,492
= 8	Moscow	Russia	7	380,530
= 8	Munich	Germany	7	485,386

Source: http://money.cnn.com/magazines/fortune/global500/2009/index.html.

Commerce Index 2008 (MasterCard Global Centers of Commerce, 2008), Seoul was ranked ninth overall. It is defined as an 'alpha'[8] city by the Globalization and World Cities Research Network (GaWC, 2008) and it was ranked tenth on the AT Kearney/ Foreign Policy and Chicago Council on Global Cities Index for 2010 (Foreign Policy, 2010).

South Korea's high-tech industries, as measured by R&D investment and technology content, have continued to concentrate in Seoul and the SMR. More than 50 per cent of all high-tech items are produced in the SMR. In terms of its share of total R&D expenditure, Seoul accounted for nearly 20 per cent in 2007 (table 6.3) and the SMR for nearly 65 per cent. South Korea was ranked fifth in 2008 amongst OECD countries in its ratio of R&D expenditure to GDP (Ministry of Science and Technology, 2008). Seoul also has the greater part of South Korea's venture capital and a heavy concentration of its IT industries.

Table 6.3. Regional distribution of R&D expenditure (2007) (in million won).

	South Korea (A)	SMR (B)	Seoul (C)	Share of A	
				B	C
R&D	313,838	201,248	61,838	64.8%	19.8%

Sources: Ministry of Science and Technology (2008) *Report on the Survey of Research and Development*, p. 108.

The development of the internet has strongly affected the creation and dissemination of information. According to the International Telecommunication Union (ITU) (2008), South Korea ranked eighth in the world at that time in its number of high-speed internet subscribers per 100 residents. By 2010 South Korea had the highest level of wireless broadband penetration amongst OECD countries with 95 subscribers per 100 residents (OECD, 2010). About 28 per cent and 53 per cent of high-speed internet users are residents of Seoul and the SMR, respectively. South Korea came first in 2010 in the United Nations 'e-government' rankings, which were based on a survey of access to information available electronically from public sector organizations (UN Public Administration Programme, 2010). In the same year, Seoul was ranked as the most advanced city in the world for municipal e-government in another UN-sponsored survey (Hicks, 2010)

The Fourth Comprehensive National Territorial Plan (2000–2020) acknowledged that, to enable South Korea to assume the role of a gateway to Northeast Asia, the nation must systematically build world-class transportation infrastructure in the form of international airports, seaports and high-speed railways (Ministry of Construction and Transportation and Korea Research Institute for Human Settlement, 2001; Baek, 2003). In 2001, the Incheon International Airport (ICN) opened. This was built on a site of 11.7 million square metres on reclaimed land approximately 50 kilometres from downtown Seoul at a cost exceeding US$5 billion. As the world's sixth-largest cargo airport, it has two runways and the capacity to handle 30 million passengers a year. The annual passenger capacity is expected to be extended to 100 million by 2020 (table 6.4) and this is predicted to exceed the capacity of the international airports at Shanghai, Hong Kong and Osaka (Kansai) by then. ICN's ability to handle 7 million tons of freight per annum in 2020 will be exceeded only by Hong Kong amongst its regional competitors. The airport has two 3.75 kilometre long runways which allow the largest planes to land and take off twenty-four hours a day. In 2010, ICN was ranked as the best airport in the world by the Airports Council International for the fifth consecutive year.[9]

Table 6.4 Increasing capacity of Incheon International Airport.

Annual Capacity	Flights	Passengers	Cargo
2002	170,000	27 million	1.7 million tons
2020	530,000	100 million	7 million tons

Source: Ministry of Finance and Economy (2003) *The Axis of Asia: Transforming Korea into Northeast Asia's Business Hub*, p. 8.

A high-speed rail link was completed between Seoul and Busan in 2010 and a second line is under construction to Mokpo. Rail freight operations also commenced in December 2007 between Munsan in South Korea and Bongdong in North Korea. While these operations are vulnerable to continuing political tensions between North and South Korea, there are ambitious longer term plans to extend them to link South Korea's rail network to the three trans-Asian railroads across China, Mongolia and Siberia. This would cut shipping times and costs of freight to Europe by up to two-thirds. It would also raise the possibility of Japanese freight being shipped via Korea, possibly through a tunnel linking South Korea and Japan. Concept plans for a trans-Korean highway have also been prepared.

The functions of a world city identified by Douglass (2000), referred to above, also include the staging of spectacular world events. Seoul hosted the 1988 Olympic Games. The planning preparations for this in the mid-1980s were significant in bringing about a reconsideration of the image of the city and in prompting a vigorous citywide programme to beautify and improve the amenity of Seoul. The 1988 Summer Olympics were also particularly important in bringing international recognition to Seoul as a basis for the further expansion of its role as a key political and economic location in the Asia-Pacific region. More recently, Seoul stole the global sporting limelight once again when South Korea and Japan co-hosted the 2002 football World Cup, the largest sporting event in the world.

Figure 6.6. Seoul's 'Hammering Man', erected in 2002.

Seoul's cultural policy now supports the strengthening of the city's image through place-making and the promotion of its cultural industries. Substantial amounts are allocated to cultural activities, public art (figure 6.6) and the operation of galleries and museums (Kim and Yoo, 2002, p. 95).

There is also increased emphasis on the protection of heritage places in Seoul, in part to enhance the city's attractiveness to tourists (see figure 6.7). Visitors from China and other nearby Asian countries are particularly targeted. The Dongaemoon Fashion Market is also an important tourist destination. Seoul was recently named 'World Design Capital 2010' by the International Council of Societies of Industrial Design and is now seeking to use this accolade to bolster its reputation as a centre of global design leadership.

There has been a remarkable growth recently in the electronic game, animation and software industries in Seoul, and also in film production and distribution. Korean pop music, television dramas, movies, fashion, food and celebrities have become popular in China, especially amongst young people, and that popularity has expanded in the past few years to Hong Kong, Taiwan, Vietnam and Japan (Dickie, 2006). The further growth of film production is being promoted through the Seoul Film Commission, while a digital media city hosting multi-media industries is being built at Sangamdong

Figure 6.7. 'Hanok' (traditional houses) as a tourist attraction in Jongno-gu.

and is expected to function as one of the primary media and entertainment hubs in Northeast Asia.

It seems clear from the evidence above that Seoul is indeed a world city. Banking, finance, insurance and other producer services are strongly concentrated in Seoul, while manufacturing has declined. Foreign direct investment continues to increase rapidly. The city is a major centre of transnational corporations, high technology and venture companies; it is a hub of high-speed internet activity; it is a major film producer; and it has hosted a series of world sporting events. However, Seoul continues to face a number of environmental challenges, while much of its urban area is low in amenity and in the quality of its public realm. The final section of this chapter considers these issues together.

Environmental Challenges and Urban Quality

Seoul's environmental problems are inextricably linked to its growth in the second half of the twentieth century and the concentration of population and activities which occurred. During the first two decades of South Korea's economic boom, there was little attention paid to the damaging effects of rapid industrialization. It was not until the 1980s that South Korea began to show some concern for the environment and the capital city continues to suffer from air and water pollution, solid waste disposal problems and a shortage of green space.

Air quality in Seoul has improved since 1988 and has generally been in compliance with relevant minimum environmental standards since 1991. During the winter, however, several areas in the city still fail to meet these standards because of pollutants discharged from household heating systems. In addition, the number of short-term violations of air quality standards is increasing, particularly in relation to pollutants emitted from mobile sources. Air quality is also poor in subway stations and malls.

Major changes to clean air policies came into effect in 2007 when higher minimum standards were adopted for nitrogen dioxide and particulate matter. A new minimum standard was also introduced for benzene in 2010, adding to longer-established standards for carbon monoxide, lead, ozone and hydrocarbons. Under Article 3 of the Clean Air Conservation Act, there is also now an improved network for monitoring air pollution across the nation, involving the Ministry of the Environment as well as lower-level governments.

As a consequence of the release of hazardous pollutants and wastewater from industrial developments, as well as increased amounts of municipal sewage, pollution of rivers and water resource conservation areas has occurred, leading to the need to introduce advanced water purification facilities. The installation of galvanized steel pipes was officially banned in 1994 but they remained in use and a recent survey conducted by the Seoul city government showed that 63 per cent of households with such pipes have experienced discoloured water from their taps. More systematic monitoring of water quality has now been introduced and, in 2009, the government began an indoor water pipe improvement project for low-income earners.

Recently the national government has taken a major initiative to tackle South Korea's environmental problems and to position the country at the forefront of 'green growth' through a new national strategy. This is described briefly below.

South Korea's National Strategy for Green Growth

South Korea's rapid industrialization and urbanization have resulted in significant pressure on its natural resources. South Korea's 'National Strategy for Green Growth', announced in 2008, and its related Five-Year Plan set out a new approach to address the problems of the past and to pursue ambitious policy goals and targets in relation to sustainable development. There is a commitment to develop new green industries and to invest substantially in new environmental infrastructure (UNEP, 2010). The Five-Year Plan outlines a set of three strategies, ten policy directions and fifty core projects. The three strategies comprise measures for addressing climate change and securing energy independence; the creation of new 'growth engines' based on green technologies; and the improvement of the quality of life (table 6.5). The Basic Law for

Green Growth came into effect on 13 January 2010. The South Korean government has committed to spending 2 per cent of GDP over 5 years, a total of 107.4 trillion won (or US$83.6 billion), on green technologies and materials, renewable energy initiatives, sustainable transport, green buildings and ecosystem restoration. This represents a very significant initiative to reorient and refocus the economy on sustainable growth. The government anticipates that this level of investment will lead to the creation of jobs in green industries for between 1.18 and 1.47 million people during the implementation of the Five-Year Plan.

Table 6.5. Korea's 5-year green growth plan: three strategies and ten policy directions.

Strategies	Policy directions
Measures for climate change and securing energy independence	Reduce carbon emissions Decrease energy dependence and enhance energy self-sufficiency Support adaptation to climate change impacts
Creation of new growth engines	Develop green technologies as future growth engines Greening of industry Develop cutting-edge industries Set up policy infrastructure for green growth
Improving quality of life and strengthening the status of the country	Green city and green transport Green revolution in lifestyle Enhance global co-operation on green growth

Source: UNEP, 2010, http://www.greengrowth.go.kr/index.do.

Figure 6.8. Jongno-ro, looking towards Chongno tower. Seoul has a fleet of more than 7,000 buses.

South Korea's Green Growth policies are reflected in Seoul in a range of initiatives including the progressive replacement of its diesel bus fleet (figure 6.8) with natural gas-powered vehicles, the creation of a green network through the city, connecting its parks and open spaces, and a number of urban design, landscape and wetland restoration projects along the Han River, designed to improve its water quality and to refocus the city on its waterfront.

Many trace the origins of Seoul's current move to embrace environmental values to an earlier riverine improvement project. The Cheonggye stream (Cheonggyecheon), which used to flow through the city centre, was enclosed in a concrete culvert in the 1950s and an elevated highway was built over it in the mid-1970s. In July 2003, the then mayor of Seoul, (and subsequently President) Lee Myung-bak, initiated the Cheonggyecheon project to remove the highway and restore the stream. It is now an attractive public space extending for more than 5 kilometres through the heart of the city (see figures 6.9 and 6.10). The restored stream is not entirely natural and substantial volumes of water have to be pumped in at certain times of year. Nevertheless, the restoration of Cheonggyecheon has clearly been important as a symbol of the desire to reintroduce nature to the city and to promote a more eco-friendly approach to urban design. Cheonggyecheon stream now provides a peaceful urban oasis running through downtown Seoul below street-level. It also provides visible evidence of the paradigm shift which is occurring in Seoul, away from an emphasis on the quantitative aspects of

Figure 6.9. View of Cheonggyecheon.

Figure 6.10. View of Cheonggyecheon.

economic development and towards a more participatory approach concerned with the quality of the urban environment (Park and Young, 2009).

Conclusions

Seoul is a dynamic city, which has experienced, and is still experiencing, remarkable economic, technological, social and urban development. Its transformation to a world city has occurred in a very short space of time, however, and its rapid growth, fuelled by high in-migration, has been accompanied by some negative environmental and social impacts, with consequences for the health and quality of life of its citizens.

South Korea developed very quickly in the 1960s and 1970s and the pace of its economic growth has been maintained over the past three decades. It recovered from the Asian financial crisis of the late 1990s to become one of the world's fastest growing economies. The global financial crisis of 2008 caused a slump in Seoul house prices, increased interest rates and a significant slow-down in South Korea's economic growth, but GDP is now recovering and the indications to date are that the South Korean government's response to the crisis has been effective. South Korea is an export-led economy and exports remain strong, particularly to China. South Korea's growth prospects seem good, therefore, with a strong commitment to green growth in the next stage of the country's development.

It is clear that the South Korean government has been centrally involved in developing the policies and providing the infrastructure, which have helped Seoul to achieve world city status and to become a leading city in East Asia. But some tension remains between these policies and the intent, also expressed by governments over many years, to contain growth and to 'spread the wealth' more equally across the nation. The centripetal forces of world city development continue to spur growth in the SMR while the proponents of regional development still seek to direct development away from the centre. Government policy is to support both, although the evidence seems to be that there has been more commitment to date to the former than the latter. A major challenge for South Korea is to reconceptualize the relationship between Seoul and the rest of the country through an approach to mega-regional planning that encourages greater economic integration between the world city and the provincial areas in a way that strengthens both (Sassen, 2011).

Notes

1. Joseon (July 1392–August 1910) was a Korean sovereign state founded by Yi Seong-gye that lasted for approximately five centuries. During most of the Joseon Dynasty, Korea was divided into eight provinces. The eight provincial boundaries remained unchanged for almost five centuries from 1413 to 1895, and form the geographic basis of today's administrative divisions, dialects and regional distinctions of the Korean Peninsula (see Choe *et al.*, 2000).
2. As well as the return of expatriate Koreans from Manchuria and Japan
3. There were five zones – Restricted Development Zone, Selective Development Zone, Development Promotion Zone, Deferred Development Zone and Conservation Zone.
4. The five zones listed in note 3 above were reduced to three at this time – a Restricted Density Zone covering Seoul and Incheon, a Restricted Growth Zone, which included the previous Development Promotion Zone, and a Deferred Development Zone, which corresponded to the previous Conservation Zone.
5. Building the new administrative capital away from Seoul had also been proposed in the 1970s under Park Jung-hee. At that time, it was proposed mainly for reasons of national defence. The location of the large Seoul agglomeration within range of North Korean artillery was regarded as a military risk. The Seoul green belt was also initially justified, in part, as providing a buffer for the city against artillery fire.
6. The Seoul Metropolitan Government administers the city of Seoul itself rather than the broader SMR
7. Hill and Kim's article led to responses from Friedmann (2001) and Sassen (2001) and to a rejoinder from Hill and Kim (2001) – see the list of references for this chapter.
8. Very important world cities that link major economic regions and states into the world economy.
9. See http://www.korea.net/detail.do?guid=46847. Accessed 23 January 2011.

References

Ahn, K.H. and Ohn, Y.T. (2001) Metropolitan growth management policies in Seoul: a critical review, in Kwon, W.Y. and Kim, K.J. (eds.) *Urban management in Seoul: policy issues and responses*. Seoul: Seoul Development Institute.

Baek, Y.H. (2003) *Can Seoul become a World City?* Seoul: Seoul Development Institute.

Choe, S.C. (2004) The thirty-year experiment with British greenbelt policy in Korea, in Richardson, H. and Bae, C.C. (eds.) *Urban Sprawl in Western Europe and the U.S.* Aldershot: Ashgate.

Choe, Y.H., Lee, P.H. and de Bary, W.T. (eds.) (2000) *Sources of Korean Tradition*. Volume II. *From the Sixteenth to the Twentieth Centuries*. New York: Columbia University Press.

Dickie, Lance (2006) The Korean wave. *Seattle Times*, 4 June. Available at http://seattletimes.nwsource.com/html/opinion/2003036091_sundaykorea04.html. Accessed 23 January 2011.

Douglass, M. (2000) Mega-urban regions and world city formation: globalism, the economic crisis and urban policy issues in Pacific Asia. *Urban Studies*, **37**(12), pp. 2315–2335.

Embassy of the Republic of Korea (2003) Capital relocation plan to be completed by 2004. *Bi-weekly Review of Korean Affairs*. Washington DC: Embassy of the Republic of Korea (13 June).

Foreign Policy (2010) *The Global Cities Index 2010*. Available at www.foreignpolicy.com/node/373401. Accessed 18 January 2011.

Friedmann, J. (1986) The world city hypothesis. *Development and Change*, **17**(1), pp. 69–84.

Friedmann, J. (1995) Where we stand: a decade of world city research, in Knox, P.L. and Taylor, P.J. (eds.) *World Cities in a World* System. Cambridge: Cambridge University Press.

Friedmann, J. (2001) World cities revisited: a comment. *Urban Studies* **38**(13), pp. 2535–2536.

Globalization and World Cities Research Network (GaWC) (2008) *The World According to GaWC 2008*. Available at http://www.lboro.ac.uk/gawc/gawcworlds.html. Accessed 30 August 2010.

Ha, S.K. (2004) New shantytowns and the urban marginalized in Seoul Metropolitan Region. *Habitat International*, **28**(1), pp. 123–141.

Hicks, R. (2010) Seoul tops global cities ranking for e-govt. *Asia-Pacific FutureGov Newsletter,* 21 July. Available at http://www.futuregov.asia/articles/2010/jul/21/seoul-tops-global-cities-ranking-e-govt/. Accessed 14 October 2010.

Hill, R.C. and Kim, J.W. (2000) Global cities and developmental states: New York, Tokyo and Seoul. *Urban Studies*, **37**(12), pp. 2167–2195.

Hill, R.C. and Kim, J.W. (2001) World cities revisited: a comment. *Urban Studies*, **38**(13), pp. 2541–2542.

International Telecommunications Union (2008) Information and Communication Technology (ICT) Statistics. Available at http://www.itu.int/ITU-D/ict/. Accessed 23 December 2009.

Joongang Ilbo (2003) *What Reasons to Change Capitals?* 4 February (Korean Daily Newspaper).

Kim, K.J. (2004) Residential rebuilding boom and planning response, in Sorensen, A., Marcotullio, P. and Grant, J. (eds.) *Towards Sustainable Cities*: *East Asian, North American and European Perspectives on Managing Urban Regions*. Aldershot: Ashgate.

Kim, K.J., Jung, H.Y., Kim, Y.R. and Yun, H.R.H (eds.) (2002) *Seoul and World Cities: Comparative Reference on Urban Context and Infrastructure*. Seoul: Seoul Development Institute.

Kim, M.H. and Jung, H.Y. (2001) Spatial patterns and policy issues of the Seoul metropolitan region, in Kwon, W.Y. and Kim, K.J. (eds.) *Urban Management in Seoul: Policy Issues and Responses*. Seoul: Seoul Development Institute.

Kim, W.B. and Yoo, J.Y. (2002) *Culture, Economy and Place: Asia-Pacific Perspectives*. Seoul: Korea Research Institute for Human Settlement.

Kim, Y.W. and Cha, M.S. (1998) *Regional Development Policy: Experiences and Issues*. Seoul: Korea Research Institute for Human Settlement.

Ko, B.H. (2003) The Strategy and Task for National Balance for Building the New Administrative Capital. Paper presented to the Forum for Building the New Administrative Capital, Administrative Capital Relocation Team, Seoul, June.

Lee, G.Y. and Kim, H.S. (1995) *Cities and Nation: Planning Issues and Policies of Korea*. Seoul: Nanam.

Leyshon, A. and Thrift, N. (1997) *Money/Space: Geographies of Monetary Transformation*. London: Routledge.

MasterCard Global Centers of Commerce (2008) *Worldwide Centers of Commerce Index 2008*. MasterCard Wordwide. Available at www.mastercard.com/us/company/en/insights/pdfs/2008/MCWW_WCoC-Report_2008.pdf. Accessed 10 October 2009.

Mawson, J. (1997) Regional policy in the Republic of Korea. *Regional Studies*, **31**(4), pp. 417–434.

McGee, T.G. (1991) The emergence of *desakota* regions in Asia: expanding a hypothesis, in Ginsberg, N., Koppel, B. and McGee, T.G. (eds.) *The Extended Metropolis*. Hawaii: University of Hawaii Press.

McGee, T.G. and Robinson, I. (eds.) (1995) *The New Southeast Asia: Managing the Mega-urban Regions*. Vancouver, BC: University of British Columbia Press.

Ministry of Construction and Transportation and Korea Research Institute for Human Settlement. (2001) *The Fourth Comprehensive National Territorial Plan for Korea (2000–2029)*. Seoul: Korea Research Institute for Human Settlement.

Ministry of Finance and Economy (2003) *The Axis of Asia: Transforming Korea into Northeast Asia's Business Hub*. Seoul: MFE.

Ministry of Science and Technology (2008) *Report on the Survey of Research and Development*. Seoul: MST.

Nam, Y.W. (2003) Evaluation of Discussion about the New Administrative Capital and Work for Urban Spatial Structure. Paper presented to the Forum for Building the New Administrative Capital, Administrative Capital Relocation Team, Seoul, June.

OECD (2010) *OECD Broadband Statistics, June 2010*. Available from http://www.oecd.ord/document/4/0,3746,en_2649_34225-42800196_1_1_1_1,00.html. Accessed 14 December 2010.

Park, J.G. and Young, A.L. (2009) The role of urban planning in the process of making livable cities in Korea, in Horita, M. and Koizumi, H. (eds.) *Innovations in Collaborative Urban Regeneration*. Tokyo: Springer.

Park, S. (1985) Regional changes in the industrial system of a newly industrializing country: the case of Korea, in Hamilton, F.(ed.) *Industrialization in Developing and Peripheral Regions*. London: Croom Helm.

Park, Y.H. and Kim, C.H. (2001) *Vision 2011 and Strategies for Balanced Regional Development*. Seoul: Korea Research Institute for Human Settlement.

Sassen, S. (1991) *The Global City: New York, London, Tokyo*. Princeton, NJ: Princeton University Press.

Sassen, S. (1995) On concentration and centrality in the global city, in Knox, P. and Taylor, P. (eds.) *World Cities in a World System*. Cambridge: Cambridge University Press.

Sassen, S. (2001) Global cities and developmentalist states: how to derail what could be an interesting debate: a response to Hill and Kim. *Urban Studies*, **38**(13), pp. 2537–2540.

Sassen, S. (2011) Novel spatial formats: megaregions and global cities, in Xu, J. and Yeh, A.G.O. (eds.) *Governance and Planning of Mega-City Regions: An International Comparative Perspective*. London: Routledge.

Seoul Metropolitan Government (2006) *Seoul Master Plan for 2020*. Seoul: Seoul Metropolitan Government.

UNEP (2010) *Change Tomorrow: Green Growth Will Change Korea's Future*. Available at http://www.greengrowth.go.kr/index.do. Accessed 10 October 2010.

UN Public Administration Programme (2010) *United Nations E-Government Survey 2010*. Available from http://www2.unpan.org/egovkb/global_reports/10report.htm. Accessed 9 June 2010.

Yokohari, M., Takeuchi, K., Watanabe, T. and Yokota, S. (2000) Beyond greenbelts and zoning: a new planning concept for the environment of Asian mega-cities. *Landscape and Urban Planning*, **47**, pp. 159–171.

Acknowledgement

The editors thank Damien Mugavin of the Seoul National University for his assistance in obtaining images for this chapter.

Chapter Seven

Hong Kong: The Turning of the Dragon Head

Anthony Yeh

Hong Kong developed from a manufacturing centre to an Asian world city in the half century after World War II. At the doorstep of China, it benefited tremendously from China's adoption of economic reform and an Open Door policy after 1978. In the early Chinese Reform Period, while most of Hong Kong's manufacturing industries moved to China and, in particular, to the nearby Pearl River Delta (PRD), its economy grew strongly because of a regional division of labour based on a 'front shop, back factories' model. However, with China's economic growth and increasing technological 'know how', Hong Kong is losing some of its competitive edge. In the early reform years, Hong Kong served as China's 'dragon head', providing the finance, marketing and office services for factory production in the PRD. More recently, however, this 'front shop, back factories' model has begun to lose its relevance. On the one hand, with further growth of manufacturing industry and service activities, the PRD has become a dynamic regional engine of growth in its own right. On the other hand, Hong Kong is starting to show signs of decline with job loss, deflation, a rising unemployment rate, declining relative income and a slowing down of infrastructure development. This chapter examines the changing nature of the competition that Hong Kong is facing and the challenges that it has in sustaining its economy and maintaining its position as the dragon head of the PRD.

Hong Kong's Historical Development

Hong Kong has a total land area of 1,068 square kilometres and is located on the southeastern coast of China, adjoining the Chinese province of Guangdong (figure

7.1). Hong Kong Island was ceded to Britain on 29 August 1842 by the Treaty of Nanking after the First Opium War. The Kowloon Peninsula and Stonecutter's Island were ceded later, on 24 October 1860, by the First Convention of Peking which followed the Second Opium War. The New Territories, comprising the area extending north of Kowloon as far as the Shenzhen River at the border with China, together with more than 200 outlying islands, were leased to Britain on 9 June 1898 for 99 years by the Second Convention of Peking. Hong Kong remained a British colony until it became a Special Administrative Region (SAR) of China on 1 July 1997. This followed the signing in Beijing of the Sino-British Joint Declaration on the Future of Hong Kong by the Chinese and British governments on 19 December 1984 and the enactment of the Basic Law of the Hong Kong SAR by the National People's Congress (NPC) of China in April 1990. In order to maintain the prosperity and stability of Hong Kong after the return of sovereignty to China, the Joint Declaration and Basic Law specified that Hong Kong would be governed differently from the rest of China ('One Country, Two Systems') for 50 years after 30 June 1997. Under this arrangement China has responsibility only for Hong Kong's defence and foreign policy while the Hong Kong SAR retains autonomy in domestic affairs. Hong Kong possesses its own executive, legislature and independent judiciary, including its own court of appeal. It is responsible for public order, continues to decide its own economic and trade policies and maintains its former capitalist system.

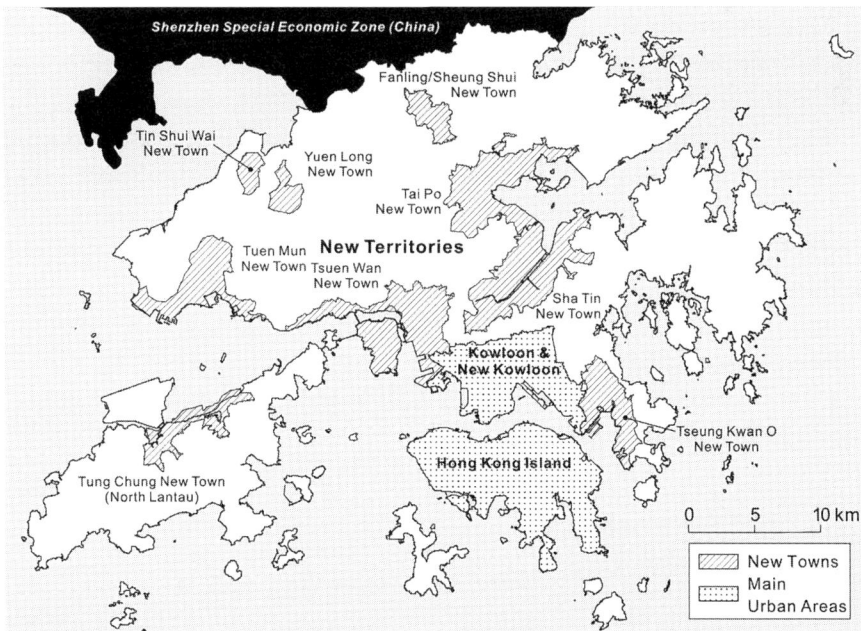

Figure 7.1. Hong Kong urban areas and new towns.

Before World War II, Hong Kong was primarily an entrepôt for Southern China. Post-war urban development and economic growth in Hong Kong were spectacular, however. Modern high-rise buildings began to appear as a result of private sector property development while the public sector embarked on a massive programme of rental housing construction. This was given particular impetus by the Christmas fire of 1953, which made more than 53,000 people homeless (Yeh and Fong, 1984). By 2006, over 3.3 million people, or 48 per cent of Hong Kong's population, were residing in subsidized public rental housing. Hong Kong is second only to Singapore in terms of the proportion of its population living in public housing. The quality of public housing has been constantly improved, especially since the introduction of the 'Ten Year Housing Program' in 1973 which was aimed particularly at raising the quality of the living environments of low-income people. Since then, there have been significant improvements in the administration, management, design and construction of public housing and the redevelopment of older estates (Yeung and Wong, 2003).

The Ten-Year Housing Program required a large amount of land which was difficult and expensive to obtain in the existing urban areas around Victoria Harbour in Kowloon, New Kowloon and on Hong Kong Island. New towns were therefore built on undeveloped sites, including some in the previously rural areas of the New Territories (Wang and Yeh, 1987; Yeh, 1987, 2003), and this changed the spatial structure of Hong Kong from a monocentric city to a polycentric city (see figure 7.1). In 1971, 81.1 per cent of the total population of Hong Kong lived in the main established urban areas. By 1981, after the new town developments of the 1970s, this had decreased to 76.8 per cent and by 2008 it was only 47 per cent. The proportion of the population living in the New Territories increased correspondingly from a mere 23.3 per cent in 1976 to 47.3 per cent in 2008.

There was considerable uncertainty in the lead up to the return of Hong Kong to China in 1997 about what this would mean for Hong Kong's future urban development and its economic relationship with the PRD (Yeh, 1985; Cheng and Lo, 1995). This uncertainty gradually diminished after 1997 and, with China's rise as a world economic power, the political and economic future of Hong Kong is now much clearer. There are significant challenges ahead, however, in dealing with continuing economic restructuring and the increasingly severe competition that Hong Kong is experiencing from cities in China and elsewhere in Asia. These are explored in some detail in the following sections of this chapter.

Economic Restructuring and Global Financial Crisis

As noted in the previous section, economic growth in Hong Kong was mainly a post-war phenomenon. In the early 1950s, the traditional entrepôt trade declined and Hong Kong received an influx of refugees from China, who provided cheap labour, capital and entrepreneurship which transformed Hong Kong rapidly into an industrial city (Szczepanik, 1958). External demand for goods also helped in developing export-led industries (Koo, 1968; Lin *et al.*, 1980). The textile and plastic industries expanded quickly at an early stage while the electrical and electronics industries started to grow in the 1960s. From the early 1970s, manufacturing industries progressively diversified and there was a shift from labour-intensive and simple goods to more technology-intensive and sophisticated products. The label 'Made in Hong Kong' became well known internationally and Hong Kong was referred to as the 'world's factory'. Manufacturing remained important to Hong Kong's economy throughout the 1970s but, by the end of the decade, it had also developed as a major regional financial centre (Chen, 1984, 1990). Significant economic restructuring occurred in the 1980s, following the adoption of economic reform by China in 1978 which, as mentioned earlier, led to the migration of Hong Kong industries to China, especially to the nearby PRD, and resulted in a marked decline in Hong Kong's manufacturing sector.

These stages in Hong Kong's economic growth and development were reflected in changes in the employment structure between 1961 and 2008 (figure 7.2). Employment in the manufacturing sector increased steadily from 43 per cent of total employment in 1961 to 47 per cent in 1971, but then decreased to 41.2 per cent in 1981, to 28.2 per cent in 1991 and to only 4.7 per cent in 2008. There was a steady growth of employment in the service sector (wholesale and retail trade, restaurants and

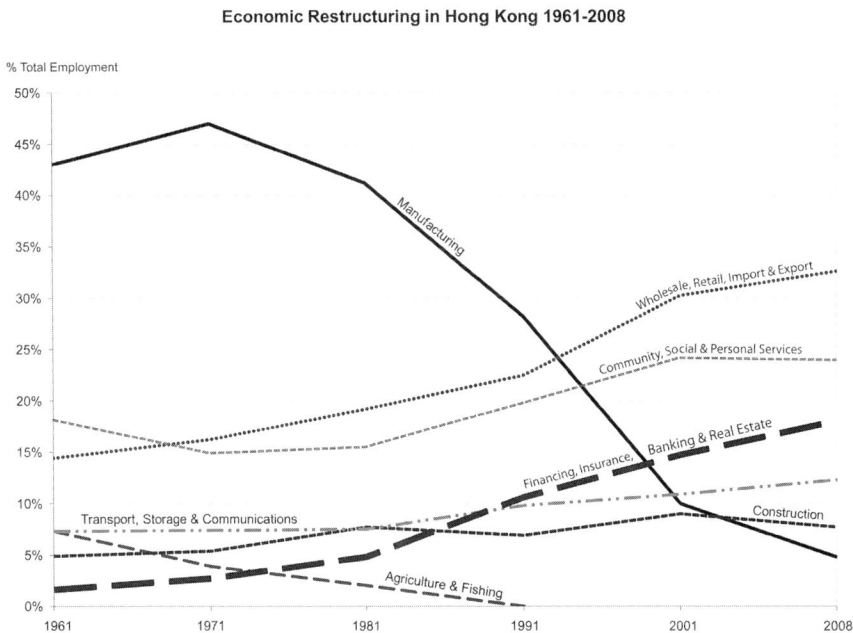

Figure 7.2. Economic restructuring in Hong Kong, 1961–2008.

hotels), from 14.4 per cent in 1961 to 32.8 per cent in 2008, and a slight increase in the professional services sector (community, social and personal services), from 18.3 per cent to 24.1 per cent over the same period. The marked increase of employment in the office sector (financing, insurance, real estate and business services) occurred mainly after 1981, rising from 4.8 per cent in 1981 to 18 per cent in 2008. A substantial decline in employment in the manufacturing sector has occurred not only in relation to its proportion of total employment but also in absolute terms. Employment in this sector fell from 0.9 million at its peak in 1981 to 0.6 million in 1991 and to 0.2 million in 2007. Manufacturing's contribution to GDP also dropped from 26.9 per cent in 1981 to 15.9 per cent in1991 and to only 2.4 per cent in 2007.

The decline in the manufacturing sector and the restructuring of Hong Kong's economy reflected the influence of both internal and external factors. Internally, serious labour shortages in the 1980s boosted wages significantly. Externally, protectionism in foreign markets forced local industrialists to upgrade the quality of their products and to diversify into new areas, as other ASEAN (Association of Southeast Asian Nations) countries became more competitive in the production of labour-intensive goods. Under the 'New International Division of Labour' (NIDL), industrialists, especially those engaged in large-scale manufacturing, sought to maximize profits by moving production processes to countries or regions with lower costs. This was the rationale for the transfer of manufacturing from Hong Kong to China in the late 1970s and 1980s, following China's establishment of new development areas to attract foreign investors, including the Shenzhen Special Economic Zone, adjacent to Hong Kong (Phillips and Yeh, 1989; Vogel, 1989). Hong Kong soon became the major source of foreign investment in China. Many Hong Kong industrialists engaged in the manufacture of labour-intensive products, such as garments, toys and electrical goods, began to set up branch plants or to undertake sub-contracting processes in the PRD where cheaper, unskilled labour was readily available. These sub-contracting arrangements often made use of personal contacts, because many people in Hong Kong came originally from the PRD and Guangdong Province (Leung, 1993). The proximity of this area also made it easier for management in Hong Kong to maintain operational control.

Although economic reform and economic development in China, and especially in the PRD, contributed to the decline in Hong Kong's manufacturing industries, they also stimulated the growth of the service and office sectors. Hong Kong industries which had moved production to the PRD still used Hong Kong as a base for research, marketing and distribution. Hong Kong also provided a 'hub' function for China, resuming its former role as an entrepôt for the rapidly growing areas of the PRD. Re-exports to and from China have increased sharply since 1990, many making use of the efficient trans-shipment services of Hong Kong's Kwai Chung container port. Hong Kong also has a very efficient banking system and provides other financial services for international firms, which find it convenient to use Hong Kong as a stepping stone for doing business with China. Despite the decline in the manufacturing sector, Hong Kong's economy has continued to grow, albeit at a slower rate, because of this counter-growth in the service and office sectors of the economy related to trade with China. Many Chinese provincial and municipal entities have also set up their own companies in Hong Kong, investing in all major economic sectors (Ho, 1992).

Since, the early 1980s, therefore, China has been increasingly important to the economic development of Hong Kong, and a close partnership has developed between Hong Kong and Guangdong province (Kwok and So, 1995). China is now, of course, a key player in the global economy and the major supplier of manufactured goods to the world. In addition, the rapid increase of China's per capita GDP, from US$226 in 1980 to US$1,148 in 2000 and US$2,494 in 2009, has made China's 1.3 billion people a huge global market for consumer goods, food, construction materials and natural resources. With growing wealth, Chinese tourists are also increasingly important to the tourist industries of many countries. In 2009, they ranked fourth (behind Germany, the United States and the United Kingdom) in international tourism expenditure, up from seventh position in 2005. This trend will continue according to the World Tourism Organization (WTO, 2010).

To capitalize on the huge market in China, Hong Kong signed the 'Closer Economic Partnership Arrangement' (CEPA) on 29 June 2003 (Trade and Industry Department, nd). This was a free trade agreement that opened up the China market to Hong Kong goods and services, enhancing the already close economic co-operation between the signatories. The agreement provided Hong Kong businesses with greater access to the Mainland China market while facilitating the use of Hong Kong as a springboard for Chinese enterprises seeking to reach out to the global market. Under CEPA, all goods of Hong Kong origin imported into Mainland China enjoy tariff free treatment and Hong Kong service suppliers receive preferential treatment. Professional bodies in Hong Kong and regulatory authorities in China have also signed a number of agreements or entered into arrangements on mutual recognition of professional qualifications.

There are now no restrictions on Hong Kong people who wish to travel to China. In 1979, soon after the adoption of the Open Door policy, in order to facilitate the movement of people and investment from Hong Kong to China, Hong Kong's Chinese citizens were granted Home Return Permits,[1] providing unlimited entry to China. However, citizens of China had to go through a lengthy process to obtain an entry permit to visit Hong Kong. These arrangements were retained after the return of Hong Kong to China in 1997 (Yeh, 1999) but, with the increasing wealth of China and to foster closer integration with Hong Kong, the application process for Chinese residents who wish to visit Hong Kong has been simplified subsequently. The Individual Visit Scheme (IVS), which allowed permanent residents of four cities in Guangdong province (Dongguan, Zhongshan, Jiangmen and Foshan) to visit Hong

Kong, was introduced soon after the signing of the CEPA in 2003 and this scheme now extends to forty-nine Mainland China cities. The IVS is the most visible indicator of the success of CEPA and has led to an increase in the number of Chinese visitors to Hong Kong from 4.4 million in 2001 to 16.9 million in 2008, surpassing the numbers of visitors from Japan, North America and Europe which used to dominate in the 1980s and 1990s (see table 7.1). The percentage of Chinese tourists was 57 per cent of all the visitors to Hong Kong in 2008. The spending of Chinese visitors is comparable to that of those from Western developed countries. In 2008, the average per capita spending of overnight mainland Chinese visitors in Hong Kong was US$728, only slightly lower than the US$768 spent by visitors from the US, Europe and Australia. It is no wonder that now, in the shopping and tourist districts of Hong Kong, most of the employees in the shops and restaurants need to speak Putonghua.[2]

Table 7.1. Changing Mix of Visitors to Hong Kong, 1981–2008.

Country	1981	%	1991	%	2001	%	2006	%	2008	%
Japan	507,960	20.0%	1,259,837	18.5%	1,336,538	9.7%	1,311,111	5.2%	1,324,797	4.5%
North America	436,588	17.2%	781,469	11.5%	1,219,103	8.9%	1,494,722	5.9%	1,525,410	5.2%
Western Europe	313,900	12.4%	578,596	8.5%	736,569	5.4%	1,039,526	4.1%	1,128,581	3.8%
Australia	201,793	8.0%	235,064	3.5%	324,156	2.4%	563,933	2.2%	643,538	2.2%
Taiwan	135,621	5.3%	1,298,039	19.1%	2,418,827	17.6%	2,177,232	8.6%	2,240,481	7.6%
China	101,790	4.0%	875,062	12.9%	4,448,583	32.4%	13,591,342	53.8%	16,862,003	57.1%
East and Southeast Asia (excluding Japan, Taiwan & China)	602,517	23.8%	1,258,294	18.5%	2,010,538	14.6%	3,084,386	12.2%	3,489,853	11.8%
Total	2,535,203	90.7%	6,795,413	92.5%	13,725,332	91.0%	25,251,124	92.1%	29,506,616	92.2%

Source: Hong Kong Annual Digest of Statistics, 1987, 2001, and 2009.

Before 1997, many people in Hong Kong were concerned that the return of the colony to China would lead to instability and would affect Hong Kong's prosperity. However, apart from the replacement of the British flag with the Chinese flag, nothing much has changed under 'One Country, Two Systems'. Hong Kong has gone through some economic difficulties in recent times, but it is not the handover of power to China that has been responsible for these. The reasons lie elsewhere. Because of the rise of globalization, Hong Kong's economy and development were inevitably affected by the Asian Financial Crisis of 1997 and the Global Financial Crisis (GFC) of 2008. Hong Kong's unemployment rate rose from 2.2 per cent in 1997 to 4.7 per cent in 1998. It reached an all time high of 7.9 per cent in 2003[3] (figure 7.3), falling to 3.6 per cent in 2008, but then rising again to 5.4 per cent in 2009 following the GFC.

To resurrect the economy in the early years of the twenty-first century, the 2002–2003 Budget Speech proposed further restructuring of Hong Kong's economy and identified financial services, trading and logistics, tourism, and producer and professional services as four important industrial sectors that could provide impetus to growth and generate employment. More recently, in response to the Global Financial

GDP and Unemployment

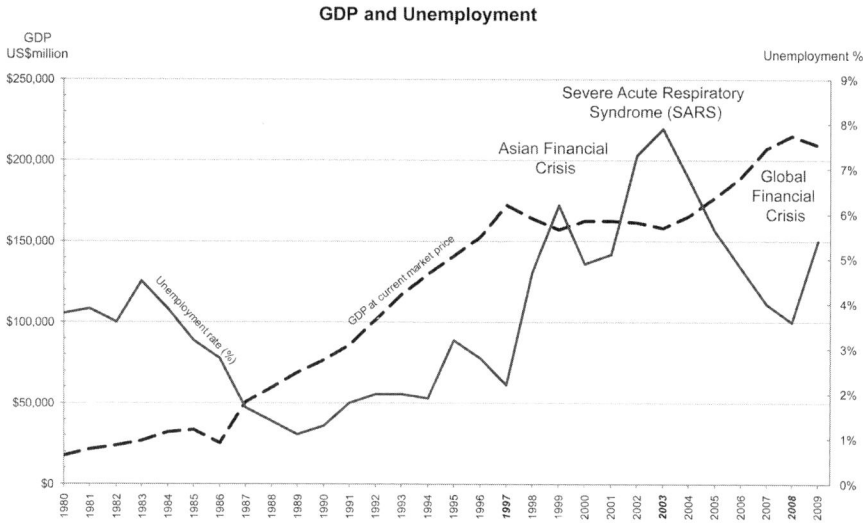

Figure 7.3. GDP and unemployment in Hong Kong, 1980–2009.

Crisis, the government established a Task Force on Economic Challenges (TFEC) in October 2008 (Hong Kong Government, 2008) which identified six economic sectors to underpin the medium- and long-term growth of the Hong Kong economy. These six sectors were educational services, medical services, testing and certification, environmental industries, innovation and technology, and cultural and creative industries. However, what happens in Hong Kong remains inextricably bound up with developments in its broader hinterland of the PRD. The changing nature of the relationship between Hong Kong and the PRD is examined in more detail in the next section.

Increasing Competition between Hong Kong and the Pearl River Delta

In the early days of economic reform in China in the 1980s, Hong Kong was more advanced than China and exported capital, technological expertise and managerial skills to the PRD, passing on also the mantle of 'world's factory'. After 30 years of development, the PRD is now much more advanced than it was in the early days of China's economic reforms and it now increasingly competes with Hong Kong (Yeh and Xu, 2006). The economy of the PRD was diversified in the 1990s to include tertiary activities and property development. Some sectors formerly monopolized by the state were opened to participation by the private sector and overseas investors. The Shenzhen Special Economic Zone (SEZ), the Zhuhai SEZ and Guangzhou and Foshan were set up by the central government as test-beds for these new policies.

Land values were re-introduced in China after 1992, creating a property market and increasing the rate of housing construction (Yeh and Wu, 1996). Local governments in the PRD quickly became active in land markets, generating funds which were subsequently invested in city-building and infrastructure, further reinforcing the attractiveness of PRD cities to foreign capital.

By the mid-1990s, Hong Kong firms had invested in all sectors opened up to competition by Chinese authorities (Cheng and Zheng, 2001). Hong Kong and the PRD had become two highly integrated and mutually dependent entities (Sit, 2001; Sit and Yang, 1997). Inherent in the evolution of the regional division of labour was a profound qualitative change for Hong Kong, as its economic dependence shifted from the West to China in general, and to the PRD in particular (Enright *et al.*, 1997; Ash, 2000).

However, the Hong Kong economy manifested increasing stress around the turn of the century. The boom of the 1990s boosted already high salary levels and led to skyrocketing land and property prices, so that Hong Kong became one of the most expensive cities in the world with high operating costs (Yeh, 2001). A high deficit (estimated at HK$80 billion in 2003) forced the government to impose higher taxes on people and businesses, despite a fairly stagnant economy and deflation. R&D investment fell and there were cuts to government programmes, to education spending and to the pay of civil servants.

In the face of these trends, a number of firms in the service industries relocated their operations elsewhere. For example, the data processing work of Cathay Pacific, Hong Kong's airline, moved to Perth in Australia, while the Hong Kong and Shanghai Bank relocated its data functions to Guangzhou. Logistics service providers (e.g. trucking and warehousing companies), ranging from large multinationals to the smallest of firms, also relocated, mostly into Guangzhou and Shenzhen, attracted by lower costs and by the growth of cargos in the delta. In these and other ways the diminished importance of Hong Kong was paralleled by the rise of its delta hinterland.

The PRD itself is now facing pressure to further upgrade its own economy in the face of a rising tide of domestic and global competition and, in particular, the intensified export activities of other provinces and countries. ASEAN countries, such as Thailand, Malaysia, the Philippines and Indonesia, which are at roughly the same level as Guangdong in terms of economic development, have similar export-based economies, focused particularly on labour-intensive industries, and compete ferociously for international markets (Cheng and Zheng, 2001). Shanghai, Jiangsu, Zhejiang and other northern Chinese provinces have also become strong competitors with Guangdong.

Against this background, local governments in the delta have been continuously upgrading their manufacturing activities to become more 'high-tech' focused so as to sustain their competitiveness (Xu and Ng, 1998; Xu, 2001). As economic restructuring accelerated after the late 1990s, new types of regional production systems emerged in the PRD. Important amongst these were the substitution of competitive industrial clusters for the former scattered factories. Also emphasized was the development of a more skilful, knowledgeable and professional workforce, to replace the mainly unskilled labour force of earlier years. New industrial districts were established, including a Science Park in Shenzhen, the Songshan Science and Industrial Park, around Songshan Lake in Dongguan, and 'Bio Island' in Guangzhou. With a strong supporting policy regime, these new industrial districts were intended to foster local integration of firms, technological leadership, an appropriate institutional framework and the development of local human capital.

As the PRD has developed its own services, its dependence on Hong Kong has diminished. The new world-class Baiyun International Airport and the Nansha Port in Guangzhou will help to boost the city's role as a freight and trans-shipment hub in the region, providing strong competition to Hong Kong's logistics facilities (Yeh, 2005; Xu and Yeh, 2003, 2005). There is also now a well-developed highway and railway network which provides Guangzhou with high accessibility to nearby cities and distant provinces. Increasingly, Guangzhou is becoming a gateway for international companies seeking to penetrate the domestic market in Southern China. With the improvement in the quality of local business services, firms in the delta no longer need to use more expensive services in Hong Kong, especially if their market is mainly in the PRD or in other parts of China.

Several influential comparative assessments recorded a fall in Hong Kong's international competiveness in the early years of the twenty-first century,[4] citing cost factors (e.g. cost of living index and electricity costs); a lack of research and development (e.g. total R&D personnel in business *per capita*); environmental degradation (e.g. pollution problems); low technological input factors (e.g. firm-level innovation, technological sophistication, and patent and copyright protection); and a lack of adequate education and manpower (e.g. secondary school enrolments, total public expenditure on education and universities) as growing comparative disadvantages for Hong Kong. Under the 'One Country, Two Systems' model, Hong Kong does still possess some advantages over its competitors in China, however. These include an open and fair market system; efficient government with a lack of corruption; a good legal system; an efficient and well-connected sea and air cargo-handling infrastructure; a high quality communications hub; and a highly regarded international financial centre. With a population only 30 per cent of that of the PRD, Hong Kong has 50 per cent of the employment in producer services of the PRD but, as table 7.2 shows, this gap is closing quickly and the quality of services in the PRD is also improving while labour costs remain much lower. Although the 'front shop, back factories' regional division of labour with the PRD saved Hong Kong's economy from declining in the 1980s and 1990s (Sit and Yang, 1997), there is some doubt as to whether the 'shop'

Table 7.2 Comparison of producer services employment in the Pearl River Delta and Hong Kong, 2002 and 2007 (in 1,000 persons).

	2002			2007		
Cities	Producer Services	Service Industries	% of Producer Services in Service Industries	Producer Services	Services Industries	% of Producer Services in Service Industries
Guangzhou	579.7	2,210.5	26.2	911.4	3,178.1	28.7
Shenzhen	241.1	1,549.6	15.6	1,041.0	3,002.2	34.7
Zhuhai	74.3	430.6	17.2	117.9	492.9	23.9
Foshan	135.9	722.6	18.8	215.2	1,280.9	16.8
Huizhou	48.1	364.6	13.2	101.3	607.9	16.7
Dongguan	75.6	381.1	19.8	127.0	554.9	22.9
Zhongshan	52.4	443.5	11.8	87.8	472.8	18.6
Jiangmen	75.5	661.2	11.4	73.9	583.8	12.7
Zhaoqing	72.8	637.5	11.4	88.7	751.3	11.8
PRD Total	*1,355.4*	*7,401.2*	*18.3*	*2,764.1*	*1,092.48*	*24.1*
Hong Kong	*1,317.7*	*2,621.2*	*50.3*	*1,441.2*	*2,985.1*	*48.3*
Hong Kong/PRD	0.97	0.35		0.52	0.27	
Hong Kong/ Shenzhen	5.47	1.69		1.38	0.99	
Hong Kong/ Guangzhou	2.27	1.19		1.58	0.94	

Sources: Statistical Bureaus of Respective PRD Cities, 2003, 2008. Hong Kong Census and Statistics Department, 2003, 2008.

function, on which the economy is now heavily dependent, can be sustained in the face of very high labour and operating costs in Hong Kong and the rapid socio-economic changes occurring in the PRD and elsewhere in China (Yeung and Sung, 1996; Yeh, 1996).

Strategic Spatial Planning in a Changing Economic Environment

In the early years of Hong Kong's industrialization, the government's economic policy was basically one of *laissez faire*. Its role in economic development was mainly through the provision of financial and business services, labour force training and an investment promotion programme to attract foreign investment for technology transfer (Yeung, 1991; Yeh, 1999). Spatial planning was mainly limited to ensuring an adequate supply of serviced industrial land and the construction of transport and communications networks (Industry Department, 1991). However, strategic spatial planning, intended to sustain key economic activities in Hong Kong, as well as meeting the requirements for land, housing, services and facilities for a growing population, commenced in the early 1980s (Hong Kong Government, 1985). This was prompted, in part, by problems arising from the lack of co-ordination between urban development and transport provision in the new towns established in the 1970s. The first strategic plan of significance was the Territorial Development Strategy (TDS), published in 1984 after 4 years of work. This was prepared at a time when there was considerable uncertainty about the outcome of negotiations between the British and Chinese governments over the return of Hong Kong to China (Yeh, 1985). At an early stage of the TDS study, the relocation of Hong Kong's international airport was recognized as a very significant factor affecting the future spatial development pattern. The existing airport had been at Kai Tak since 1925 and a new airport was clearly needed, but the uncertain political future made it difficult to make progress with relocation proposals. Two alternative long-term growth patterns were put forward, therefore, using a 'common component' approach, which identified projects common to both options, thus allowing detailed planning and associated works to proceed on these projects. These included further expansion and development of the existing urban area at West Kowloon in Yau Ma Tei; at West Hong Kong, near Green Island, for public housing; and at Central for expansion of business offices and cultural activities (Pryor, 1985). Other common components were the construction of a second cross harbour tunnel and passenger rail systems to link Sha Tin, the North East New Territories and West Kowloon with Tsuen Wan, and West Hong Kong with Green Island.

In the later part of the 1980s, further studies were undertaken of future port and airport requirements. They sought to meet the forecast port and air traffic growth of Hong Kong up to 2011 and to ensure that all new port, airport, associated industrial and residential facilities, transport links and other infrastructure, would be incrementally provided according to an integrated and cohesive plan (Hong Kong Government, 1989). By the time the preferred development strategy was announced in October 1989 (*Ibid.*) agreement had been reached to relocate Kai Tak airport to Chek Lap Kok, on the north coast of Lantau Island, with future port expansion planned to be around the existing Kwai Chung container port in the first instance, followed by later developments on the north-eastern tip of Lantau.

After the adoption of the Territorial Development Strategy (TDS) in1984, there had been a sharp increase in economic interaction with China, for reasons described earlier, and this was accompanied by a major increase in cross-border traffic, including truck traffic between Hong Kong and Shenzhen. The government therefore decided that it needed to have a major review of the TDS in 1990. A consultation paper, 'Territorial Development Strategy Review (TDSR) – Development Options', was published in 1993 for public comment (Planning Department, 1993*a*, 1993*b*). This review was also required to address the broader spatial implications of the key decisions taken in 1989 on the location of the new airport and ports. A long-term planning framework for the provision of land and infrastructure that would also take into consideration the increase in cross border traffic was a priority (Planning Department, 1993*c*). The review began under British rule and was eventually approved by the Executive Council of the Hong

Kong SAR government on 24 February 1998. It sought to provide a blueprint for Hong Kong's urban development into the twenty-first century (Yeh, 1999) and to lay the foundation for a longer-term vision of Hong Kong's changing role in a wider regional context, especially with regard to development trends in the PRD and other parts of South China (Planning, Environment and Lands Bureau, 1998). Under the proposals of the TDSR, the Metro Area (focused on the old core of Hong Kong Island and Kowloon) was to remain the key employment centre, but new nodes of employment were also to be developed in the New Territories at major transport interchanges, at sites close to the new Chek Lap Kok airport, and in a north-south 'technology corridor' along the Kowloon-Canton Railway. Much emphasis was given to further rail projects, especially for developing the strategic growth areas in the north-west New Territories. The TDSR proposed a number of new highway links and extensions that would help create a denser web of north-south and east-west high-capacity expressways within Hong Kong. In addition, investigations were proposed into new cross-boundary highways to provide additional connections to the Shenzhen and Zhuhai Special Economic Zones.

In 1997 the Planning Department initiated a 'Study on Sustainable Development for the 21st Century' (SUSDEV21) (Planning Department, 2000) to give more emphasis to environmental and social, as well as economic, concerns in planning for a more sustainable future for Hong Kong. This study popularized the concept of sustainable development and also introduced new methods to allow people to participate in the decision-making process for government projects. The outputs of SUSDEV 21 included a definition of sustainable development appropriate to the Hong Kong context and a set of guiding principles for moving towards sustainability. A decision-making tool, known as 'CASET' (Computer Aided Sustainable Evaluation Tool), was also developed to assess the implications of strategic policies and projects against a set of sustainability and social indicators. The study further suggested institutional changes, including the setting up of a Sustainable Development Unit reporting directly to the Chief Executive, and the establishment of a Council for Sustainable Development. Members of this Council were to be appointed by the Chief Executive and drawn from the environmental, social and business sectors. Senior government officials were also appointed to the Council, including the Secretary for the Environment, the Secretary for Transport and Housing, the Secretary for Development, and the Secretary for Home Affairs (Council for Sustainable Development, 2003).

A consequence for planners of the economic restructuring described earlier was the need to deal with an over-supply of industrial land and buildings because of the decline of manufacturing industry. New mixed-use industrial and office building provisions had been introduced into the Hong Kong planning system as far back as in 1989 (Yeh, 1997), but the rezoning of surplus and suitable industrial land for mixed uses, including non-polluting industrial uses, was further facilitated by reforms adopted by the Town Planning Board more recently in 2001 (LegCo, 2009). Improvements in the quality of Hong Kong's urban environments through the redevelopment, amongst other things, of older industrial sites are central to its most recent strategic spatial plans. These also have a strong focus on sustainability, and place increased importance on the need to develop a planning framework which can integrate the future growth of Hong Kong and the PRD. The latest of these plans are considered in the final part of this section, below.

Hong Kong 2030 and Integrated Planning for the Guangdong, Hong Kong and Macao World City Region

'Hong Kong 2030' (HK 2030) was released in October 2007. The overarching goal of this long-term conceptual plan was 'to adhere to the principles of sustainable development to balance social, economic and environmental needs to achieve better quality of life for present and future generations' (Planning Department, 2008). Its more detailed planning objectives were as follows:

(i)   Providing a good living environment to its residents, paying special attention to its environmental carrying capacity, enhancing its townships, and regenerating old urban areas;

(ii)   Conserving natural landscapes, preserving important ecological, geological, scientifically significant sites, and preserving cultural heritage;

(iii)   Enhancing Hong Kong's hub functions by setting aside land reserves to meet the changing needs of commerce and industry, strengthening its role as a global and regional financial and business centre, international trading, transport and logistics centre, and further development as an innovative and technology centre for the region;

(iv)   Meeting housing and community needs via adequate land and infrastructure, development of housing and community facilities;

(v)   Providing a framework to develop a safe, efficient, economically viable and environmentally friendly transport system;

(vi)   Promoting arts, culture and tourism to ensure HK continues to be a world-class destination;

(vii)   Strengthening links with the Mainland to cope with the rapid growth of cross-boundary interaction.

In HK 2030 development areas and axes were proposed in order to better distribute Hong Kong's growing population and to provide its residents with an improved living environment. The four major development areas and axes were:

1. *Metro Development Core*: intensive commercial/ business zones and housing for urban living;

2. *Central Development Axis*: community-type housing and education/ knowledge-building facilities;

3. *Southern Development Axis*: logistics and major tourism facilities; and

4. *Northern Development Axis*: non-intensive technology and business zones and other uses that capitalize on the strategic advantage of the boundary location.

New development areas were identified, located mostly near the boundary with China, away from the busy existing urban areas, so as to allow for a more balanced territorial development pattern and a less congested environment. Proposed new areas included Kwu Tung North, Fanling North and Ping Che/Ta Kwu Ling. A rail-based development strategy was to remain central to Hong Kong's development as a low-carbon city. *Hong Kong 2030* (figure 7.4) proposed a 'two-hour living' circle, allowing residents of Hong Kong and the PRD to move easily between the cities of the region through a planned rapid railway and highway system. Cross-border transport infrastructure, intended to facilitate business and industrial development between Hong Kong and the PRD, included the Hong Kong-Zhuhai-Macao Bridge (see figure 7.5).

Further proposals to strengthen the integration with the PRD were contained in the *Planning Study on the Co-ordinated Development of the Greater Pearl River Delta Townships*, initiated in 2006 between the Hong Kong and Macao Affairs Office of the State

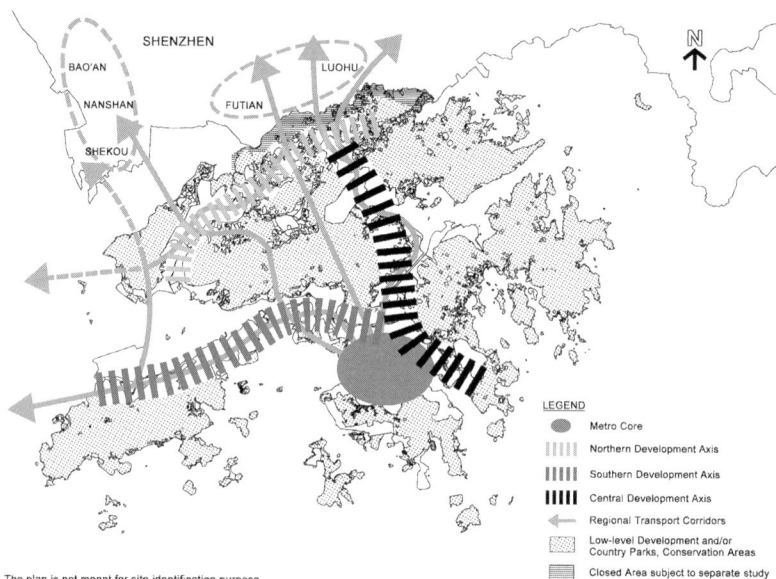

Remark : The plan is not meant for site identification purpose

Figure 7.4. Hong Kong 2030: preferred development pattern.

Council and the governments of Guangdong Province, Hong Kong SAR and Macao SAR (Planning Department, 2009). This was the first strategic planning study carried out collaboratively by the separate jurisdictions in the Greater Pearl River Delta and it gave explicit recognition to the fact that Guangdong, Hong Kong and Macao together now represent an important world city-region. This study proposed collaborative efforts to build global competitiveness and influence across the city region, creating: a world-class advanced manufacturing base with innovative abilities; a world-class centre of modern services; a world-class domestic and international transportation hub; a cultural centre of global influence; and a quality living area that is affluent, civilized, harmonious and liveable. To give effect to these objectives, work has commenced on a further series of plans and studies, including spatial co-ordination master plans; ecological and environmental protection plans; a cross-boundary co-operative development plan; and plans for the collaborative development of transportation (see figure 7.5).

The Hong Kong-Zhuhai-Macao Bridge, referred to above, is planned for completion in 2015–2016 and will have the potential to reduce the travelling time between the three cities to less than one hour. Guangdong Province and the Chinese

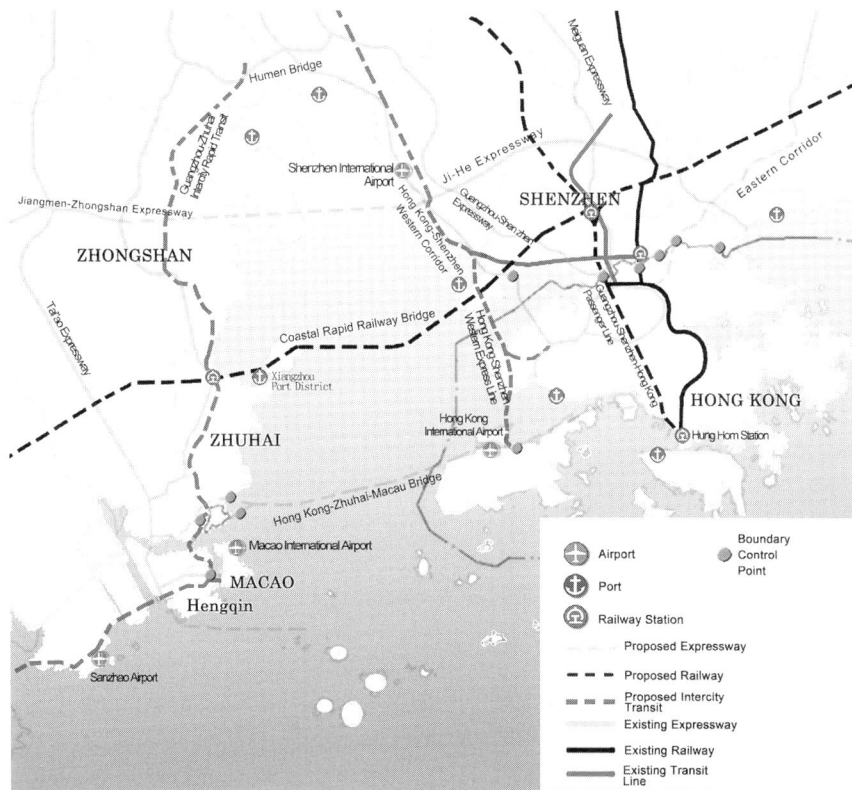

Figure 7.5. Transportion network for the Greater Pearl River Delta.

central government are contributing US$1 billion towards this project, while Hong Kong and Macao are contributing US$0.96 billion and US$0.28 billion, respectively. Another cross-boundary transport project is the 142 kilometre Guangzhou-Shenzhen-Hong Kong Express Rail Link that will be completed in 2016. The expected travel times from the West Kowloon Terminus are 12 minutes to Shenzhen's Futian station and 48 minutes to Guangzhou South station in Shibi. Once this link is completed, Hong Kong will be connected with the Wuguang (Wuhan-Guangzhou) High-Speed Railway at Guangzhou South Station and with the Xiashen (Xiamen-Shenzhen) High-Speed Railway at Shenzhen North Station. It will thus become an important gateway for local and international passengers wishing to travel to other major cities of Mainland China by high-speed rail.

Conclusions

In the period since decolonization, China has exercised less obvious political control over Hong Kong than the British government did, but Hong Kong's economy has become much more dependent on China than it was on the UK (Yeung, 2007). As this chapter has shown, Hong Kong's former role as the 'dragon head' of the Pearl River Delta has changed as the PRD has been transformed from an agricultural region to a major global manufacturing region and service provider. Hong Kong now faces severe competition on a number of fronts with the cities of the PRD and other booming Chinese cities (Yeh, 2005, 2006).

An important element of Hong Kong's recent attempts to reposition itself in the face of growing regional and global competition has been its emphasis on the qualitative aspects of urban development. Since the period after the Asian Financial Crisis in 1997, Hong Kong has been taking active steps to revitalize its economy and attractiveness to visitors through a conscious strategy of 'place-making' (Kotler et al., 1993, 2001). In Hong Kong, there has been an emphasis on a branding strategy which seeks to promote Hong Kong, variously, as 'the city of life' and as 'a fusion of east and west' (Hong Kong Government, nd). 'Brand Hong Kong' was launched at a FORTUNE Global Forum in May 2001 with a flying dragon logo and the slogan of 'Asia's world city'.

A project of particular interest in the light of this strategy has been the proposal to develop the West Kowloon Cultural District on 40 hectares of reclaimed land on the western waterfront of Kowloon. This project was first announced in 1999 as an explicit attempt to redress Hong Kong's perceived lack of cultural interest by creating an iconic area of cultural and entertainment facilities which would attract more tourists to Hong Kong. A high-level international design competition was organized in April 2001 with a 'dragon-like' gigantic canopy design scheme, submitted by Foster and Partners, as the winner. Unfortunately, the project stalled in 2006 due to doubts about its feasibility. The West Kowloon Cultural District Authority was set up in 2008 to revive and manage this megaproject, with a greater emphasis the second time around on the economic and cultural benefits for the local population as well as for international visitors and tourists. Public consultation on three new options came to an end in November 2010. Construction is now expected to start in 2011 with completion of the first phase by 2015.

The saga of the West Kowloon Cultural District is perhaps indicative of a particular area of competitive risk facing Hong Kong – delay in the decision-making process and in the lead time for major projects. It is now more than a decade since the West Kowloon Cultural District was first announced. Similarly, Hong Kong's airport was relocated from Kai Tak to Chek Lap Kok in 1998, but the land on which the old airport was located, close to the heart of the city, remains vacant, with plans for its future use still to be confirmed. Cities in China, by contrast, seem able to bring major projects to fruition much more quickly and are making very effective use of 'mega-events' for marketing and place-making purposes – for example, the Beijing Olympics in August 2008, the Shanghai World Expo in May to October 2010 and the Asia Games, planned for Guangzhou in November 2011. These cities are now attracting more international attention than Hong Kong as a consequence of these megaprojects.

Hong Kong's producer services are well regarded by the international business community and continue to operate in one of the world's best business environments. However, their longer-term future, and the future of Hong Kong more generally, seem inevitably to lie in closer integration with the cities of the PRD, as the infrastructural and policy barriers to integration continue to be removed. A combination of Hong Kong's efficient, reliable, high-quality financial and business services with the increasingly sophisticated and high-tech manufacturing industries of the PRD has the potential to make the Guangdong-Hong Kong-Macao region a major world economic region, able to compete for capital, skilled labour and central government resources with other fast growing cities in China and the wider region.

Notes

1. The full name is 'Mainland Travel Permit for Hong Kong and Macau Residents'.
2. 'Putonghua' is the standard form of modern Chinese.
3. At the time of the SARS epidemic.
4. For example, Hong Kong's ranking in the Global Competitiveness Report, published by the World Economic Forum, dropped from second in 1998 to seventeenth by 2002. Hong Kong has recovered somewhat in recent years and is ranked eleventh in the 2010–2011 Report, available from www3.weforum.org/docs/WEF_GCR_Highlights_2010-11.pdf. Accessed 15 December 2010.

References

Ash, Robert F. (2000) Like fish finding water: economic relations between Hong Kong and China, in Ash, R., Ferdinand, P., Hook, B. and Porter, R. (eds.) *Hong Kong in Transition: the Handover Years*. Basingstoke: Macmillan.

Chen, Edward K.Y. (1984) The economic setting, in Lethbridge, David G. (ed.) *The Business Environment in Hong Kong*, 2nd ed. Hong Kong: Oxford University Press.

Chen, Edward K.Y. (1990) The Hong Kong economy in a changing international economic environment, in Kulessa, M. (ed.) *The Newly Industrializing Economies of Asia*. Berlin: Springer Verlag.

Cheng, J.Y.S. and Lo, S.S.H. (eds.) (1995) *From Colony to SAR – Hong Kong's Challenges Ahead*. Hong Kong: The Chinese University Press.

Cheng, J.Y.S. and Zheng, Peiyu (2001) High-tech industries in Hong Kong and the Pearl River Delta: development trends in industrial cooperation. *Asian Survey*, **41**(4), pp. 584–610.

Council for Sustainable Development (2003) *Terms of Reference of the Council and Establishment of Sub-committees*. Available at http://www.susdev.gov.hk/html/en/council/Paper01-03e.pdf. Accessed on 10 July 2010.

Enright, M.J., Scott, E.E. and Dodwell, D. (1997) *The Hong Kong Advantage*. Hong Kong: Oxford University Press.

Ho, Y.P. (1992) *Trade, Industrial Restructuring and Development in Hong Kong*. London: Macmillan.

Hong Kong Government (1985) *Planning for Growth*. Hong Kong: Hong Kong Government.

Hong Kong Government (1989) *Gateway to New Opportunities: Hong Kong's Port and Airport Development Strategy*. Hong Kong: Hong Kong Government.

Hong Kong Government (2008) *CE appoints members of Task Force on Economic Challenges*. Available at www.info.gov.hk/gia/general/200810/28/P2008102801135.htm. Accessed 19 October 2010.

Hong Kong Government (n.d.) *Brand Hong Kong*. Available at http://www.brandhk.gov.hk/en/#/ Accessed 10 July 2010.

Industry Department (1991) *Hong Kong's Manufacturing Industries 1990*. Hong Kong: Hong Kong Government.

Kotler, P., Haider, D.H. and Rein, I. (1993) *Marketing Places: Attracting Investment, Industry, and Tourism to Cities, States, and Nations*. New York: Free Press.

Kotler, P., Hamlin, M. A., Rein, I. and Haider, D. H. (2001) *Marketing Asian Places: Attracting Investment, Industry, and Tourism to Cities, States, and Nations*. New York: Wiley.

Koo, S.E. (1968) The role of export expansion in Hong Kong's economic growth. *Asian Survey*, **8**(6), pp. 499–515.

Kwok, R.Y.W. and So, Alvin Y. (eds.) (1995) *The Hong Kong-Guangdong Link – Partnership in Flux*. Armonk: Sharpe.

LegCo (2009) *Optimizing the Use of Industrial Buildings to Meet Hong Kong's Changing Economic and Social Needs*. Available at http://www.devb.gov.hk/industrialbuildings/filemanager/press_release_publication/eng/legco_brief_091015.pdf. Accessed 10 July 2010.

Leung, C.K. (1993) Personal contacts, subcontracting linkages and development in the Hong Kong-Zhujian Delta Region. *Annals of the Association of American Geographers*, **83**(2), pp. 272–302.

Lin, T.B., Mok, V. and Ho, Y.P. (1980) *Manufactured Exports and Employment in Hong Kong*. Hong Kong: The Chinese University Press.

Phillips, D.R. and Yeh, A.G.O. (1989) Special Economic Zones, in Goodman, David (ed.) *China's Regional Development*. London: Routledge.

Planning, Environment and Lands Bureau (1998) *Territorial Development Strategy Review – A Response to Change and Challenges: Final Executive Report*. Hong Kong: Hong Kong SAR Government.

Planning Department (1993*a*) *Territorial Development Strategy Review: Development Options*. Hong Kong: Hong Kong Government

Planning Department (1993*b*) *Territorial Development Strategy Review – Foundation Report*. Hong Kong: Hong Kong Government

Planning Department (1993*c*) *Assessment of Land Requirement and Supply 1993*. Hong Kong: Hong Kong Government.

Planning Department (2000) *Sustainable Development for the 21st Century*. Hong Kong: Hong Kong Government.

Planning Department (2008) *Hong Kong 2030*. Available at http://www.pland.gov.hk/pland_en/p_study/comp_s/hk2030/. Accessed 10 July 2010.

Planning Department (2009) *Planning Study on the Coordinated Development of the Greater Pearl River Delta Townships*. Available at http://www.pland.gov.hk/pland_en/misc/great_prd/gprd_e.htm. Accessed

10 July 2010.

Pryor, E.G. (1985) An Overview of Territorial Development Strategy Studies in Hong Kong. *Planning and Development (Journal of the Hong Kong Institute of Planners)*, **1**(1), pp. 8–20.

Sit, V.F.S. (2001) Economic Integration of Guangdong Province and Hong Kong: Implications for China's Opening and Its Accession to the WTO. *Regional Development Studies*, **7**, pp. 129–142.

Sit, V.F.S. and Yang, C. (1997) Foreign investment-induced exo-urbanization in the Pearl River Delta, China. *Urban Studies*, **34**(4), pp. 647–677.

Szczepanik, E. (1958) *The Economic Growth of Hong Kong*. London: Oxford University Press.

Trade and Industry Department (n.d.) Mainland and Hong Kong Closer Economic Partnership Arrangement (CEPA), Hong Kong SAR Government. Available at http://www.tid.gov.hk/english/cepa/index.html. Accessed 10 July 2010.

Vogel, E.F. (1989) *One Step Ahead in China: Guangdong under Reform*. Cambridge, MA: Harvard University Press.

Wang, L.H. and Yeh, A.G.O. (1987) Public Housing-Led New Town Development: Hong Kong and Singapore. *Third World Planning Review*, **9**(1), pp. 41–63.

World Tourism Organization (2010) *UNWTO World Tourism Barometer: Interim Update*, April, pp. 1–3.

Wu, F. (2000*a*) Place promotion in Shanghai. *Cities*, **17**(5), pp. 349–361.

Wu, F. (2000*b*) The global and local dimensions of place-making: remaking Shanghai as a world city. *Urban Studies*, **37**(8), pp. 1359–1377.

Xu, J. (2001) The role of land use planning in the land development process in China: the case of Guangzhou. *Third World Planning Review*, **(23)**3, pp. 229–248.

Xu, Jiang and Ng, Mee Kam (1998) Socialist urban planning in transition: the case study of Guangzhou, China. *Third World Planning Review*, **20**(1), pp. 35–51.

Xu, J. and Yeh, A.G.O. (2003) Guangzhou. *Cities*, **20**(5), pp. 361–374.

Xu, J. and Yeh, A.G.O. (2005) City repositioning and competitiveness building in regional development: new development strategies in Guangzhou, China. *International Journal of Urban and Regional Research*, **29**(2), pp. 283–308.

Yeh, A.G.O. (1985) Planning for uncertainty – Hong Kong's urban development in the 1990s. *Built Environment*, **11**(4), pp. 252–267.

Yeh, A.G.O. (1987) Spatial Impacts of new town development in Hong Kong, in Phillips, D.R. and Yeh, A.G.O. (eds.) *New Towns in East and Southeast Asia – Planning and Development*. Hong Kong: Oxford University Press.

Yeh, A.G.O. (1995) Planning and management of Hong Kong's border, in Cheng, J.Y.S. and Lo, S.S.H. (eds.) *From Colony to SAR – Hong Kong's Challenges Ahead*. Hong Kong: The Chinese University Press.

Yeh, A.G.O. (1996) Pudong – remaking Shanghai as a world city, in Yeung, Y.M. and Sung, Y.W. (eds.) *Shanghai: Transformation and Modernization under China's Open Policy*. Hong Kong: The Chinese University Press.

Yeh, A.G.O. (1997) Economic restructuring and land use planning in Hong Kong. *Land Use Policy*, **14**(1), pp. 25–39.

Yeh, A.G.O. (1999) Land and infrastructure development before and after the handover, in Chow, L.C. and Fan, Y. (eds.) *The Other Hong Kong Report*. Hong Kong: The Chinese University of Hong Kong Press.

Yeh, A.G.O. (2001) Hong Kong and the Pearl River Delta: competition or cooperation? *Built Environment*, **27**(2), pp. 129–145.

Yeh, A.G.O. (2002) Further cooperation between Hong Kong and the Pearl River Delta in creating a more competitive region, in Yeh, A.G.O., Lee, Y.S., Lee, T. and Sze, N.D. (eds.) *Building a Competitive Pearl River Delta Region: Cooperation, Co-ordination and Planning*. Hong Kong: Centre of Urban Planning and Environmental Management, University of Hong Kong.

Yeh, A.G.O. (2003) Public housing and new town development, in Yeung, Yue Man and Wong, Timothy K.Y. (eds.) *Fifty Years of Public Housing in Hong Kong: A Golden Jubilee Review and Appraisal*. Hong Kong: The Chinese University Press.

Yeh, A.G.O. (2005) Producer services and industrial linkages in the Hong Kong-Pearl River Delta Region, in Daniels, P.W., Ho, K.C. and Hutton, T.A. (eds.) *Service Industries and Asia-Pacific Cities: New Development Trajectories*. London: Routledge.

Yeh, A.G.O. (2006) Hong Kong's producer services linkages with the Pearl River Delta, in Yeh, A.G.O., Sit, V.F.S., Chen, G. and Zhou, Y. (eds.) *Developing a Competitive Pearl River Delta in South China Under One Country-Two Systems*. Hong Kong: Hong Kong University Press.

Yeh, A.G.O. and Fong, P.K.W. (1984) Public housing and urban development in Hong Kong. *Third World Planning Review*, **6**(1), pp. 79-94.

Yeh, A.G.O. and Wu, Fulong (1996) The new land development process and urban development in Chinese cities. *International Journal of Urban and Regional Research*, **20**(2), pp. 330–353.

Yeh, A.G.O. and Xu, J. (2006) Turning of the dragon head: changing role of Hong Kong in the regional development of the Pearl River Delta, in Yeh, A.G.O., Sit, V.F.S., Chen, G. and Zhou, Y. (eds.), *Developing a Competitive Pearl River Delta in South China Under One Country-Two Systems*. Hong Kong: Hong Kong University Press.

Yeung, K.Y. (1991) The role of Hong Kong Government in industrial development, in Chen, Edward, Nyaw, Mee-Kau and Wong, Teresa Y.C. (eds.) *Industrial and Trade Development in Hong Kong*. Hong Kong: Centre of Asian Studies, University of Hong Kong.

Yeung, Yue-man (ed.) (2007) *The First Decade: The Hong Kong SAR in Retrospective and Introspective Perspectives*. Hong Kong: Chinese University Press.

Yeung, Y.M. and Sung, Y.W. (eds.) (1996) *Shanghai: Transformation and Modernization under China's Open Policy*. Hong Kong: The Chinese University Press.

Yeung, Yue Man and Wong, Timothy K.Y. (eds.) (2003) *Fifty Years of Public Housing in Hong Kong: A Golden Jubilee Review and Appraisal*. Hong Kong: The Chinese University Press.

Chapter Eight

Singapore: Planning for More with Less

Belinda Yuen

Singapore's prominence in the global marketplace has been increasingly recognized in recent years. The Globalization and World Cities Research Network (GaWC) ranked Singapore as an 'alpha plus'[1] world city in 2008 (GaWC, 2008). Singapore is the world's principal logistics hub, the third largest oil-refining centre, the fifth largest foreign exchange trading centre and one of South-East Asia's major cities for transportation, aviation, financial services and investment (Hoong, 2009). It was also the 'most globalized country' on the A.T.Kearney/Foreign Policy globalization index, which compared sixty-two countries across four dimensions of economic integration, personal contact, technological connectivity and political engagement (Foreign Policy, 2006). Singapore is ranked fourth after London, New York and Tokyo on the Worldwide Centers of Commerce Index, 2008, a comparative ranking of seventy-five of the world's leading global cities on seven dimensions: legal and political framework, economic stability, ease of doing business, financial flow, business centre, knowledge creation and information flow, and liveability (Mastercard Worldwide, 2008). Since 2008, Singapore has been the world's easiest place to do business (World Bank, 2010) and in 2010 it was ranked first worldwide amongst fifty-eight countries in the World Competitiveness Yearbook (World Competitiveness Center, 2010). The list goes on.

Singapore also has a comprehensive pro-immigration policy, which seeks to raise overall immigration, integrate foreign migrants and encourage the return of Singapore citizens from abroad. The result has been a substantial increase in the number of foreigners, from 1 per cent of the population in 1970 to more than 20 per cent by the 1990s. Likewise, the number of visitors to the city has been expanding, with 9.7 million tourists arriving in 2009. Singapore has sought to brand itself as an Asian destination with a unique culture, built on its multi-ethnic population comprising Chinese (74.2 per cent), Malays (13.4 per cent), Indians (9.2 per cent), Eurasians and others (3.2 per cent).

This chapter seeks to explain Singapore's aspirations and strategy for dealing with the rise of economic globalization in the post-industrial era and with the competition which it faces from other emerging world cities in Asia. It also traces Singapore's impressive responses to the challenges of sustainability, distinctive in this city-state with a limited land area and few resources. Singapore is a city that has been successful in orchestrating economic growth, responding to globalization and transforming constraints into opportunities. As the future is influenced by the present and past, the chapter will briefly consider the influence of colonial planning ideas, and a well-established planning system, on the present form and functions of the city. It will then set out the characteristics of Singapore's world city role, examine the risks and challenges associated with this, and describe the innovative policies that the city is seeking to develop in the face of those risks and challenges.

The Rise of Singapore as a Global City

Compared to most of the world's other global cities, Singapore is small, with a land area of only 700 square kilometres and a population of 5 million. Located 1 degree north of the equator, Singapore has been described as 'a little red dot' (Koh and Chang, 2005) but one that increasingly 'punches above its weight' in terms of innovative urban development. Under a largely state-directed policy of economic and urban growth, Singapore has been transformed from a Third World to a First World city in a remarkably short time (Lee, 2000). Under this transformation, the city's landscape has changed from that of a largely low-rise, British colonial trading post[2] of congested slums and squatter settlements to a predominantly high-rise, modern post-industrial garden city-state. More than 80 per cent of the current housing stock has been built by the government's Housing Development Board (figure 8.1). Since the mid-1960s Singapore has sought to encourage home ownership through a variety of subsidies and loans. Nearly 90 per cent of households in Singapore are now owner -occupiers (Singapore Department of Statistics, 2010).

Singapore's ability to provide housing, manage traffic congestion, develop a green city and, above all, turn plans into reality, has attracted the attention of many urban scholars and global policy-makers (see, for example, Wong and Yeh, 1985; Yeung, 1987; Castells *et al.*,1991; Ooi and Yuen, 2009; UN-HABITAT, 2009). Building on Singapore's expertise in developing workable solutions to urban problems, the World Bank and Singapore set up a 'World Bank-Singapore Urban Hub' in 2009 to provide advice and technical assistance to developing countries, particularly on urban solutions for city management, financing, urban design and climate change. While some have

Figure 8.1. High-rise apartment living is common in Singapore.

admired and commended Singapore on its transformation into one of the world's most liveable cities, others have criticized the city for becoming too ordered, efficient, sanitized or 'dull and drab' (Koolhaas and Mau, 1995; *New Straits Times*, 2003). Despite these differing perspectives, Singapore continues to have big plans for the future and is attempting to reinvent itself, this time with a serious focus on sustainable development and design.

Singapore's rise as a global city is the story of a small island 'planning for more with less'. Planning in Singapore is about making the most effective use of land to create a 'distinctive, dynamic and delightful city' (Urban Redevelopment Authority, 2001). Land is its biggest constraint as all the needs of the population and the nation – housing, business, social and recreational needs, defence, port, airport, utilities, water catchment and other elements – must be accommodated within its limited land area. While reclamation from the sea has helped to increase the amount of land – about 70 square kilometres have been reclaimed since 1960 – the scope to do more of this is limited at present by cost. There are also limitations of current technology, which only allows reclamation from water of 15 metres or less in depth, as well as territorial constraints, given Singapore's proximity to Malaysia and Indonesia.[3]

This land scarcity calls for trade-offs to be made between different land uses while retaining the city's competitive advantage. Competition from other emerging Asian world cities has threatened Singapore's position in recent times. The growth of Shanghai, for example, has enticed a number of regional headquarters of multinational corporations to move away from Singapore, leading to declining membership in Singapore's American Chamber of Commerce (Lawrence, 2002). For its economic survival, Singapore has to stay ahead of the pack by offering leading edge urban infrastructure and services, responsive to market demands.

Foreign investment has long been one of the key drivers of Singapore's economy (Mirza, 1986; Rodan, 1989). During its early phase of industrialization, generous incentive schemes were offered to foreign firms to compensate for the lack of competitive advantages from locating in Singapore – for example, through the Pioneer Industries Ordinance of 1959 and the Economic Expansion Incentives Act of 1967. The establishment of the Economic Development Board in 1961 and the Jurong Town Corporation in 1968 provided further institutional support for the foreign investment development strategy. Singapore now has the highest *per capita* foreign direct investment inflow among South-East Asian countries. Moreover, following its 'Regionalization 2000' strategy to build an 'external wing' to the economy (Okposin, 1999), Singapore has itself become a major source of foreign direct investment and one of the largest foreign investors in other Asian economies, through projects such as the Indonesia-Malaysia-Singapore growth triangle, the Singapore-Suzhou Industrial Park and the Singapore-Tianjin Eco-city. Private sector entrepreneurs are expected to bear the primary risks and take on majority stakes in most overseas projects, with government-linked companies taking the lead only in major infrastructural projects of strategic significance (Ministry of Finance, 1993; Pereira, 2007). By projects of this sort Singapore is seeking to address its lack of a hinterland by globalizing beyond its national territory, developing an 'external economy' (Seetoh and Ong, 2008). This innovative economic strategy is explained as follows:

… no matter what Singapore does in terms of business promotion policy, it is an unavoidable fact of modern economic life that Singapore will face keener competition from its neighbors as a center for regional manufacturing or service industry operations. So it would seem better for Singapore to promote the outward regional expansion of its own private sector, and in the process capture for itself some of the benefits of the region's dynamic development. (Kanai, 1993, p. 41)

The pathway to an 'external economy', while allowing Singapore capital to escape the country's spatial limits and high costs, is not without risks as it often involves long gestation and payback periods. Financial market instability and political problems, arising from the complex geographies of power among countries involved in regional co-operation, are ever-present threats (Ooi, 1995; Henderson, 2001). Singapore's involvement in the Indonesia-Malaysia-Singapore growth triangle also opened it to criticism as a latter-day colonial power, imposing new forms of tenure and development that marginalized the local indigenous population of the Riau archipelago. According to Chou and Wee (2003, p. 332):

... within the framework of the Growth Triangle as an economic zone oriented towards the global market, the tribal indigenes of Riau have been completely invisible and disenfranchised. Their existence, livelihood needs, and resource rights are totally ignored. Their livelihood resources are appropriated without compensation or even acknowledgement.[4]

Nevertheless, for the reasons outlined above, Singapore has no real option but to accept the risks inherent in internationalization, modern globalization and the vulnerabilities demonstrated by the Asian financial crisis. Thus, since the early years of this century, Singapore has pursued an economic vision to become a hub of talent, enterprise and innovation, and a leading global city by 2018 – a 'key basing point' for transnational corporate headquarters, both facilitating and attracting outward and inward investment flows and providing services for the global economy. According to the Economic Review Committee (2003, p. 3), which was responsible for fundamentally rethinking Singapore's economic strategies, the vision is to make Singapore:

A *globalized* economy, a key node in the global network, linked to all major economies and emerging regions;

A *creative and entrepreneurial* city, willing to take risks to create fresh businesses and start new paths to success; and

A *diversified* economy, powered by the twin engines of manufacturing and services, where Singapore companies complement multinational corporations and new startups co-exist with traditional businesses exploiting technology, new and innovative ideas.

The government believes in being pro-active in its thinking about the future. In this regard, the Economic Review Committee has identified six key areas that are critical to realizing its vision. These are expanding external ties; improving micro-economic competitiveness and flexibility; nurturing the spirit of entrepreneurship, innovation and creativity; strengthening the twin growth engines of manufacturing and services, which will become ever-more innovation and knowledge driven; developing people and talent; and restructuring and responding flexibly to change. The intent is to make Singapore a vibrant and distinctive global city, the most open and cosmopolitan city in Asia and one of the best places to live and work.

The Importance of Place

Like many other global cities, Singapore has also found itself re-examining its priorities for local place making in order to retain its own citizens and to attract the world's top talents and hyper-mobile capital. As the Prime Minister said during the 1999 National Day Rally Speech,

Singapore should be a fun place to live. People laugh at us for promoting fun so seriously. But having fun is important. If Singapore is a dull, boring place, not only will talent not come here, but even Singaporeans will begin to feel restless. (*Straits Times*, 1999)

An important assumption is that if Singapore can become a more exciting city, then its citizens are less likely to want to emigrate and expatriate Singaporeans may be more likely to return. As a result, an increasing emphasis is directed to making Singapore a global city economy driven by creativity and lifestyle. The government's interest in developing creativity seems to be as much about developing cultural capital as it is about advancing an additional sector of economic growth, although creative industries have been observed to be among the fastest growing sectors in most advanced countries and in world trade (Howkins, 2001; Florida, 2002; and UNCTAD, 2008). In the Singapore tradition of 'whole-of-government' effort, its planning agency, the Urban Redevelopment Authority, has included heritage, identity and leisure plans in its recent land-use planning. The Ministry of Information, Communications and the Arts has also launched the 'Creative Community Singapore' initiative to provide opportunities for the community to express its creativity and to unleash the economic potential of its arts and cultural sector. The overall aim is to reposition Singapore as a 'Renaissance City' – an innovative global city for arts and culture, redressing its highly planned 'boring city' image (Ministry of Trade and Industry, 2002; Tng, 2008). The way in which recent plans have sought to achieve this aim is examined in some more detail below.

Ordered Development *Par Excellence* – Master Plans and Concept Plans

Singapore is a city that has sought to create order in its urban landscape since its early years as an independent country. A comprehensive plan for the control of urban development and growth was prepared and implemented at an early stage of its development. The first statutory master plan, approved in 1958 for a period extending to 1972, sought primarily to provide a basic standard of living and amenity for the population on the British model of urban containment and new towns (Motha and Yuen, 1999). It followed a traditional blueprint planning approach that emphasized the rational use of land through density controls and an all-encompassing zoning scheme, providing transparency and certainty for investment decisions. As has been demonstrated elsewhere (Motha and Yuen, 1999; UN-HABITAT, 2009), the master plan approach is not well suited to dealing with the challenges of a dynamic urban setting. Thus, as the plan period of the first master plan came towards an end, a new type of strategic plan, the long-range Concept Plan, was formulated with UNDP assistance and adopted in 1971 to guide the massive public investment required for Singapore's further national development. Since the adoption of the Concept Plan, the role of the master plan has been redefined as a short-term (five-year) detailed statutory local development guide.

The Concept Plan has guided the systematic and comprehensive development of industrial estates, housing estates and new towns; the airport, port, roads, expressways, mass rapid transit; and green areas, the last as part of Singapore's deliberate transformation from a city of slums to a green city. Despite the increase in population between 1970 and 1990 from 2 million to 4 million, the quality of Singapore's urban fabric improved in a number of ways. For example, between 1970 and 1990, the proportion of the population living in public housing increased from 32 per cent to 86 per cent and the numbers living in squatter settlements decreased to less than 1 per cent; commercial floor space underwent an eightfold expansion; and the area available for open space and recreation increased from 700 hectares to 4,000 hectares. A continuing increase in the proportion of land available as green and recreation spaces remains a high priority in Singapore's current and future urban landscape planning (see table 8.1).

Table 8. 1. Proposed land use distribution for Singapore in Year X*.

Parameters	1967	1988	Year X
Population	2 million	2.6 million	5.5 million
Number of households	0.34 million	0.6 million	1.8 million
Land-use distribution (hectares)			
Working space	17,900	13,400	14,000 (19 per cent)
Living space	7,600	8,000	14,000 (19 per cent)
Community uses (includes recreation space)	2,800	5,900	16,900 (23 per cent)
Infrastructure	2,500	7,400	15,500 (21 per cent)
Others	27,500	27,600	13,200 (18 per cent)
Total land supply	58,300	62,300	73,600

* Year X refers to an unknown date in future at which the population would reach 5.5 million.

Source: URA Concept Plan, various years.

In the earlier years after independence, functional considerations were uppermost in the minds of Singapore's planners. As Liu, the former Chief Planner (1991, p. 253) explained:

we wanted the City to function efficiently in order to support the needs of the population and business community; we wanted the City to have the inner vitality to generate ample jobs in order to ensure its long-term survival; and the city had to be a friendly and secure community for all residents.

As indicated above, however, the emphasis has now changed and the main planning goals embodied in the Concept Plan are to develop a city that is distinctive in its identity, dynamic for work and business, good to live in and full of energy and excitement (Urban Redevelopment Authority, 2001). There is now a better understanding of the importance of image and imagination in planning (see, for example, Lynch, 1960; Hillier, 1996; Mazzola, 2009). Participatory planning is also seen as increasingly important, with individuals engaged in improving the quality of places and their local environments, thereby enhancing the chance of a 'good fit' between built form and the social life it is intended to support and sustain (Healey, 1997; Smith, 2007). Contrary to earlier planning norms in Singapore, when the style was 'top-down', an extensive public consultation process was started early in the plan-making process for the 2001 Concept Plan (the currently adopted plan) (Soh and Yuen, 2006). As the Minister for National Development stated at the launch of the 2001 Plan, the development of Singapore 'will require the collective efforts of many Singaporeans to contribute ideas and make those ideas work' (Urban Redevelopment Authority, 2001, p. 49).

Aside from reflecting global trends in the urban planning process, this move towards a more consultative and participatory approach reflects broader shifts in the conditions for economic competitiveness. In the age of globalization, as Ng and Hills (2003) have argued, the state can no longer monopolize decision-making: new partnerships between government, business and civil society are the embodiment of enlightened governance. Like Hong Kong, Singapore has long been considered an 'administrative state' with an efficient administration, a vibrant market and an ability to deliver public goods (Cheung 2008, p. 121) but its ability to demonstrate sensitivity to public expectations will be an increasingly important source of its political legitimacy in the twenty-first century (Chua, 1997).

As well as the involvement of the community, the preparation of the Concept Plan is a collaborative effort involving many government ministries and agencies. Incorporating the land-use requirements of various agencies into the same plan provides an opportunity for overview, reflection and conflict resolution, with balance and trade-offs between development objectives addressed at a strategic level. A co-ordinated, comprehensive land-use planning approach is a hallmark of Singapore's urban development. Singapore is, of course, unique as it has a single level of government, which greatly facilitates the co-ordination process.

Thus, Singapore, having solved its acute housing shortage, traffic congestion and urban redevelopment challenges, is now giving much more attention to developing a better living and working environment and a better quality of life for its population. The general thrust of its development plans is now to compensate with quality for what Singapore lacks in size. More and more, the challenge is to provide a place where its citizens will continue to dwell despite the greater mobility afforded by globalization. The 2008 Master Plan set out a vision of Singapore as 'a Global City, a Vibrant

Playground and an Endearing Home'. The image of home conjures up feelings of affection for most people. Home is where people feel a sense of belonging and security. It implies emotional solidarity with those inside the 'home'. Thus, whether it is home building at the individual or family level or the more collective spatial activity of place making, there are processes of community formation that contribute to a sense of home. In Singapore this rhetoric is also intended to provide a sense of belonging to those in transnational communities (see Kastoryano, 1998; Gupta and Ferguson, 1997). Immigration has transformed demographic structure and influenced notions of identity. As alluded to earlier, more than 20 per cent of Singapore's population is now made up of foreigners, but this immigrant population is not homogeneous. In particular, there are two groups, commonly referred to as 'foreign talents' and 'foreign workers'. The first group comprises the transnational elite workers that the government is keen to attract. Since 2004, the criteria for citizenship have been broadened to offer suitably qualified foreigners permanent residency or citizenship. By contrast, at the opposite end of the labour market, there are policies to ensure that the second group remains a transient presence in Singapore. The image of 'home' does not extend to all.

Another emerging trend is the growth of the middle class and the differentiated lifestyles of segments of the local population who are economically better off, well-travelled and with international social networks. They have consumption and social practices converging with Western standards that are different from the rest of the local population. Economic restructuring and immigration are thus changing the social structure, creating larger high- and low-income segments and a less homogeneous middle-income group. Despite increases in average wages fostered by the economic growth, the incomes of the bottom 30 per cent of Singaporean households have fallen since 2000 (*International Herald Tribune*, 2007; Reuters, 2006). The *gini coefficient* has been rising annually over the same period to reach 47.2 (Singapore Department of Statistics, 2007), indicating that income inequality has increased. The Merrill Lynch and Capgemini World Wealth Report (2007) reported Singapore as having the world's highest growth of 'high net worth individuals' (defined as holding more than US$1 million in financial assets). The number of these individuals had grown by 21.2 per cent in the previous year, giving a total of 67,000 millionaires.

The process of production restructuring, population redistribution and splintering has been widely discussed in urban theory, especially in Western Europe (see, for example, Wachter, 1974; Fields, 2007). The spatial consequence is typically an 'urban-cosmopolitan' and 'suburban-heartland' divide (Howkins, 2001; Champion and Hugo, 2004). The cosmopolitan urban spaces are generally places where modern locals and foreigners come together. They are places of consumption and novelty, often signified by names such as Gucci, Prada and other international brands. By contrast, the heartland spaces represent the core of the nation, places of domesticity and local practices. However, this dichotomy may not fully capture the fluidity and layers of physical and metaphorical space that exist. For instance, an individual with a cosmopolitan profile can choose to inhabit the heartland space. While some of the international migrants would circulate mostly in cosmopolitan spaces, others, especially the executive level workers and students, may filter into the heartland spaces, attracted by lower costs and other considerations. Whatever their spatial preferences, the increase in the number of different social groups presents challenges to harmonious living, including issues of the assimilation of migrants into the wider national community. As Putnam (2007, p. 15) observed, 'at least in the short run immigration and ethnic diversity challenge social solidarity and inhibit social capital'.

Thus, the latest version of the Concept Plan, scheduled for completion in 2011, emphasizes in its terms of reference the identification of the key attributes for a good quality of life and proposes strategies for enhancing quality of life, even as Singapore develops and its population becomes larger and more diverse. Singapore now aspires to become a leading global Asian city in which people of different cultural backgrounds can flourish and celebrate their diversity. According to the planning authority, 'as Singapore develops, we want to retain a sense of identity in our physical landscape and encourage a sense of rootedness to our country' (Urban Redevelopment Authority, 2010).

To build and nurture place identity, a better quality of life for all and investment in the nation, the latest Concept Plan proposes, amongst other things, the provision of new dwellings in familiar places, higher-quality homes and a greater degree of housing choice. More schools, hospitals, polyclinics, libraries, retail and entertainment centres, culture and arts facilities are also to be provided. There will be a broader range of leisure options: more parks, resorts, marinas, beaches, sports facilities, entertainment complexes and theme parks, building on the Master Plan of 2008 which, in its Leisure Plan, took stock of the different recreation opportunities available in Singapore and proposed ways to enhance leisure experiences and promote the 'clean and green' theme.

Another strategic thrust is to introduce opportunities for recreation on and around Singapore's canals, rivers, reservoirs and other water bodies which, prior to the 'Active, Beautiful and Clean' water programme of 2008, served only utilitarian functions. Under this water programme, more than 800 hectares of reservoirs and 90 kilometres of inland waterways will be developed and made accessible to the public for water-based activities such as canoeing and sailing. In addition, four nature reserves have been gazetted to safeguard diverse indigenous ecosystems, such as tropical rainforests and coastal mangroves, while another eighteen nature areas have been identified and are to be kept 'for as long as possible' (see figure 8. 2).

Some of these nature areas will be integrated with parks through a park connector network. This network, to be developed by 2020, will have a total length of 360 kilometres, including a 150 kilometre route around the island. The combined effect of

Figure 8.2. Parks and nature reserves form an important aspect of Singapore's urban landscape.

all these proposals is to enhance further the image of 'a city in a garden' (Yuen, 1996).

As part of the process of selling a positive image of the city to both residents and visitors, Singapore's city centre has been reconsidered recently. The emphasis now is on repositioning and branding the city centre with sites which provide both internationally renowned business facilities and consumption opportunities. A major initiative in this regard is the development of a new downtown on 360 hectares of reclaimed land at Marina Bay, anticipated to become the 'new iconic signature image' for Singapore (*Straits Times*, 2005). As indicated in the 'Downtown@Marina Bay' plan, a contemporary repertoire of 'post-industrial mega-flagship projects' will progressively emerge in the next 15–20 years (see figure 8.3). These projects include Singapore's first integrated resort and casino associated with a 'high-end' 55-storey hotel (2500 hotel rooms and 110,000 square metres of meeting and convention facilities); theatres; exhibition centres; museums; trendy restaurants; luxury retail outlets; the 3.55 hectare Marina Bay Financial Centre, with more than 250,000 square metres of prime office space; 450 super-luxurious, high-rise, 55–66-storey apartment units; and 100 hectares of gardens by the bay.

The development proposals at Marina Bay seek to embrace international consumption-driven urban development practices, in support of the goal of recreating Singapore's urban image as a 'fun 24-hour city'. This has led to a reversal of Singapore's long-standing ban on casino development. As Singapore Minister Mentor Lee Kuan Yew explained,

Weighing against that balance, we decided yes, we may have (*social*) casualties, but against those casualties, the gain in gaining these big shows, getting these big conventions to come to Singapore, which goes with the casinos which is part of the Integrated Resorts is enormous… Is it cost-free (*in terms of social costs*)? I don't think it will be completely cost-free. Is it cost-containable? Yes. (*Straits Times*, 2006) (Words in parenthesis added by author)

In this decision there is a pragmatic weighing of the economic benefits of casinos against the adverse social consequences. The government has imposed gambling controls on the local population, including an entrance fee for all Singapore citizens and permanent residents. Individuals can also be excluded at the request of their family members. Nevertheless, casinos will be provided as part of Singapore's development of a new urban landscape of consumption spaces. The first casino only opened in February 2010 so its consequences remain to be seen.

The Downtown@Marina Bay plan also includes provisions for the area to develop as a major gathering point in the city, including as a venue for national events and celebrations (such as National Day and New Year celebrations). The aim is to create a place that inspires celebration, pride and community 'bonding', a place to 'explore, exchange and entertain' (Kuek, 2007).

Figure 8.3. Marina Bay Sands development.

Heritage and Identity

As mentioned earlier, paradoxically globalization has reinforced an interest in locality and place, on the premise that the more special, distinctive and unique a city is, the more chances it has to succeed (Kearns and Philo, 1993; Jacobs and Fincher, 1998).

As a result, heritage and identity are now important 'place products' in Singapore's efforts to construct a global city in which tradition and familiar Asian charm persist (Committee of Heritage, 1988). Since the inclusion of conservation in the Planning Act in 1989, there has been a reversal of the *tabula rasa* approach to development in the city and a greater effort to reinforce and integrate past heritage with current developments in Singapore.

Figure 8.4. Old shophouses are now conserved as part of Singapore's architectural heritage.

More than 7,000 buildings and structures have now been conserved (figure 8.4). As well as monuments and historic buildings, local vernacular architecture and places of collective memory are increasingly being identified and protected (Urban Redevelopment Authority, 2001). Façade restoration guidelines and conservation manuals assist in educating building owners and ensuring that they comply with conservation policies so as not to 'compromise … the authenticity of the historic districts' (Urban Redevelopment Authority, 1988, p. 86). Building owners are encouraged under the 'Conservation Initiated by Private Owners Scheme' (started in 1991) to volunteer their buildings of architectural and historic interest for conservation. To aid the conservation process, concessions are given to building owners, including waivers of development charges, car parking requirements and car parking deficiency charges. Heritage conservation as practised in Singapore encourages adaptation towards what is perceived as an improved environment rather than just the simple preservation of entire but lifeless old areas (Liu, 1990). There is a desire to make sure that familiar trades and places continue to be relevant to people, young and old and of varied backgrounds and ethnicity (Boey, 1998; Yuen, 2006). As Bachtiar argued (2002, p. 13)

These days, more than 200,000 Singaporeans work overseas. And many more travel frequently, laying their heads down to sleep in distant lands, gazing at novel views from their windows … its people have come to assimilate global influences. Today's populace likes having a Starbucks around the corner and glittering megamalls. They want the cinema multiplexes, and offices of glass and chrome. But these are the hallmarks of a generic upscale town. If these are the only features of our landscape, we would be possessing only the typical structures of a high-end MacCity. What makes this city uniquely Singapore, distinct and separate from so many others, are the buildings of our heritage.

The general approach in Singapore is about maintaining the past to serve the present and the future. Under the new participatory planning regime, this gives rise occasionally to tensions and contesting voices. The 2001 Concept Plan review consultation process reinforced the need to give more prominence to conservation issues on the planning agenda. These issues are certain to continue to feature prominently in the city's future urban transformation, especially as the 2011 Concept Plan review has, as a key focus, planning for sustainable growth and identity (*Straits Times*, 2010).

Singapore's Brand of Sustainable Development

Singapore's CO_2 emissions have increased by 83 per cent from 21.8 metric tonnes to 39.9 metric tonnes *per capita* between 1990 and 2007, constituting in total 0.2 per cent of global CO_2 emissions. Undoubtedly, effective management of the environment and climate change are key challenges facing Singapore. Like other parts of South-East Asia, Singapore faces particular risks from sea level rise, with most of its business spaces – airports, ports and business districts – less than 2 metres above sea level.

Planning for sustainable growth and climate resilience is critical to Singapore, as an island which imports most of its resources. UN-HABITAT has observed on many occasions that urban planning is important in managing climate change, because well-planned cities provide a better foundation for sustainable development (Tibaijuka, 2007). Fortunately, Singapore has a tradition of effective planning, and sustainable development therefore means continuing to achieve 'more with less'. Singapore's key guiding strategic principles in this regard are the adoption of long-term, integrated, pragmatic and cost effective planning approaches to make the best use of limited resources, while remaining adaptable and flexible to changes in technology and the global economy.

Where land resources are concerned, with Singapore's limits on horizontal expansion, its urban form has become more compact and high-rise over time. The tallest commercial building is presently 66 storeys high, while the tallest public and private housing developments are 50 and 70 storeys, respectively. Further building upwards is constrained at present by the requirements of the airports and air bases, so Singapore is also developing an underground land-use master plan to maximize the use of its subterranean space.

Singapore has started to pursue climate change-related research and development, especially since signing the Kyoto Accord in late 2006. It has planned its first eco-precinct in a public housing township to promote sustainable green living, and this will be built by 2011. As mentioned earlier, it is also planning and developing an eco-city in China in partnership with the Chinese government. It is promoting water technology and renewable energy by investing in research and development, including the establishment of research institutes and the provision of funding and test-bed platforms to improve performance and cost-effectiveness. Experimental projects are under way in thirty public housing precincts. Just as the United States has its 'Leadership in Energy and Environmental Design' (LEED) scheme, Singapore established a Green Mark Scheme in 2005 to move Singapore's construction industry towards more environmentally-friendly buildings, with a target for 80 per cent of existing buildings to be 'Green Mark' certified by 2030.

A significant spin-off from this R&D is that Singapore has already reduced its dependence on water imports by almost half and has diversified its water sources to include desalinated water and 'NEWater', produced through purifying used water with membrane filtration technology. In parallel with moves towards greater self-sufficiency in water is the investment in a deep tunnel sewerage system with 80 kilometres of tunnels for used water collection, treatment and disposal. Using gravity to carry sewage deep underground to centralized water reclamation plants, the deep tunnel system has also freed up about 1000 hectares of land, which, in the past, were occupied by 130 pumping stations and above ground sewage treatment facilities and their buffer zones. These water initiatives demonstrate how Singapore innovates and turns constraints to its advantage. In the face of issues of water shortage and water scarcity typical of most developing countries, Singapore has turned its water vulnerability into an asset and a growth industry. The ambition now is to become a global environment and water hub for business, investment, research and technology. This development strategy is easy to comprehend when viewed against the size of the South-East Asian region's water market, which has an estimated value of US$120 billion a year, with the desalination and water treatment sectors growing annually at 12–15 per cent (*New York Times*, 2008).

As a city committed to planning and achieving long-term economic growth, Singapore set up a high level Inter-Ministerial Committee on Sustainable Development in 2008, co-chaired by the ministers for national development, environment and water resources, and with participation by the ministers for finance, transport, trade and industry. The committee's 'Sustainable Singapore Blueprint', released in 2009, sets targets and funds initiatives for continued sustainable development for the next 20 years. These include programmes to achieve a 35 per cent reduction in energy intensity from 2005 levels by 2030; to reduce the volume of waste dramatically by behaviour change programmes and by converting waste to energy; and to reduce water consumption per person to 140 litres a day. Proposed enhancements to the physical environment include raising park provision to 0.8 hectares per 1,000 persons and increasing the amount of 'sky-rise greenery', including green roofs and gardens on the upper floors of buildings, by 50 hectares by 2030.

The government has committed S$1 billion over the next 5 years to achieve the goals of the Sustainable Singapore Blueprint. The emphasis is on a 'whole-of-nation' effort that includes community participation, emphasizing the need for Singapore's people to own the programme and actively to embrace more responsible practices, habits and lifestyles, recognizing that the collaboration of government and stakeholders is often a key critical success factor in environmental management programmes (Fischer, 2003).

Conclusion

The ability to make long-term decisions about urban development is of fundamental importance for the sustainable development of society and the environment. In Singapore, that ability is found primarily in the established system of government-led planning. The Singapore development plan system and, in particular, the Concept and Master Plans, offer a framework for long-term planning that ensures that the liveability and sustainability of Singapore will not be left to chance. Transparent plans and orderly development are now the hallmarks of Singapore planning. These provide certainty to the real estate market and to foreign investors. But centralized planning is often criticized as being overly rigid and unable to create places with character, identity and vibrancy. These criticisms are not without validity when reviewing Singapore's early development under plans that emphasized 'quantity' over 'quality', leading to largely monotonous and standardized developments – functional and efficient but somewhat 'dull and drab'.

A careful balance needs to be struck and the task is never easy. There are many possible strategies for managing rapid urban growth: privatization of urban services, national urbanization strategies, the development of new towns, large city developments and more. The strategy chosen by any one country depends in part on the nature and magnitude of its urbanization pressures, and in part on its ability to respond to its own peculiar circumstance and challenges. For Singapore, the strategy has been a pragmatic one, with the preferred urban growth direction evolving to solve emerging problems. There has been a strong commitment to planning to develop a city that is orderly and purposeful, with an initial priority on satisfying basic needs and solving core economic and housing issues, before moving on to meeting the higher aspirations of the people and enhancing their quality of life with social, cultural and recreational amenities. The key guiding principles have been the adoption of long-term integrated planning, a cost-effective approach and flexibility to changes in technology and in the global economy.

Building on the solid foundation of an integrated urban planning framework, the Concept Plan has progressively refined its planning approach to reach out and actively engage the community and government agencies in the tasks of making Singapore more appealing, attractive and exciting for its residents and visitors in an increasingly global world. Participatory planning is now a key strategy in the making of a distinctive and sustainable city and further illustrates Singapore's ability to reinvent itself as it seeks to become a global city that aims to balance economic, social and environmental considerations in a situation of resource constraints and limited land. In the process, Singapore has evolved a practice of flexible and sustainable city development that achieves 'more with less'.

Notes

1.   In this classification scheme, London and New York stand out clearly as alpha ++ cities at the highest level of integration. Alpha + cities, at the next level down, are also highly integrated cities filling advanced service needs.
2.   Singapore was a British colony from 1819, granted internal self-rule in 1959, joined the Federation of Malaysia in 1963 and became independent in 1965.
3.   See The Straits Times Interactive (2002) *Jakarta says it wants to stop sand exports to Singapore.* 9 February; The Straits Times Interactive (2003) *KL calls for immediate halt to Singapore's land reclamation.* 14 January; The Straits Times Interactive (2005) *Singapore, KL to settle land reclamation dispute.* 13 January.
4.   See also Mack (2004) and Smith (1996) on 'revanchist cities'. ('Revanchist cities' are cities which, under neo-liberalism, tended to exclude or repress minorities who were not part of the dominant economic vision).

References

Bachtiar, I. (2002) The conservation story. *URA Skyline*, 2002 Commemorative Edition. Singapore: Urban Redevelopment Authority.

Boey, Y.M. (1998) Urban conservation in Singapore, in Yuen, B. (ed) *Planning Singapore: From Plan to Implementation.* Singapore: Singapore Institute of Planners.

Castells, M., Goh, L. and Kwok, R. (1991) *Shek Kip Mei Syndrome: Economic Development and Public Housing in Hong Kong and Singapore.* London: Pion.

Champion, T. and Hugo, G. (eds.) (2004) *New Forms of Urbanization: Beyond the Urban-Rural Dichotomy.* Aldershot: Ashgate.

Cheung, A.B.L. (2008) The story of two administrative states: state capacity in Hong Kong and Singapore. *Pacific Review*, **21**(2), pp. 121–145.

Chou, C. and Wee, V. (2003) Tribality and globalization: the Orang Suku Laut and the 'Growth Triangle' in a contested environment, in Benjamin, G. and Chou, C. (eds.) *Tribal Communities in the Malay World: Historical, Cultural and Social Perspectives.* Singapore: Institute for Southeast Asian Studies.

Chua, B.H. (1997) *Political Legitimacy and Housing.* London: Routledge.

Committee of Heritage(1988) *The Committee of Heritage Report 1988.* Singapore: The Committee of Heritage.

Economic Review Committee (2003) *New Challenges, Fresh Goals – Towards a Dynamic Global City.* Singapore: Ministry of Trade and Industry.

Fields, G.S. (2007) Dual Economy. *ILR Collection Working Paper 17*, School of Industrial and Labor Relations, Cornell University, Ithaca, New York.

Fischer, T.B. (2003) Strategic environmental assessment in post-modern times. *Environmental Impact Assessment Review*, **23**(2), pp. 155–170.

Florida, R. (2002) *The Rise of the Creative Class and How it's Transforming Work, Leisure and Everyday Life.* New York: Basic Books.

Foreign Policy (2006) The Global Top 20. *Foreign Policy*, November/December, pp. 74–81.

GaWC (Globalization and World Cities Research Network) (2008) *The World According to GaWC 2008.* Available at http://www.lboro.ac.uk/gawc/gawcworlds.html. Accessed 19 October 2010.

Gupta, A. and Ferguson, J. (1997) *Culture, Power, Place.* Durham, N.C.,USA: Duke University Press.

Healey, P. (1997) *Collaborative Planning.* London: Macmillan Press.

Henderson, J. C. (2001) Regionalisation and tourism: the Indonesia-Malaysia-Singapore Growth Triangle. *Current Issues in Tourism*, **4**(2), pp. 78–93.

Hillier, B. (1996) *Space is the Machine.* Cambridge: Cambridge University Press.

Hoong, V. (2009) *Leading Singapore on growth path. GIS Development*, August, pp. 18–33. Available at http://www.gisdevelopment.net/magazine/global/2009/august/18_1.htm. Accessed October 2010.

Howkins, J (2001) *The Creative Economy: How People make Money from Ideas.* London: Penguin.

International Herald Tribune (2007) Singapore tries to redress income gap. 15 February.

Jacobs, J. M. and Fincher, R. (eds.) (1998) *Cities of Difference.* New York: Guildford Press.

Kanai, T. (1993) Singapore's new focus on regional business expansion. *Nomura Research Institute Quarterly*, **2**(3), pp.18–41.

Kastoryano, R. (1998) Transnational Participation and Citizenship: Immigrants in the European Union. Working Paper Transnational Communities Project (WPTC-98-12), Centre for Educational Research and Innovation, Paris. Available at http://www.transcomm.ox.ac.uk/working_papers.htm.

Kearns, G. and Philo, C. (eds.) (1993) *Selling Places: The City as Cultural Capital, Past and Present.* Oxford: Pergamon Press.

Koh, T. and Chang, L.L. (eds.) (2005) *The Little Red Dot: Reflections by Singapore's Diplomats.* Singapore: World Scientific.

Koolhaas, R. and Mau, B. (1995) *S, M, L, XL.* New York: Monacelli Press.

Kuek, C.L. (2007) A countdown with a difference at Marina Bay. *URA Skyline*, January/February. Singapore: Urban Redevelopment Authority.

Lawrence, A. (2002) Hi-tech's promised land. *Topics*, **31**(10), pp. 15–22.

Lee, K.Y. (2000) *From Third World to First.* Singapore: Times Academic Press.

Liu, T. K. (1990) Singapore's experience in conservation. Paper for International Symposium on Preservation and Modernisation of Historic Cities, Beijing, 15–18 August.

Liu, T.K. (1991) Improving the living environment of Singapore. *Environmental Monitoring and Assessment*, **19**(1–3), pp. 251–259.

Lynch, K. (1960) *The Image of the City.* Cambridge, MA: MIT Press.

Mack, J. S. (2004) Inhabiting the imaginary: Factory women at home on Batam Island. *Singapore Journal of Tropical Geography*, **25**(2), pp. 156–179.

Mastercard Worldwide (2008) *Worldwide Centers of Commerce Index 2008.* New York: Mastercard Worldwide.

Mazzola, E.T. (2009) The importance of local spirit and sense of place: side effects of the underestimation in the modernist town planning. *The IUP Journal of Architecture*, **1**(1), pp. 7–15.

Merrill Lynch and Capgemini (2007) *World Wealth Report.* Available at www.ml.com/media/79882.pdf. Accessed 14 June 2009.

Ministry of Finance (1993) *Interim Report of the Committee to Promote Enterprise Overseas.* Singapore: Ministry of Finance.

Ministry of Trade and Industry (2002) *Report of the Economic Review Committee (ERC) Services Sub-committee Workgroup on Creative Industries.* Singapore: Ministry of Trade and Industry.

Mirza, H. (1986) *Multinationals and the Growth of the Singapore Economy.* London: Croom Helm.

Motha, P. and Yuen, B. (1999) *Singapore Real Property Guide.* Singapore: Singapore University Press.

New Straits Times (2003) On turning 'dull and drab' Singapore into a vibrant 'wow' city. 20 March.

New York Times (2008) Asia holds promise of big profits for water industry. 1 July.

Ng, M. K. and Hills, P. (2003) World cities or great cities? A comparative study of five Asian metropolises. *Cities*, **20**(3), pp. 151–165.

Okposin, S.B. (1999) *The Extent of Singapore's Investments Abroad*. Aldershot: Ashgate.

Ooi, G.L. (1995) The Indonesia-Malaysia-Singapore Growth Triangle: sub-regional cooperation and integration. *Geojournal*, **36**(4), pp. 337–344.

Ooi, G.L. and Yuen, B. (eds.) (2009) *World Cities: Achieving Liveability and Vibrancy*. Singapore: World Scientific.

Pereira, A. (2007) Transnational state entrepreneurship? Assessing Singapore's Suzhou Industrial Park project (1994–2004). *Asia Pacific Viewpoint*, **48**(3), pp. 287–298.

Putnam, R.D. (2007) E pluribus unum: diversity and community in the twenty-first century. *Scandinavian Political Studies*, **30**(2), pp. 137–174.

Reuters (2006) Income gap tears at Singapore social fabric.18 December.

Rodan, G. (1989) *The Political Economy of Singapore's Industralization: National State and International Capital*. London: Macmillan.

Seetoh, K.C. and Ong, A.H.F. (2008) Achieving sustainable industrial development through a system of strategic planning and implementation: the Singapore model, in Wong, T.C., Yuen, B. and Goldblum, C. (eds.) *Spatial Planning for a Sustainable Singapore*. Berlin: Springer.

Singapore Department of Statistics (2007) *Yearbook of Statistics*. Singapore: Department of Statistics.

Singapore Department of Statistics (2010) *Population Trends 2010*. Singapore: Ministry of Trade and Industry.

Smith, N. (1996) *The New Urban Frontier: Gentrification and the Revanchist City*. London: Routledge.

Smith, R. (2007) *The Vision: Inclusive Planning*. London: RTPI. Available at http://www.rtpi.org.uk/item/300/23/5/3. Accessed 24 March 2010.

Soh, E.Y. and Yuen, B. (2006) Government-aided participation in planning Singapore. *Cities*, **23**(1), pp. 30–43.

Straits Times (1999) *Prime Minister's National Day Rally Speech*. 22 August.

Straits Times (2005) Singapore set to be a sparkling jewel in ten years time. 20 August

Straits Times (2006) Government takes decisions that benefit the majority. 4 November

Straits Times (2010) Concept Plan 2011 feedback. 23 January.

Tibaijuka, A. (2007) Climate Change Mitigation through Urban Planning and Development. Paper presented at UN-HABITAT mitigation workshop in Bonn, Germany (26th sessions of the Subsidiary Body for Scientific and Technological Advice (SBSTA) and the Subsidiary Body for Implementation (SBI) of the United Nations Framework Convention on Climate Change).

Tng, S. (2008) Get set for more fun! *URA Skyline* March. Singapore: Urban Redevelopment Authority.

United Nations Conference on Trade and Development (UNCTAD) (2008) *Creative Economy Report*. Geneva: UNCTAD.

UN-HABITAT (2009) *State of the World's Cities 2008/2009*. Nairobi: UN-HABITAT.

Urban Redevelopment Authority (1988) *A Manual for Chinatown Conservation Area*. Singapore: Urban Redevelopment Authority.

Urban Redevelopment Authority (2001) *Concept Plan 2001*. Singapore: Urban Redevelopment Authority.

Urban Redevelopment Authority (2010). *Identity*. http://www.ura.gov.sg/conceptplan2001/identity.html. Accessed 27 March 2010.

Wachter, M (1974) Primary and secondary labor markets: a critique of the dual approach. *Brookings Papers on Economic Activity*, **3**, pp. 637–680.

Wong, A. and Yeh, S.H.K. (1985) *Housing A Nation*. Singapore: Maruzen Asia.

World Bank (2010) *Doing Business*. Washington DC: The World Bank.

World Competitiveness Center (2010) *World Competitiveness Yearbook 2010*. Available at http://www.imd.org/research/publications/wcy/World-Competitiveness-Yearbook-Results/#/wcy-2010-rankings/. Accessed on 12 December 2010.

Yeung, Y.M. (1987) Cities that work: Hong Kong and Singapore, in Fuchs, R.J., Jones, G.W. and Pernia, E.M. (eds.) *Urbanisation and Urban Policies in Pacific Asia*. Boulder, CO: Westview Press.

Yuen, B (1996) Creating the garden city: the Singapore experience. *Urban Studies*, **33**(6), pp. 955–970.

Yuen, B. (2006) Reclaiming cultural heritage in Singapore. *Urban Affairs Review*, **41**(6), pp. 830–854.

Chapter Nine

Going Global: Development, Risks and Responses in Kuala Lumpur and Putrajaya

Sirat Morshidi and Asyirah Abdul Rahim

When Malaya gained its independence in 1957, Kuala Lumpur was chosen as the capital of the newly independent nation. In 1963 it became the political and administrative capital of the Federation of Malaysia. Since the late 1990s, while Kuala Lumpur has remained pre-eminent as both a national and regional centre for commerce, consumption and finance, its capital city function has been gradually transferred to a new federal administration centre at Putrajaya, about 25 kilometres to the south. Kuala Lumpur and Putrajaya are now 'twin-cities' with complementary functions. City marketing and image making are now important elements in the competition between cities on the regional and global scale. The Kuala Lumpur-Putrajaya corridor forms the spine of the developing Kuala Lumpur city region (see figure 9.1) and is at the core, therefore, of the federal government's attempts to create a 'world class city' (Kuala Lumpur City Hall, 2008), which also presents a progressive image of a modern Malaysia to the world.

Cities in developing countries, which have been successful in attracting enterprises playing key roles in the increasingly globalized world economy, have typically made substantial investments in the quality of their city centres and infrastructure (see UNCHS, 1996, p. 263). 'World class' airports and seaports, retail-entertainment complexes along city waterfronts and symbolic skyscrapers of post-modern design are common manifestations of globalization-related developments. Increasingly, it seems that the main vehicles required to project a city's image to the world are megaprojects that require huge capital investments (Olds, 2001). The United Nations Centre

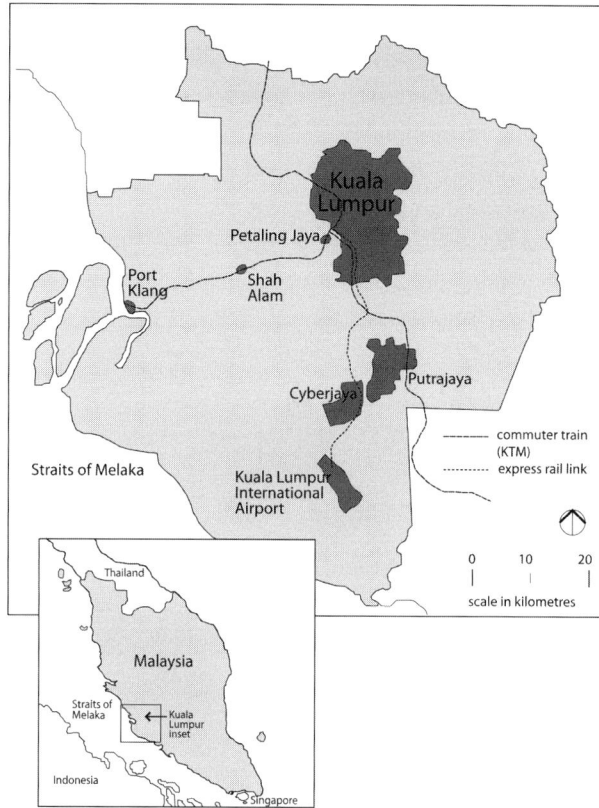

Figure 9.1. Malaysia and Kuala Lumpur.

for Human Settlements, in its 2001 report entitled 'Cities in a Globalizing World', listed the potential positive consequences of globalization as diffusion of knowledge; spread of norms of democratic governance, environmental justice and human rights; increased city-to-city exchanges of knowledge, experiences and best practices; and increased awareness on the part of both citizens and city managers of the potentials for peer-to-peer learning (UNCHS, 2001, p. vi). While acknowledging the unsustainable state of many urban environments at the start of the twenty-first century, UNCHS was optimistic about the prospects of transforming these into better and safer places to live in the new age of globalization. However, while megaprojects may succeed in attracting mobile global capital and investment for the benefit of city powerbrokers, their consequences for local people are not always positive. These projects can be empowering but they can also be harmful and disruptive. There is a need, in analyzing the growth of world cities, to pay careful attention to these 'local imprints of globalization' (Pulsipher and Pulsipher, 2003). It is also increasingly obvious that the environmental risks of urban megaprojects need to be mitigated and that a failure to do so will threaten the economic benefits of globalization, undermining the aims and aspirations of places like Kuala Lumpur to attain 'world class city' status.

This chapter describes the recent development of Kuala Lumpur and Putrajaya. It then explores climate change and sustainability issues against the broader backcloth of Malaysia's global aspirations and their spatial impacts. Risks are identified from the continuing intensification of development in central Kuala Lumpur, particularly in regard to 'heat island' effects, flooding and increasing traffic congestion. The central strand running through the chapter is that of the tension between the longer-term sustainable development of the Kuala Lumpur region as a polycentric metropolis and the continuing pressure for higher-density development at the centre as part of the aspiration to be a 'world class' city.

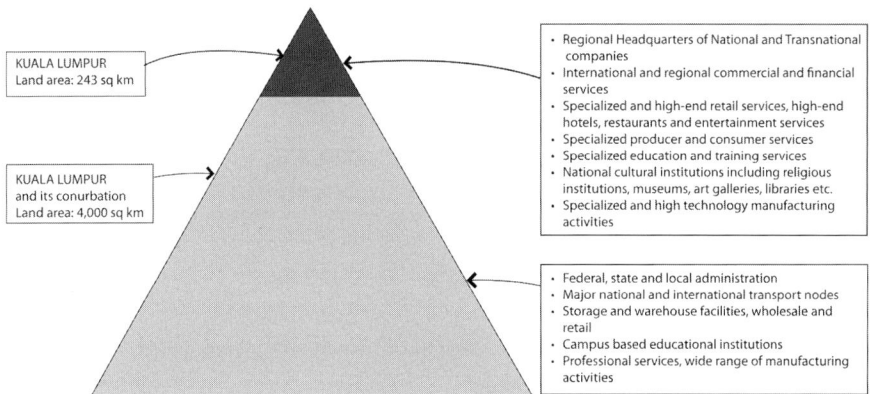

Figure 9.2. Functional pyramid of Kuala Lumpur and the wider conurbation. (*Source*: adapted from Kuala Lumpur Structure Plan 2020 (Kuala Lumpur City Hall, 2004))

Kuala Lumpur City – Going Global

Kuala Lumpur lies just north of the equator and is a hot and humid city, prone to heavy tropical downpours. It is situated at the confluence of the Klang and Gombak Rivers in the state of Selangor and traces its history back to a tin mining settlement established in the mid-nineteenth century. In the latter part of that century it became the seat of British colonial power in the Malay States (Morshidi, 2006, p. 72) and it remained the centre of political power and decision-making after independence was achieved in 1957. Thereafter, Kuala Lumpur began to emerge as a centre of consumption for foreigners and locals alike and, with the onset of economic globalization and rapid growth from the late 1980s, it became a capital that functioned not only as 'a special site for the concentration, administration and representation of political power' (Campbell, 2000, p. 2), but also as an important centre for international business and commerce.

Kuala Lumpur saw a remarkable spurt of development in the 1990s, with the completion of a number of megaprojects including the Petronas Twin Towers, the tallest buildings in the world at the time. Sardar, writing in 2000, described Kuala

Lumpur as 'a postmodern city writ large, a city that, within the short span of a decade, has been transformed from a sleepy capital into a technological marvel with a thriving, diverse and affluent cultural life'.[1]

The transformation of Kuala Lumpur to a modern, globally-networked city must be understood in the context of Malaysia's explicit desire to reposition its capital in the global hierarchy of cities (see Morshidi and Suriati, 1999; Morshidi, 2001; and Bunnell *et al.*, 2002). Kuala Lumpur's development in the last decade of the twentieth century was particularly influenced by its powerful 'urban director' and 'master planner' – the former Prime Minister of Malaysia, Mahathir Mohamad (Morshidi and Pandian, 2007; Colombijn, 2004). Mahathir's vision was that Malaysia should be a developed country by 2020 with a global city at its heart. Sardar (2000, p. 231) observed that information technology could easily have allowed the decentralization of many activities away from Kuala Lumpur's city centre, but its city planners chose to concentrate economic activities, office spaces and shopping complexes in a very small upmarket area of the city known as the 'Golden Triangle'. The iconic Twin Towers in the Golden Triangle area (see figure 9.3), the KL Sentral megaproject embracing the central station and central city air terminal, and a number of prestigious office and residential towers, in or close to the city centre, have now become landmark developments, but they have also seriously compounded existing problems of over-development in the central area.

Figure 9.3. The Kuala Lumpur Monorail in the Golden Triangle, with the Petronas Twin Towers in the background.

By the time that Abdullah Ahmad Badawi took over as Prime Minister from Mahathir in 2003, the transformation of Kuala Lumpur's centre was almost complete. Following Mahathir's departure, the federal government saw it as desirable to slow the pace of Kuala Lumpur's growth somewhat and to begin to deal with the environmental consequences of the decisions of the Mahathir period. However, in recent years, and particularly since 2009, under Badawi's successor as Prime Minister, Najib Tun Razak, there has been a renewed emphasis on the redevelopment of central Kuala Lumpur, and especially of Kampung Baru, a surviving low-rise area retaining some of the qualities of a traditional Malay settlement[2] which lies within the shadow of the Twin Towers (see figure 9.4).

The federal cabinet has approved in principle the comprehensive redevelopment of this area of over 300 hectares of prime land.[3] Given the land values, this is likely to mean modern high-rise commercial and residential buildings, at least for part of the area. The probable consequence will be the displacement of many of the 5,000 small lot owners. The redevelopment of Kampung Baru at higher density, located as it is at the fringe of the Golden Triangle, is very likely to exacerbate further the environmental and traffic congestion problems of the central city.

Figure 9.4. The Twin Towers viewed from Kampung Baru.

Putrajaya – A Malaysian City for the World

It was certainly one argument for building a new federal capital at Putrajaya that it would relieve the serious and growing congestion in the centre of Kuala Lumpur. The relocation of government departments would also free up land in the centre for global city office functions while providing more modern facilities for Malaysia's federal administration at the new site. However, this was much more than just a physical relocation. It was also a discarding of old colonial legacies and mindsets (Mahathir, 1998, pp. 10–11). To Mahathir Mohamad, Putrajaya was to be a city that embodied the spirit of Malaysia and its global aspirations in the twenty-first century (Morshidi, 2006). To reinforce this, Mahathir demanded architecture for the new capital replete with symbols representing pan-Islamic modernity. King (2007) has noted the clear influence of architectural forms from the Middle East – the 'well-spring of the Islamic world' – in the more significant buildings at Putrajaya, while the main Putra bridge is 'designed and constructed in accordance with Islamic architectural principles to resemble the Khaju Bridge in Isfahan, Iran' (Castor, 2003, p. 91).

Work on Putrajaya began effectively in December 1993 when the federal government invited a number of local consulting firms and government departments to come up with concept plans for the new capital. A preferred proposal, based on 'garden city' principles, was identified in 1994 and further work was undertaken to refine this. The final plan for Putrajaya was notable for its low overall density. The city has an area of 4,931 hectares, of which over 1,800 hectares or 37 per cent is reserved as green or open space. The key elements of the plan are:

◆   A large man-made lake and artificial wetlands, fed by the two small rivers that flow through the area;

◆   A 38 kilometre long waterfront area formed by the creation of the lake;

◆   A total of 20 precincts with the 'core employment and commercial precincts' (Core Area) located on an island surrounded by the man-made lake and peripheral precincts planned on neighbourhood planning principles to accommodate a mix of residential and local commercial uses, as well as public amenities;

◆   A projected residential population of 330,000, with 67,000 housing units;

◆   A 4.2 kilometre long axial boulevard which forms the central spine of the city.

(John, 2006) (see figure 9.5)

More important buildings, including the imposing Prime Minister's office building (figure 9.6) and the Putra mosque, are sited on the island within the lake. These buildings face on to a vast circular plaza, the Dataran Putra, which is used for

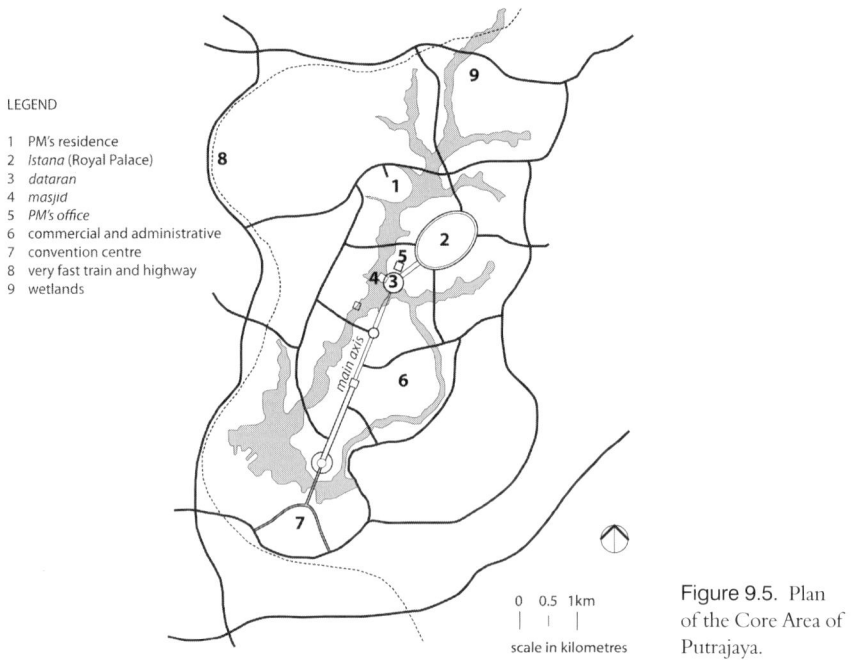

LEGEND

1 PM's residence
2 *Istana* (Royal Palace)
3 *dataran*
4 *masjid*
5 PM's office
6 commercial and administrative
7 convention centre
8 very fast train and highway
9 wetlands

0   0.5 1km
scale in kilometres

Figure 9.5. Plan of the Core Area of Putrajaya.

major events and ceremonies. The formal central axis, which runs along the island, ends significantly at the Prime Minister's building. Areas with a mix of residential and local commercial uses are found in the more peripheral precincts of Putrajaya, away from the central island. The population in 2010 was about 60,000. When completed, it is anticipated that slightly more than half of Putrajaya's residents will be federal government officers and their dependents.

Figure 9.6. Prime Minister's office building at Putrajaya.

Apart from its symbolic and administrative purposes, Putrajaya also forms part of a more ambitious vision developed in the 1990s to restructure the Kuala Lumpur metropolitan area. A decision had previously been taken to replace Kuala Lumpur's Subang Airport, opened in the 1960s, with the new Kuala Lumpur International Airport (KLIA) some 70 kilometres to the south of the city centre. The chosen site for Putrajaya was roughly midway between KLIA and the city. Its principal former use was as an oil palm plantation and it had the advantage of having only a small number of existing owners.[4] A very fast train, the KLIA Ekspres, and a new motorway were planned to connect KLIA with Kuala Lumpur and these were able to serve Putrajaya as well. Another new city, called Cyberjaya, with a target population of about 250,000 people, was also planned just to the west of Putrajaya. This formed part of Malaysia's attempt in the 1990s to establish itself as a leading international player in the information technology and computing industries. As part of these aspirations, the corridor itself was given the name of the 'Multimedia Super Corridor'. Overall, as a consequence of these well-integrated projects, the metropolitan area has acquired a new major north-south development axis, more or less at right angles to the established urban area which extends from Kuala Lumpur to Port Klang on the west coast of Malaysia, via Shah Alam and the industrialized Klang Valley.

The development of Putrajaya has been achieved remarkably quickly. King (2007) attributes this to 'the iron will of Mahathir' and also to its timing. Putrajaya's early development coincided with the Asian financial crisis after 1997 and its construction became a major counter-cyclical investment, creating work for a substantial part of Malaysia's building industry and playing a significant role in assisting the country to ride out the crisis.

Putrajaya had lofty aspirations. It was to represent nothing less than a new direction for Malaysia's planning – a paradigm shift based on the integration of moral and spiritual values and focused on quality of life and sustainability (Omar, 2004). Apart from its importance as a symbol of a modern, independent nation, it was also intended to exemplify sustainability principles in its wetlands, its mixed-use development, its use of energy efficient building materials and designs, and its public transport. The contribution that Putrajaya may make to a more sustainable metropolis is considered later, following the next part of the chapter, which explores the broader environmental challenges currently faced by Kuala Lumpur and its extended region.

Kuala Lumpur and Putrajaya – Environmental Challenges

Cities have a complex relationship with their climate and understanding this relationship is becoming more urgent as concerns grow about the effects of climate change. The concentration of economic activities in cities, and the high levels of energy consumption and emissions which result, mean that cities have profound effects on both regional weather and global climate. Some urban communities are particularly vulnerable to specific impacts of climate change (Alam and Rabbani, 2007; Roy, 2008) and Kuala Lumpur seems to be especially at risk from an increased 'heat island' effect and a higher incidence of extreme weather events, leading to flooding and landslides.

Heat Island Effects

Kuala Lumpur city has a total land area of 243 square kilometres and a current population of 1.7 million, giving a population density of about 7,000 persons per square kilometre. This population is expected to rise to 2.2 million people by 2020, with employment reaching 1.4 million by that time. The Kuala Lumpur-Putrajaya corridor will then be the primary spine of an expanding metropolitan region, which also embraces the established and growing centres of Petaling Jaya and Shah Alam as well as other settlements in the Klang Valley. This larger area has a population of about 6 million people at present and the most recent predictions suggest a population of between 8.5 and 10 million by 2020. Kuala Lumpur at present is not as large a city as most other Asian capitals and its further growth is still seen as desirable by the federal government in pursuit of the objective of increasing its importance as a node in the global city network – a 'world class city' (Kuala Lumpur City Hall, 2008). The growth of the city in recent years has not occurred without some attention to the public realm. There is, for example, a splendid park of some 20 hectares at the Kuala Lumpur City Centre (KLCC) adjacent to the Petronas Twin Towers.[5] However, to achieve the growth projected for 2020, further intensification will be required and open space in central Kuala Lumpur will be at a premium. The likely consequence of the next phase of growth will be an increase in the density of built form in the already highly developed city centre, with a corresponding increase in the severity of the 'heat island' effect.

Information on the heat island effect in Kuala Lumpur has been documented by Sham (2008) over a 25 year period of observations. These studies show that, in Kuala Lumpur and the Klang Valley, the urban centres are usually warmer by several degrees Celsius than the surrounding countryside. On average, the annual mean temperature difference between Kuala Lumpur city and Subang Airport over the observation period was approximately 1–2°C, but on calm nights the urban-rural temperature difference could be as much as 4–5°C. The higher temperatures in Kuala Lumpur are a consequence primarily of the density of buildings and paved surfaces, which retain heat, together with the extensive use of air-conditioning units in the core commercial and financial areas, such as the Golden Triangle and KL Sentral.

Sham (2008) has further observed that, while the horizontal temperature variations in the Kuala Lumpur and Klang Valley conurbations have been well documented, little work has been undertaken to assess the vertical temperature changes associated with tall buildings in the central area. There is some evidence, however, that temperature distributions within the heat island inhibit the upward dispersal of pollutants, resulting in a greater concentration of these at lower levels within the central area. In 1990, there were only about twenty-five buildings in Kuala Lumpur with twenty-six or more storeys (Morshidi and Suriati, 1999, p. 41). Since then, the number of high-rise structures has increased tenfold and there are more than fifty buildings taller than 100 metres, with several taller than 200 metres including, of course, the Petronas Twin Towers which reach a height of 450 metres. The effects of tall buildings on temperature are likely to become even more significant as more megastructures are proposed for central Kuala Lumpur.

Flash Floods

Flash flooding has long been a regular occurrence in Kuala Lumpur, given its location in a river basin and its tropical climate. There has been recent evidence to suggest that changes in rainfall patterns, attributable to global warming, are occurring. These changes relate to both the amount of rain that falls on particular areas and its distribution across the year. Annual rainfall data for two recording stations located within Kuala Lumpur and the Klang Valley illustrate this. The first station is at Subang Airport and the second is at Petaling Jaya, an urban sub-centre to the west of Kuala Lumpur. Annual rainfall data for both stations showed somewhat erratic changes in rainfall patterns between 1980 and 2008 (figure 9.7). However, both also showed

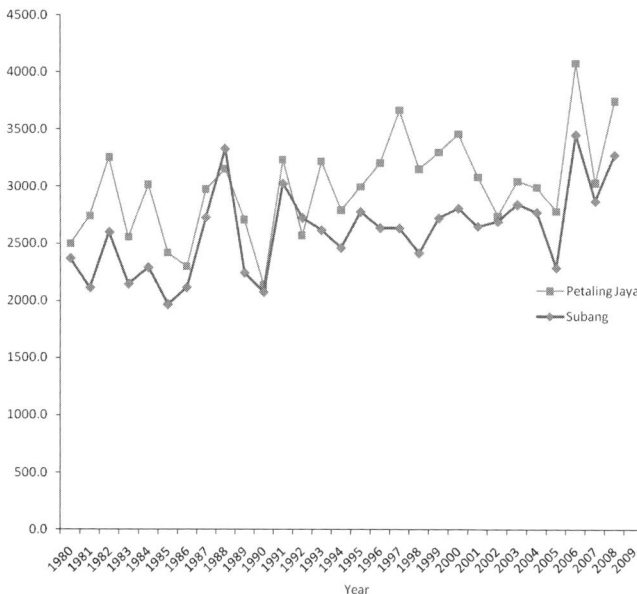

Figure 9.7. Annual rainfall patterns for Petaling Jaya and Subang (in millimetrers).

significant overall increases in the amount of precipitation. Increases in rainfall, associated with increased intensity and duration, pose serious threats to Kuala Lumpur with its susceptibility to devastating landslides after periods of heavy rain.

The natural drainage system of Kuala Lumpur includes three main rivers with a total length of 32.8 kilometres and eight other rivers with a combined length of 40.6 kilometres (Kuala Lumpur City Hall, 2008). A study prepared by the Japan International Co-operation Agency (JICA, 1989) reported that, over the period 1949–1988, eleven occurrences of significant flooding of the Klang River had been recorded at Sulaiman Bridge in the heart of Kuala Lumpur. The most severe flood, lasting for five days, was recorded in January 1971. The JICA study identified the main factors contributing to flooding as being the increase in stormwater run-off, related to the growing amount of urban development, and inadequate drainage facilities.

Flood alleviation projects in Kuala Lumpur were initiated as a result of the recommendations of the JICA study. A master plan for the mitigation of floods in the Klang River basin was prepared and subsequently acted upon by the federal Department of Irrigation and Drainage (DID). Implementation was spread over 15 years at a total cost of RM[6] 619 million. The Kuala Lumpur Structure Plan for 2020 (Kuala Lumpur City Hall, 2004) identified a number of further measures with respect to drainage, to be undertaken jointly between the city authorities and DID as part of the Kuala Lumpur Drainage Master Plan. As noted earlier, Kuala Lumpur began as a tin mining town and many of the remaining mining ponds have potential to serve as

Figure 9.8. The diversion channel (under construction) of the Kuala Lumpur Flood Mitigation project in 2006.

retention ponds for flood mitigation. The redirection of storm water to these ponds was also an important outcome of the Kuala Lumpur Flood Mitigation (KLFM) project (see figure 9.8), as was the installation of a modern rain gauge system as part of an early warning system.

A particularly innovative recent flood mitigation project was the Stormwater Management and Road Tunnel (SMART) along the Klang River, upstream of the historically significant Masjid Jamek mosque. This tunnel, opened in May 2007, involved the construction of a 3 kilometre long twin-deck toll road within a 9.7 kilometre long stormwater diversion tunnel (figure 9.9).

Figure 9.9. Stormwater Management and Road Tunnel (SMART).

Major stormwater works of this sort are managed by DID but Kuala Lumpur City Council also has important responsibilities, including the removal of silt and rubbish from the rivers and the regular maintenance of a network of drains over 2,400 kilometres in length. Table 9.1 shows how the cost of regular maintenance of rivers, borne by the city council, rose from about RM 3 million in the year 2000 to RM 9 million in 2005. A significant part of this cost is incurred in clearing rubbish from the rivers. Kuala Lumpur's waste management systems are not effective and there is a pressing need to educate the inhabitants of Kuala Lumpur in the proper disposal of their rubbish.

Table 9.1. Regular maintenance cost of rivers in Kuala Lumpur City, 2000–2005.

Year	De-silting and Operation of Rubbish Traps (RM)	Mopping-up and Cleaning of Rubbish (RM)	Total (RM)
2000	1,886,611.98	1,315,719.48	3,202,331.46
2001	1,943,941.93	1,926,160.02	3,870,101.95
2002	2,558,908.70	3,472,606.89	6,031,515.59
2003	2,067,688.47	3,521,358.33	5,589,046.80
2004	3,650,765.91	3,862,019.88	7,512,785.79
2005	4,082,947.90	4,947,464.53	9,057,412.43

Source: Adapted from Kuala Lumpur City Hall, 2007.

The draft Kuala Lumpur City Plan for 2020 (Kuala Lumpur City Hall, 2008) included a number of further initiatives to achieve best practice in the sustainable management of urban stormwater, including the adoption of water-sensitive urban design principles in new and retrofit projects. Other environmentally-friendly stormwater management practices were gathered together in an Urban Stormwater Management Manual (Department of Irrigation and Drainage, 2009). These recent initiatives now pay attention to the aesthetics of stormwater drainage projects as well as their engineering design.

Pereira (2008) has emphasized the importance generally of achieving collaborative action between all levels of government, as well as with stakeholders in the private and community sectors, in responding to climate change. There is much to be admired in the recent collaborative initiatives on flood mitigation undertaken by federal and city authorities in Kuala Lumpur. Nevertheless, there is evidence that the propensity for flood events to occur has increased since the 1990s. Substantial floods followed rainstorms in the north-east catchments of the Ampang and Klang rivers in April and October 2001, with drainage channels having insufficient capacity to contain the flood waters. The Gombak river and the Batu river (see figure 9.10) require further flood mitigation works fairly urgently and floods still occur regularly near the confluence of the Gombak and Klang rivers. On 4 March 2009 thousands of motorists were stuck in massive traffic jams for several hours after a storm caused flash floods up to a depth of 2 metres on Jalan Ipoh and in the Kampung Baru area (Menon, 2009). Floods of a similar severity were also experienced in December of that year (Menon and Yi, 2009). There are also other areas, which are still very susceptible to flooding in low-lying areas along the Klang River, including Kampung Datok Keramat, Jalan Yap Kwan Seng, Jalan Dang Wangi and the Jalan Tun Perak-Jalan Melaka area of the city centre. Given the physical configuration of the river valleys and the indications of a trend towards increased rainfall, further growth in the intensity of urban development in the Kuala Lumpur city centre seems likely to lead to heightened flood risks in future.

Figure 9.10. The confluence of the Gombak and Batu rivers near Putra World Trade Centre.

Increased Traffic Volumes and Road Congestion

As the major centre of employment and commerce in the country, Kuala Lumpur has a large number of daily vehicle movements and, as noted earlier, traffic congestion has become a serious problem. In 1996 there were 860,000 vehicles registered within the city, which then had a population of 1.2 million. At that time Kuala Lumpur already had a ratio of 620 vehicles per kilometre of road, compared to 220 vehicles per kilometre for Singapore (Mazlan, 1996). By 2007 there were about 3.8 million vehicles registered in the federal territory (comprised primarily of Kuala Lumpur and Putrajay).[7] Over 1.3 million private vehicles enter and leave Kuala Lumpur city centre each day during the peak periods. About 70 per cent of these vehicles have only a single occupant. The number of private cars (and also motorcycles) has grown rapidly with the increased affluence of Malaysians and there seems little prospect of the automobile dependence of the metropolitan region's inhabitants reducing significantly in the foreseeable future (Barter, 2004; Ramadas, 2010).

Kuala Lumpur's transport infrastructure is already under great strain with its present population of 1.7 million (Ramadas, 2010) and conditions are likely to get worse as the population increases towards its forecast target of 2.2 million in 2020. In order to deal with the inefficiencies in current travel patterns, the draft Kuala Lumpur City Plan aspires to reduce the length of trips and to influence modes of travel through co-ordinated land-use planning arrangements which encourage major employers and other traffic generating uses to locate close to public transport (Kuala Lumpur City Hall, 2008). Significant efforts have been made to improve the urban public transportation system in Kuala Lumpur in recent years, but it remains overcrowded and unreliable. There are currently ambitious targets for increasing public transport ridership in the city centre to 60 per cent of all trips by 2020 (Kuala Lumpur City Hall, 2008) with a shorter term aim of increasing ridership in the Klang Valley from its present 10–12 per cent of all trips to 25 per cent by 2012 (Foong, 2010). A major barrier, however, to increasing public transport ridership in Kuala Lumpur in the short-term, and certainly by these heroic amounts, is that only about 10 per cent of the population has convenient access to a station at present (Ramadas, 2010) and only about 60 per cent lives within reasonable walking distance of a bus stop (Transit Malaysia, 2010).

The Kuala Lumpur Sentral megaproject is, in part, an attempt to achieve better integration between a number of different public transport services. It is a large mixed-use and commercial development project built around the central station, Kuala Lumpur's major transport hub, which provides direct links by high-speed train to KLIA and Putrajaya. It also serves as an interchange for intercity, commuter and light rail lines. Two light rail lines were built in the 1990s to ease traffic congestion in the city (see Morshidi and Suriati, 1999) and a monorail system was also opened in 2003, running for 8.6 kilometres from Kuala Lumpur Sentral through the Golden Triangle

Figure 9.11. Parking problems in Putrajaya.

to Titiwangsa. These new lines are popular but much more substantial improvements in frequency of service, network coverage and ease of interchange between different public transport providers are required before public transport usage can increase to a level where it begins to have some significant impacts on automobile dependency. The new capital of Putrajaya, despite its sustainability aspirations, is unlikely to make much difference in the short term either. It is mainly served by buses at present. A monorail system is planned, although its introduction has been much delayed because of concerns about its viability (Transit Malaysia, 2009). While the generous areas of open space at the new capital certainly have environmental and aesthetic value, they contribute to an urban form which is fairly low in density. Putrajaya seems likely to be an automobile city for the foreseeable future and car parking spaces are already at a premium (see figure 9.11).

Conclusion

Kuala Lumpur's centre has been dramatically transformed in the past two decades as part of an explicit globalization policy, driven by the federal government, and by Mahathir Mohamad in particular, to attract mobile international capital and investment to the city. The modernization of Kuala Lumpur's urban fabric through megaprojects in the late 1980s and early 1990s paid little regard to the environmental impacts of such development. The severe flash floods and landslides, which are still regular occurrences, bear testimony to some of the consequences of poorly conceived developments in the recent past. Since the late 1990s, both the federal government and the city council have begun to pay more attention to the language and imperatives of sustainable development and to incorporate environmental principles into development proposals for the planning period to 2020 and beyond. However, in a city which for many years has been developed at higher and higher densities on steep hill slopes, floodplains and other ecologically sensitive areas, new discourses on sustainability and its importance to the future image of the city will take time to be fully embraced by administrators, developers and the wider citizenry. This is all the more so because, in Kuala Lumpur, the development machinery of the Mahathir era still appears to be intact. The current draft Kuala Lumpur City Plan contains proposals for an additional 100,000 people to be living in the city centre by 2020 at a density of nearly 14,000 people per square kilometre. This is similar to the sorts of inner area densities experienced in Tokyo and Seoul.

Putrajaya's future trajectory is hard to predict with confidence. As described earlier, it was developed explicitly as a city to project a new modern image of Malaysia to the world in an era when images and symbols have become all-important to the global positioning of cities. Its residential areas are well-planned and much of its architecture is stunning. But Putrajaya was also conceived as a sustainable city, with co-existence with nature as part of its underlying design philosophy. The large wetland areas are undoubtedly ecologically important, but some have questioned whether Putrajaya's 'green' credentials stand up to close scrutiny in other respects. For example, Moser (2010, p.293) suggests that opportunities were lost to advance 'microclimatic design' by creating a city cooled passively through lush planting and with the narrow, shady streets of older Middle Eastern cities rather than the wide formal avenues, large air-conditioned buildings and vast plazas of Mahathir's capital. Critics of the new city's social sustainability, meanwhile, have also commented on Putrajaya's failure to recreate the rich multicultural diversity of older, denser parts of Kuala Lumpur (King, 2007). However, its low overall density is the main reason why Putrajaya's sustainability is questioned. It appears well on the way to becoming an automobile dependent city, although judgement on this should perhaps be suspended until its proposed monorail system and other sustainable transport proposals are fully implemented. At the metropolitan scale, more time is also needed to see whether Putrajaya can assist in changing Kuala Lumpur's structure to that of a more polycentric city, or whether, without a substantial and strongly-enforced buffer, it will gradually be absorbed by the rapid outward spread of development which is already occurring along the Multimedia Super Corridor.

The demand to increase the city's global city functions will continue to drive changes to Kuala Lumpur's urban fabric. The general tendency on the part of Kuala Lumpur's city fathers to date has been to meet these demands by allocating more space in the centre for commercial development, thereby pushing urban residents further out to the suburbs. There has been only limited priority given to the sustainability of development hitherto and the lack of attention to effective environmental protection raises questions about Kuala Lumpur's future resilience in the face of the risks it clearly faces.

The future growth of Kuala Lumpur must be planned effectively through a strategic framework which encompasses the whole of the emerging metropolitan region, including Putrajaya and other expanding cities like Petaling Jaya and Shah Alam.

Kuala Lumpur will inevitably remain as the globally focused core of this region but regional balance, environmental sensitivity and an integrated framework for responding to climate change must be high priorities for the planning of this broader area. One key to the future sustainability of the region must be a drastically improved urban public transportation system. The future of the Kampung Baru area remains in question at the time of writing, but there seems every chance, based on previous experience, that this area too will be developed at a higher intensity for the benefit of transnational firms and for those who can afford high-end property prices. Some existing residents would inevitably be displaced by this to areas at the urban fringe, adding to the number of commuters and the daily traffic congestion. There are those who express the view that, in order to enhance its attractiveness to footloose global investors, Kuala Lumpur should be protecting heritage areas like Kampung Baru which also provide housing for the low- and middle-income workers needed in the city. It does not seem likely at present, however, that this alternative view will prevail.

Postscript

Just as this chapter was being completed in late 2010, the federal government announced a new economic development strategy for Greater Kuala Lumpur and the Klang Valley. This new planning region includes Kuala Lumpur and Putrajaya as well as eight other metropolitan municipalities. The government's 'Economic Transformation Programme' seeks to drive further rapid economic growth by attracting '100 of the world's most dynamic companies' to the Greater Kuala Lumpur region. A new high-speed rail line to Singapore is also promised (Pemandu, 2010). Through these and related measures, the intention is to raise the contribution of Greater Kuala Lumpur and the Klang Valley to gross national income from RM 258 billion to RM 650 billion a year by 2020.

The aspiration now is for Kuala Lumpur to be ranked within the top twenty world cities by 2020. The government acknowledges, however, that Greater Kuala Lumpur faces 'stiff competition from neighbouring cities because liveability lags compared to many other Asian cities (and) public transport remains inadequate' (Pemandu, 2010). There are commitments, therefore, to building an integrated urban mass rapid transit system (with construction commencing in 2011), opening up the Klang River waterfront as a heritage and commercial centre, creating 'iconic places and attractions' and ensuring that every resident has access to sufficient green space. An estimated RM 58 billion has been identified as the required public sector investment in these projects and a Greater Kuala Lumpur and Klang Valley development corporation, responsible ultimately to the prime minister, has been foreshadowed to provide firm leadership and direction.

The analysis of these new proposals is a matter for another day. The commitment to planning at the level of the extended metropolitan region is encouraging, as is the projected investment in a mass rapid transit system. It remains to be seen how the balance between sufficient open space and increased urban density can be struck or what the new emphasis on liveability will mean for the residents of Kampung Baru.

Notes

1. This quotation appears on the inside front cover of Sardar (2000).
2. *Kampung* can be loosely translated as 'village'. Over a century ago British colonial administrators designated Kampung Baru as an agricultural settlement where Malays could retain a village lifestyle within the city.
3. 'Bill paves way for Kg Baru development', *The Star Online*, 20 December 2010.

4.   Although it was not quite the *tabula rasa* that it is often said to have been, since some 2,400 plantation workers were displaced according to Bunnell, 2002.
5.   Designed by the celebrated Brazilian landscape architect Roberto Burle Marx.
6.   Malaysian Ringgit.
7.   The Federal territory also includes the island of Labuan.

References

Alam, M. and Rabbani, M.G. (2007) Vulnerabilities and responses to climate change for Dhaka. *Environment and Urbanization*, **19**(1), pp. 81–97.

Barter, P.A. (2004) Transport, urban structure and 'lock-in' in the Kuala Lumpur Metropolitan Area. *International Development Planning Review*, **25**(1), pp. 1–24.

Bunnell, T., Barter, P.A. and Morshidi, S. (2002) City profile. Kuala Lumpur Metropolitan Area: a globalizing city-region. *Cities*, **19**, pp. 357–370.

Bunnell, T. (2002) Multimedia utopia? A geographic critique of high-tech development in Malaysia's multimedia super-corridor. *Antipode*, **34**(2), pp. 265–295.

Campbell, S. (2000) *Cold War Metropolis: The Fall and Rise of Berlin as a World City*. Minneapolis, MN: University of Minnesota Press.

Castor, R.T. (2003) Putrajaya Malaysia. *Lookeast*. Bangkok: Tourism Authority of Thailand, pp. 90–91.

Colombijin, F. (2004) High hopes versus a low profile: the urban development of Kuala Lumpur and Singapore, in Nas, P.J. (ed.) *Directors of Urban Change in Asia*. London: Routledge.

Department of Irrigation and Drainage (2009) *Manual Saliran Mesra Alam*. Available at http://www.water.gov.my/index.php?option=com_content&task=view&id=167&Itemid=362.   Accessed 17 July 2009.

Foong, J. (2010) Big strides in urban public transport. *The Star*, 12 May, p. N10.

JICA (Japan International Cooperation Agency) (1989) *The Study on the Flood Mitigation of the Klang River Basin Main Report*. Kuala Lumpur: JICA.

John, J.I. (2006) *Creating the Essence of Cities: The Planning and Development of Malaysia's New Federal Administrative Capital, Putrajaya*. Available at http://info.worldbank.org/etools/docs/library/235906/s5_p22paper.pdf. Accessed 1 January 2010.

King, R. (2007) Rewriting the city: Putrajaya as representation. *Journal of Urban Design*, **12**(1), pp. 117–138.

Kuala Lumpur City Hall (2004) *Kuala Lumpur Structure Plan 2020*. Kuala Lumpur: Kuala Lumpur City Hall.

Kuala Lumpur City Hall (2007) *Annual Report, Drainage and River Management Department*. Kuala Lumpur: Kuala Lumpur City Hall.

Kuala Lumpur City Hall (2008) *Kuala Lumpur City Plan 2020*. Kuala Lumpur: Kuala Lumpur City Hall.

Mahathir, M. (1998) *Excerpts from the Speeches of Mahathir Mohamad on the Multimedia Super Corridor*. Kuala Lumpur: Pelanduk.

Mazlan, A. (1996) Managing Kuala Lumpur: the environmental constraints, in Komoo, I. (ed.) *Roundtable Dialogues: Urban Planning Within the Context of the Environment: The Kuala Lumpur Experience*. Bangi: LESTARI.

Menon, P. (2009) Thousands caught unawares as 2m-high flash floods hit KL. *The Star*, 4 March. Available at http://thestar.com.my/news/story.asp?file=/2009/3/4/nation/3397378&sec=nation. Accessed 1 April 2010.

Menon, P. and Yi, Tho Xin (2009) Flash floods wreak havoc on roads in the Klang Valley. *The Star*, 4 December. Available at http://thestar.com.my/metro/story.asp?file=/2009/12/4/central/5232798&sec=central. Accessed 1 April 2010.

Morshidi, S. (2001) Kuala Lumpur, globalization and urban competitiveness: an unfinished agenda. *Built Environment*, **27**(2), pp. 96–111.

Morshidi, S. (2006) Kuala Lumpur: primacy, urban system and rationalizing capital city functions, in Ho, K.C. and Hsiao, M.H.-H. (eds.) *Capital Cities in Asia-Pacific: Primacy and Diversity*. Taipei: Center for Asia-Pacific Area Studies, Research Centre for Humanities and Social Sciences, Academia Sinica.

Morshidi, S. and Suriati, G. (1999) *Globalisation of Economic Activity and Third World Cities: a case study of Kuala Lumpur*. Kuala Lumpur: Utusan Publications and Distributors Sdn Bhd.

Morshidi, S. and Pandian, S. (2007) Globalizing Kuala Lumpur: competitive city strategies under Dr. Mahathir Mohamed. *Asian Profile*, **35**(6), pp. 489–504.

Moser, S. (2010) Putrajaya: Malaysia's new federal administrative capital. *Cities*, **27**(4), pp. 285–297.

Olds, K. (2001) *Globalization and Urban Change. Capital, Culture, and Pacific Rim Mega-Projects.* Oxford: Oxford University Press.

Omar, D.B. (2004) The total planning doctrine and Putrajaya development, in Marchettini, N., Brebbia, C.A., Tiezzi, E. and Wadhwa, L.C. *The Sustainable City III: Urban Regeneration and Sustainability.* Southampton: Wessex Institute Press.

Pemandu (2010) *Economic Transformation Programme- A Roadmap for Malaysia:Greater Kuala Lumpur/ Klang Valley.* Available at http://etp.pemandu.gov.my/Overview_of_NKEAs_-@-Greater_Kuala_Lumpur-s-Klang_Valley.aspx. Accessed 12 December 2010.

Pereira, J.J. (2008) Stakeholder participation, in Raja Zaharaton, R.Z., Pereira, J.J., Koh, P.F. and Tan, C.T. (eds.) *Round Table Dialogue 17: A New Approach to Climate Change: Balancing Adaptation and Mitigation.* Bangi: LESTARI.

Pulsipher, L.M. and Pulsipher, A. (2003) *World Regional Geography. Global Patterns, Local Lives*, 2nd ed. New York: W.H. Freeman.

Ramadas, M. L. (2010) KL under pressure. *The Sun*, 24 May, p. 9.

Roy, M. (2008) Planning for sustainable urbanisation in fast growing cities: mitigation and adaptation issues addressed in Dhaka, Bangladesh. *Habitat International*, **33**(3), pp. 276–286.

Sardar, Z. (2000) *The Consumption of Kuala Lumpur.* London: Reaktion Books.

Sham, S. (2008) The urban heat island: planning and management implications, in Abdul Samad, H., Ahmad Fariz, M., Shaharudin, I., Abdul Hadi, H. S. and Norlida, M. H. (eds.) *Round Table Dialogues18: Urbanization and the Climate for Change: the livable city in Malaysia.* Bangi: LESTARI.

Transit Malaysia (2009) *The Curious Case of the Putrajaya Monorail.* 29 July 2009. Available at http://transitmy.org/2009/07/29/the-curious-case-of-the-putrajaya-monorail/. Accessed 21 December 2010.

Transit Malaysia (2010) PEMANDU Lab Highlights: Urban Public Transport. Available at http://transitmy.org/ideas/. Accessed 14 December 2010.

UNCHS (United Nations Centre for Human Settlements) (HABITAT) (1996) *An Urbanizing World: Global Report on Human Settlements, 1996.* Oxford: Oxford University Press.

UNCHS (2001) *Cities in a Globalizing World. Global Report on Human Settlements 2001.* London: Earthscan.

Chapter Ten

Governing the Jakarta City-Region: History, Challenges, Risks and Strategies

Wilmar Salim and Tommy Firman

Asian cities are confronting common challenges related to environmental sustainability, the consequences of international financial crises and the impacts of climate change. Jakarta has been dealing with multi-faceted problems of environmental degradation for several decades; it was severely affected by the Asian monetary crisis in the late 1990s; and it is one of the most vulnerable cities in Southeast Asia to flooding, which is likely to be exacerbated by climate change (Yusuf and Fransisco, 2009). To these critical threats must be added unresolved problems of population pressure that reinforce urban agglomeration and drive the continuing expansion of the Jakarta city-region. A range of approaches and strategies has been adopted over the last 20 years or so to address Jakarta's problems and anticipated threats, but good and effective governance for the city-region still seems hard to attain. This chapter assesses the achievements of these past efforts and explores Jakarta's vulnerability to a range of current risks, including those related to climate change. It begins with a description of the Jakarta city-region and of the historical governance arrangements and development issues that have shaped the city as it is today. The authors then review current demographic, traffic, environmental and economic challenges facing Jakarta and discuss emerging strategic responses to date. The main theme of the chapter is the continuing inadequacy of governance and planning arrangements for the metropolitan region and the need to address this as a matter of great urgency if Jakarta's high level of vulnerability to flooding and other environmental risks is to be reduced.

The Jakarta City-Region

Jakarta's city-region is comprised of the city of Jakarta at the core and a number of surrounding municipalities and districts. The extended city-region is generally referred to by the acronym 'Jabodetabek',[1] since it encompasses the cities of *Ja*karta, *Bo*gor, *De*pok, *Ta*ngerang, and *Bek*asi (see figure 10.1). It is an extensive urban agglomeration which also includes sizeable rural areas – the archetypal *desa kota* – including irrigated agricultural lands and environmentally sensitive zones. The total area of the city of Jakarta is around 650 square kilometres, but the greater Jabodetabek metropolitan area has an area of 6,175 square kilometres, about half of which is urbanized. The population of the city of Jakarta in 2005 was around 8.8 million, but the extended Jabodetabek metropolitan region had a population at that time of 23.6 million, of whom over 20 million lived in urban settlements.[2]

Until the mid-1990s, this extended area consisted of only two municipalities or cities (*kota*) – Jakarta and Bogor – and three districts (*kabupaten*) – Bogor, Tangerang and Bekasi. However, with the physical expansion of Jakarta city into its neighbouring districts, more urban centres were created near the borders and the three new municipalities of Depok, Tangerang and Bekasi were founded. To complicate the institutional setting, Jabodetabek is also administered by governments at a number of different levels. As the capital city, Jakarta is where the Indonesian national government sits, but Jakarta itself has a provincial-level government[3] headed by a governor. Parts of the urbanized area also fall under the jurisdiction of two other provinces, West Java and Banten. Another relevant factor is that, for the past decade or so, Indonesia has

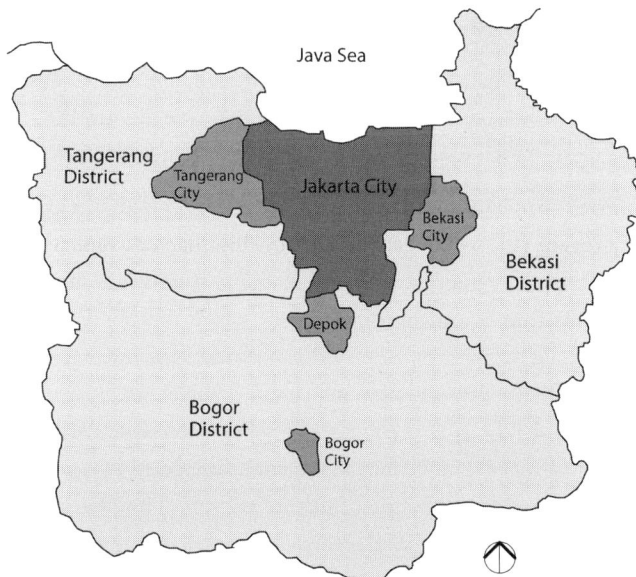

Figure 10.1. The Jabodetabek city-region. (*Source*: Adapted from Rustiadi, 2007)

been pursuing a decentralization policy that gives more autonomy to the municipal and district levels. The institutional arrangements for managing the Jabodetabek mega-region are thus fairly complex and there are conflicting interests between the different jurisdictions, which are discussed later.

Governing and Planning in the Past: From Batavia to Jakarta

Historically, Jakarta grew gradually from the small pre-colonial port of Sunda Kelapa (later renamed as Jayakarta) into the major trading post of Batavia, the centre of the first and largest multinational company of the Dutch colonial era, the *Vereenigde Oost-Indische Compagnie* (VOC).[4] The city served as the capital of the colonial Dutch East Indies until World War II and the attainment of Indonesian Independence which followed soon after the war.

The Kingdom of The Netherlands took direct responsibility for its eastern colonies away from the VOC in the early 1800s and a number of separate urban developments were built around Batavia. A suburban area south of Batavia known as Weltevreden was developed as a site for the colonial administration and as a settlement for Europeans. Governor-General H.W. Daendels, who ruled from 1808 to 1811, also established a cantonment at Meester Cornelis, south of Weltevreden. Later, in 1870, a new harbour was built at Tanjung Priok to the east of Batavia. Following the adoption by the Dutch of their 'Ethical Policy'[5] in 1901, a formal plan of Batavia was issued in 1910. At that time Batavia comprised the old town, Weltevreden and Tanjung Priok. In the 1930s Meester Cornelis was formally annexed to the city and another plan of Batavia was issued in 1930. However, as Silver (2011, p. 83) asserts, none of these plans for Batavia was effective since:

Batavia apparently lacked powerful, wealthy and enthusiastic planning sponsors to back and possibly even fund preparation of a plan. Those who were the leading advocates for better planning in Batavia were a handful of architects, and not the business elite, and these professionals were in many cases constrained in their advocacy by the need to make a living.

Batavia was renamed Jakarta after Indonesia's independence in 1945 but the planning practices of the earlier Batavia era continued – unilateral proposals by certain groups within or outside government which did not attract much support from the major powerbrokers of the city. Planning ideas formulated for Jakarta remained mainly on paper, with little attention in practice to environmental and social issues. A Concept Plan of 1952 for Jakarta proposed a pattern of concentric rings with Merdeka Square (in the former Weltevreden) at the core, supported by a series of ring highways and a greenbelt to define the outer limits of the city. This was followed by an Outline Plan in 1957 that stressed that the focus of new development should be in satellite towns surrounding Jakarta – Bogor, Tangerang and Bekasi. However, the main elements of the 1952 and 1957 plans were not implemented. The greenbelt, intended to separate Jakarta from Bogor, Tangerang and Bekasi, is non-existent, with Jakarta's development sprawling across it rather than leaping over to the satellite towns. Only the series of ring highways was built according to the prescriptions of the Concept Plan. Part of the reason why some of these early plans were not implemented may well have been that they were outstripped by the dynamism of urban growth, but it also indicates that plans have not generally been regarded as important by Jakarta's powerbrokers. There has been a continuing gap between plans and their implementation and the shape of Jakarta has been influenced more by the rulers than by the rules. Pratiwo and Nas (2005) have suggested that the development, which has taken place in Jakarta since Independence, can be attributed mainly to a number of 'directors' whom they identify as follows:

♦ The first post-war president, Soekarno, who displayed a hunger for symbolism and large-scale projects in seeking to impose a degree of uniformity on the dispersed constituent parts of Jakarta.

♦ Governor Ali Sadikin who reformed the administrative structure and put forward the Master Plan of 1965–1985, proposing the expansion of Jakarta to include an area larger even than its existing area today.

♦ Hendropranoto Suselo and his Department of Public Works which instituted the policy of deconcentrated urban development and put forward the idea of a greater Jabotabek region (now subsumed in Jabodetabek).

♦ President Soeharto who, during his three 'New Order' decades in power, pursued the development of broad avenues, highways, electric railway lines, high-rise buildings, megamalls, luxurious housing estates, golf courses and industrial estates within and outside Jakarta as symbols of development and modernization.

♦ CiPutra, a major real estate entrepreneur who translated Soeharto's concepts into condominiums, superblocks and Western-style new town developments on a large scale.

The interests and visions, occasionally conflicting, of the national leadership, the governor of the city, the head of the principal technical department and the real estate developers, of whom CiPutra was the most influential, shaped the way that Jakarta was governed, planned and developed during much of Indonesia's post-Independence era. Over this period Jakarta grew from a harbour town and trading centre to the national capital of the third largest developing country in the world. However, the interests of its 'directors' were not necessarily in the best interests of Jakarta and its people. As will be discussed below, there are many longstanding challenges to Jakarta's sustainable development that remain unresolved, despite recent changes to the regimes and elites that have shaped the city's growth. Appropriate leadership and new governance arrangements remain key requirements for formulating and implementing plans for Jakarta's future.

Challenges – Past, Present and Future

The problems facing the Jakarta city-region have been documented over the past two decades or so and there has been a rich set of critical contributions to the literature since the early 1990s by, for example, Douglass (1991, 2010), Firman (1992, 1996, 2004), Dharmapatni and Firman (1995), Forbes (1996), and Silver (2011). A number of challenges that Jakarta has struggled with for decades, including population pressure, traffic congestion and flooding, have been well described but they remain daunting and, to date, there have been no comprehensive measures taken to resolve them. More recently, Jakarta was badly affected by the regional financial crisis that hit countries in Asia in 1997 and the city is also becoming increasingly aware of its vulnerability to the consequences of climate change and global warming. The following sections discuss these challenges in turn.

Population Pressure

The root cause of land-use change and congestion problems in Jakarta is population pressure. This issue was identified in Governor Ali Sadikin's era (1966–1977) and led in the early 1970s to the declaration of Jakarta as a city 'closed' to all except those who had residence permits. However, this policy could not be enforced and was ultimately ineffective (Firman, 1999a). Jakarta acts as a magnet for migrants from all over Indonesia and, from the time of Independence, the population growth of Jakarta has always been the highest in the country, due mainly to in-migration. Figure 10.2 below shows the population of Jakarta and its surrounding cities from the time before Independence until the latest intra-census population survey in 2005. It can be seen that the population increased from around 600,000 in 1930 to 3 million in 1961, with much of the additional population coming from elsewhere in the country after the Independence War ended in 1949. The population then more than doubled to 6.5 million over the next 20 years, with economic opportunities as the main reason for the influx of people from other regions during this period.

Megaprojects continue to be pursued in the city core (including the much-delayed Jakarta Tower which, if eventually completed, will be considerably taller than the Petronas Towers in Kuala Lumpur – see Chapter 9 in this book). The residential population of parts of Central and South Jakarta has been declining for some time,

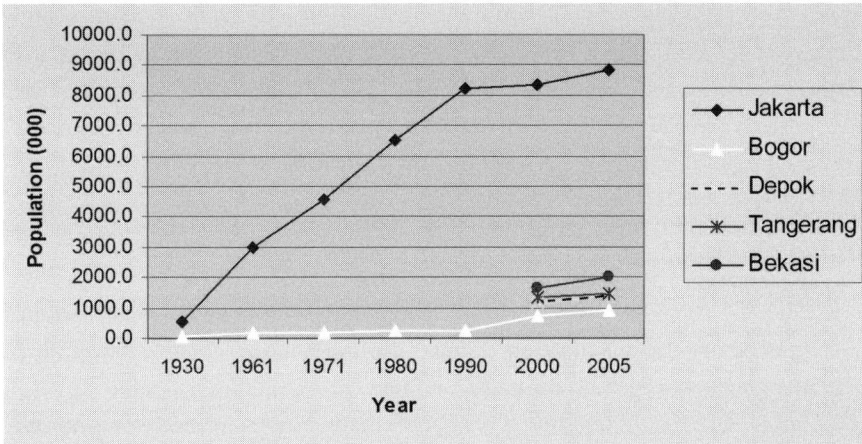

Figure 10.2. Population of Jakarta and surrounding cities, 1930–2005. (*Source*: Various BPS data)

however, and it is estimated that, between 1995 and 2000, 111,000 people moved out each year from Jakarta to neighbouring cities – Tangerang, Bekasi and Depok and also Bogor, Serpong and Karawang (a district to the east of Bekasi). The recent pattern of population growth has continued to be one of a declining population in the core of the city and increasing population in the surrounding areas (Jones, 2002; Mamas and Komalasari, 2008). As can be seen from figure 10.2, the three new municipalities of Tangerang, Bekasi and Depok each had a population of more than a million people by the year 2000. The population of each of these three cities is now higher than that of the much older city of Bogor, 60 kilometres to the south of Jakarta.

These movements of people have created a mega urban region consisting of large urban centres within a mosaic of rural areas in the districts surrounding Jakarta. This region had a population of more than 23 million in 2005 (see table 10.1) and it is probably as high as 25 million today. Many people who have moved out from

Table 10.1. Population of the Jakarta city-region in 2005.

	Urban Population	Rural Population	Total Population
Municipality			
DKI Jakarta	8,839,247	0	8,839,247
Bogor	891,467	0	891,467
Depok	1,339,263	35,640	1,374,903
Tangerang	1,451,595	0	1,451,595
District			
Bekasi	1,940,308	53,170	1,993,478
Bogor	2,180,910	1,648,143	3,829,053
Tangerang	2,292,672	966,391	3,259,063
Bekasi	1,272,550	711,265	1,983,815
Total	20,208,012	3,414,609	23,622,621

Source: BPS, 2007.

Jakarta to surrounding areas still commute back into the city centre to work, leading to heavy traffic and daily congestion on the toll roads which connect Jakarta with Bogor, Depok, Serpong, Tangerang and Bekasi. Without an effective comprehensive plan to deconcentrate employment from Jakarta to other cities, the environmental consequences of this traffic congestion are only likely to get worse in the future.

Traffic Congestion

Jakarta's traffic congestion is now chronic and is estimated to give rise to an annual economic cost of 43 trillion Indonesian rupiah (Rp), an amount double the government of Jakarta's budget in 2007 (Kompas, 2007a). This chronic traffic congestion is a result of the poor level of public transport service provision and the rapidly increasing number of private vehicles on the roads. It is estimated that between 2002 and 2007 the number of private cars increased by 269 every day while motorcycle numbers increased by as many as 1,235 a day (figure 10.3 below). Table 10.2 shows that, by contrast, the length and size of roads in Jakarta decreased between 2002 and 2006. Taken together these

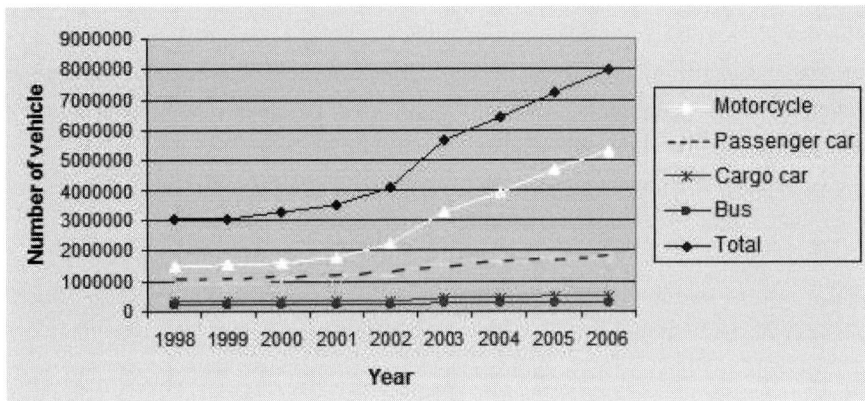

Figure 10.3. Growth of vehicle numbers in DKI Jakarta 1998–2006. (*Source*: BPS DKI Jakarta, 2003, 2007)

Table 10.2. Length, area and status of road by type in DKI Jakarta, 2003 and 2006.

Road Type	*Length* (km) 2003	2006	*Area* (ha) 2003	2006	*Status*
Toll	94.18	112.96	207.83	247.27	Enterprise
Primary arterial	102.14	112.15	214.00	214.01	National
Primary collector	55.13	51.63	86.07	67.14	National
Secondary arterial	514.01	502.64	855.38	829.91	Province
Secondary collector	966.60	823.91	822.94	697.09	Province
Local	5884.20	4936.93	2590.61	2098.81	Municipal
Total	7616.27	6540.22	4776.83	4154.23	

Source: BPS DKI Jakarta, 2003, 2007.

figures suggest that, without substantial investment in new transport infrastructure, Jakarta is approaching gridlock.

The problem of traffic congestion in Jakarta nowadays has its roots in the lack of urgency on the part of decision-makers to implement the visionary ideas found in the Master Plan 1965–1985 (Pemerintah DKI Djakarta, 1977). This Master Plan projected that the number of peak-hour passengers carried by city transportation would increase from 1.25 million in 1965 to 3.2 million in 1985. It therefore recommended the construction of a new rail-based rapid transit system (including a subway). It was proposed to build the new system by 1980, by which time an equal number of passengers were to be carried by bus and train, and by 1985 it was anticipated that rail passengers would be in the majority. However, this rapid transit system was not built.

The General Spatial Plan 1985–2005 (Pemerintah DKI Jakarta, 1987), which officially replaced the Master Plan 1965–1985 in 1987, proposed a rather more modest extension to the bus system by adding double-decker buses on a dedicated lane along main roads. There were no proposals for rail-based mass transit. The subsequent Spatial Plan for Jakarta 2010 made some vague references to a 'Mass Public Transit System' without specifying the proposed passenger shares of different modes of transportation or the date when this proposed system would come into operation.

There was a further study of a possible mass transit system to relieve Jakarta's congestion in the mid-1990s, resulting in a proposal to build a subway system connecting South Jakarta with the main business districts in Central Jakarta. This involved a 15 kilometre underground railway running beneath the main boulevards of Jalan Sudirman and Jalan Thamrin. However, the monetary crisis that hit Asia in 1997 halted this multi-million dollar project. The development of an inner ring of tollways has not been effective as an alternative. These also have become heavily congested during peak hours.

One innovative public transport development that has been introduced is a busway system which commenced operation in 2004. Based on the successful 'TransMilenio' system in Bogotá, Colombia, this makes use of a designated bus lane along major arterial roads. This is discussed further later in this chapter.

Environmental Issues

Douglass (1991) and Hadiwinoto and Leitmann (1994) identified serious environmental problems in Jakarta and its surrounding areas about 20 years ago, related especially to the impacts of development on the water cycle and on water and air pollution generally. Flooding is a problem that has existed since the Dutch colonial era. Batavia was situated in a low-lying area, crossed by major rivers and their tributaries, and the city was always flood-prone, with the first major flood recorded in 1670 (Caljouw et al., 2005). Flooding also leads to widespread disease, and impacts severely in other ways on the welfare of poor people who are usually worst hit by floods because they tend to live in the most vulnerable areas. Over the last 10 years the frequency of flooding has become progressively higher. Levels of inundation formerly associated with a 5 or 10 year flood have become annual events since 2002. The worst flood on record in Jakarta occurred in early February 2007, affecting 60 per cent of the Jakarta city area. Hundreds of thousands were forced to leave their homes, thousands of houses were destroyed and forty-eight lives were lost (Nurbianto, 2007).

A prime cause of flooding in Jakarta was identified decades ago as the environmental degradation of the water catchment areas of the Ciliwung and Cisadane rivers and the condition of the rivers themselves. Puncak, a mountainous area to the south of Jakarta, has been a popular tourist area for locals for many years and it has been highly developed since the 1980s. As the land cover has changed from vegetation to buildings, erosion has increased in the upper catchments of Jakarta's principal rivers. The district government of Bogor has been blamed for issuing an excessive number of development permits, allowing vegetated areas to be built over, but it is the case that many of these permit requests came from residents of Jakarta, as well as from government agencies and private developers from the city. Thus, the environmental degradation in Puncak, which gives rise, in part, to the flooding in Jakarta, is caused substantially by the behaviour of the people of Jakarta themselves.

There are actually other, more direct causes of severe flooding in Jakarta that relate to the condition of the rivers and to land-use changes within Jakarta itself. For example, in several places the Ciliwung river has become much narrower because of illegal structures that occupy its banks as well as an accumulation of rubbish dumped in the river. Land-use changes, from open space or parks to housing and commercial buildings, have also played a part in creating rainwater run-off which cannot be absorbed by the soil. There appears to be an inadequate understanding of the importance to the urban ecosystem of maintaining open spaces and river channel ecosystems, not least on the part of the city administration which, in its zeal to promote development, tends to neglect the environmental consequences.

Climate Change

Indonesia has only recently begun to focus on the possible impacts of climate change. Increased variability of rainfall will affect many parts of Indonesia in the future and sea level rise will be a significant threat. Using IPCC modelling, it is predicted that sea level will rise between 0.7 to 0.8 centimetres per year, while the El Nino Southern Oscillation (ENSO) will occur more frequently, from once in 3 to 7 years to once every 2 years (Sofian, 2009). A further study suggests that many areas on the northern coast of Java and the eastern coast of Sumatera will be inundated by the year 2100. This is a very alarming forecast as 48 per cent of Java's 130 million people inhabit the northern coastal areas (Suroso and Sofian, 2009). With a projection of land subsidence of 3 centimetres per year (Hirose *et al.*, 2001),[6] a sea level rise of 1 centimetre per year and impacts of tidal waves up to 80 centimetres, seawater inundation could submerge areas of Jakarta up to 10 kilometres from the coast (Susandi, 2010) (see figure 10.4 below). This would include all parts of North Jakarta from Muara Angke to Grogol, including vital infrastructure such as the Tanjung Priok harbour and the toll road to the Soekarno-Hatta International Airport. Many parts of East and West Jakarta could also be inundated, as well as areas around Menteng, Setiabudi and Gambir in Central Jakarta. The flood of February 2007 (referred to earlier), which occurred during the rainy season, is indicative of how vulnerable the low-lying parts of Jakarta are. With approximately 40 per cent of its area below sea level and predictions of increased intensity of flooding during the rainy season, this vulnerability can only increase as a consequence of climate change.

Figure 10.4. Projected area subject to inundation by sea level rise in Jakarta in 2100. (*Source*: Adapted from Suroso and Sofian, 2009)

Economic Adjustments

The Asian monetary crisis hit several countries in 1997, but its impacts were probably more serious in Indonesia than anywhere else, with the value of the Indonesian Rupiah falling significantly. As the centre of the Indonesian economy, Jakarta was particularly affected. 12,000 out of 30,000 private firms, including small and medium-sized businesses, went bankrupt. Domestic investment decreased from Rp 14,395.5 billion in 1996 to Rp 8,553.5 billion in 1997 and the incidence of poverty recorded in the city increased significantly.[7] It is also estimated that income *per capita* decreased from Rp

7.4 million in 1997 to Rp 6.0 million in 1998 (Firman, 1999*a*). In the opinion of one of the authors of this paper, the crisis changed Jakarta from a 'global city' into a 'city of crisis' (Firman, 1999*b*). The adverse impacts were not only on the economy, but also political. Price increases for almost every item in the market, but especially basic food items such as rice, sugar and cooking oil, as well as fuel price increases, triggered social unrest in many cities in Indonesia, culminating in the demand for President Soeharto to go. After months of protests and riots in Jakarta, Soeharto stepped down on 14 May 1998 after three decades in power.

Thereafter, Indonesia embarked on a period of wide-ranging political reforms, which included extensive fiscal decentralization. These reforms appear to have assisted with economic recovery, especially in the regions (Salim and Kombaitan, 2009). The growth of gross regional domestic product (GRDP) occurred at a faster rate in several regional cities than it did in Jakarta after the implementation of new decentralization laws in 2001 (see table 10.3 below). Growth was also strong in the three new cities adjacent to Jakarta, with Depok showing a faster GRDP growth than both Jakarta and Surabaya in 2004. As table 10.3 indicates, Jakarta continues to dominate the national economy in absolute terms, with its GDRP in the early years of the twenty-first century being about five times that of the second largest city economy, that of Surabaya.

Table 10.3. Gross Regional Domestic Product of five largest Indonesian cities, 2001 and 2004.

Cities	GRDP (Rp Billion)		GRDP Growth (%)	
	2001	2004	2001	2004
Jakarta	236541	275937	4.32	5.75
Surabaya	48947	56020	4.25	5.78
Medan	19828	23623	4.66	7.29
Bandung	16079	19874	7.54	7.49
Semarang	13624	15509	3.40	4.76
Cities in Jakarta City-Region				
Tangerang	16762	19766	3.43	4.10
Bekasi	9531	11111	5.09	5.36
Depok	3694	4433	5.89	6.41

Source: BPS, 2006.

There are two factors that can help to explain why Jakarta was able to recover strongly from the Asian economic crisis. Figure 10.5 below details trends in investment in DKI Jakarta before and after the crisis. It shows that foreign direct investment has accelerated since 2003 and is likely to reach the same level as before 1996 in the next few years. However, the recovery of domestic investment after the crisis was not as strong, perhaps reflecting the effects of decentralization in stimulating domestic investment in the regions. Figure 10.6 shows transactions on the Jakarta Stock Exchange (JSX) between 1993 and 2006. These fell in 1998 and remained constant until 2002, but since 2003 there has been a strong increase in both the volume and value of stock trading. Thus, 5 years after the economic crisis and the implementation of new fiscal decentralization measures, Jakarta's economy was back to its pre-crisis state. There is little doubt that the increase in foreign direct investment and capital formation was influenced to a considerable extent by the political stability in the country which was eventually achieved following Soeharto's removal. The political reforms since 1999 seem to have been the main drivers of economic recovery.

The impact on the Indonesian economy of the global financial crisis, which started in the United States in 2008, appears to have been more limited so far than that of

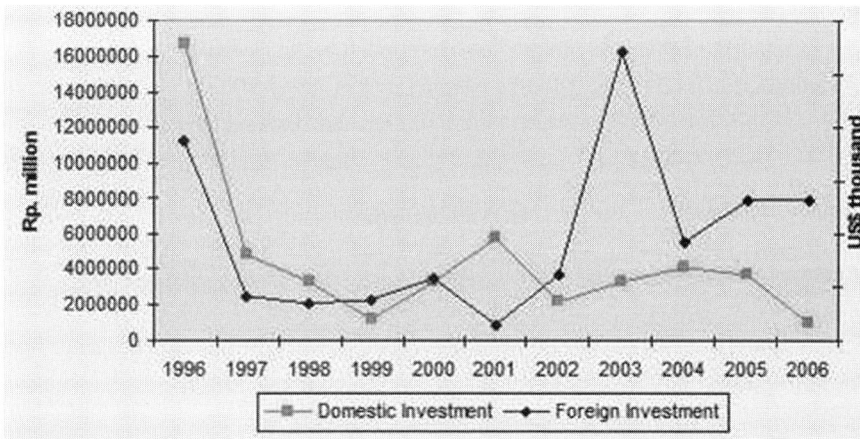

Figure 10.5. Investment trends in DKI Jakarta 1996–2006. (Source: BPS DKI Jakarta, 2003, 2007)

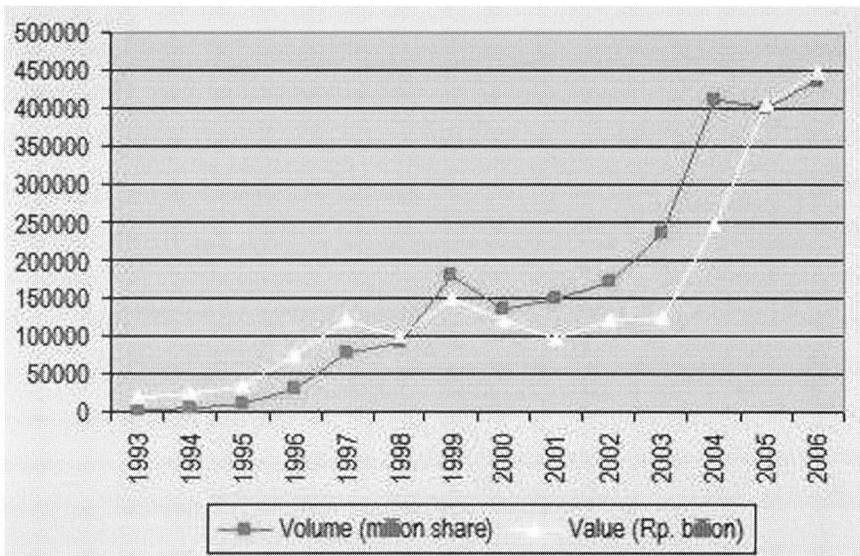

Figure 10.6. Stock trading on the Jakarta Stock Exchange 1993–2006. (Source: BPS DKI Jakarta, 2003, 2007)

the monetary crisis of the late 1990s. The latest economic update on Indonesia by the World Bank (2010, p. 4) suggests that 'Indonesian equities have been relatively resilient and the currency and the fixed income markets continue to perform strongly' despite global turbulence. The contribution from net exports has declined due to the slowing down of the global market, but buoyant domestic demand is expected to offset that decline. There has been some downturn in manufacturing, however, with industrial activities in Bekasi and Tangerang experiencing a fall in demand for their products.

National and City Strategies to Address Jakarta's Problems

The national and DKI governments have responded to the challenges faced by Jakarta in a number of ways. The measures adopted can be grouped into two categories, as either 'inward-looking' or 'outward-looking'. These are discussed below.

Inward-Looking, City-Based Strategies

Recent programmes or projects executed by DKI Jakarta and the government of Indonesia have sought to address Jakarta's flooding and traffic congestion problems. There has also been a continuing focus by the national government since the economic crisis in 1997 on poverty alleviation programmes throughout the country, including Jakarta. The following section discusses flood prevention programmes and, in particular, the East Flood Canal; a number of transport improvement projects, including the Jakarta Busway; and the nature of poverty alleviation issues in the city.

It has been acknowledged for many years that the colonial network of flood canals and drains is old and inadequate to cope with water levels during floods. The Jakarta General Spatial Plan 1985–2005 prioritized three areas for flood management as follows (Pemerintah DKI Jakarta, 1987): the construction of the Depok Dam to cope with a 100-year flood of the Ciliwung River; the completion of the Cengkareng Flood Canal to the west; and the construction of the East Flood Canal. Twenty years later, only the Cengkareng Flood Canal has been completed. The Depok Dam was not constructed and is unlikely to be. Following the relocation of the University of Indonesia to Depok in the late 1980s, urban development encroached rapidly into the area designated for the dam and, as indicated earlier, Depok has now become a city in its own right. Meanwhile, the construction of the 23 kilometre East Flood Canal did not commence until 2003, following a major 50-year flood in January 2002 which inundated about 16,000 hectares in total of Jakarta, including the area of the presidential palace in Merdeka Square (Firman *et al.*, 2011). The completion of this project has gone slowly, held up in part by land clearing issues in areas which have been occupied by squatter settlements for the past 20 years. Although national President Susilo Bambang Yudhoyono and the current Governor of Jakarta, Fauzi Bowo, have indicated that this canal will be completed in the near future, this seems unlikely. In general, the pace at which flood prevention measures are being implemented seems much too slow, given the high vulnerability of Jakarta to flood risks. Ali (2010) has identified a number of actions needed immediately to reduce this vulnerability and to safeguard Jakarta's water resources – rehabilitation of drainage systems; the building of new storage capacity to control stormwater run-off; stronger requirements for private firms to treat their wastewater; and the revitalization of rivers, ponds and dams, which provide domestic water supplies, so as to contribute to improved drinking water supplies and to reduce pressure on groundwater and the aquifers from bores.

In order to address traffic congestion in Jakarta, the national and provincial governments have implemented several recent measures, but with limited results. Inner and outer ring roads have been completed, increasing the length of toll roads between 2003 and 2006 by about 20 per cent, as indicated in table 10.2 above. There was also a slightly smaller increase in the length of primary arterial roads. However, as noted earlier, the total length of roads in Jakarta decreased overall in the same period as a consequence of the loss through development of secondary arterial, collector and local roads. It is clear that Jakarta is not progressing at present towards an adequate and integrated road hierarchy and network.

Within the central area a system of car-pooling has been introduced along Jalan Sudirman in peak hours. This so-called 'three-in-one' system limits private vehicle use of the road to vehicles with at least three passengers. The policy is considered to have had some effect in reducing traffic congestion during peak hours. Commuters who are not car-pooling either take different routes to go to their destination or time their journeys to take place before or after the operating hours of the 'three-in-one' restrictions. However, the effectiveness of the system has been undermined to some extent by the practice of paying 'three-in-one jockeys' – students or street children who wait at the roadside at the start of the restricted zone and who are paid a small amount to ride in cars which would otherwise have a single occupant.

The busway system, referred to earlier, was initiated by Governor Sutiyoso in 2004. This is the only mass transit system running in the city, but it has attracted a good deal of criticism since its inception. Of the fifteen corridors planned, eight are now operational (see figures 10.7 and 10.8). There is a single flat fare of Rp 3,500 (about 40 US cents). This is reduced to Rp 2,000 before seven o'clock in the morning or after ten at night. The busway system appears to be affordable for those who normally commute by other modes of public transportation, such as regular buses, micro buses or the public vans that are a very visible part of Jakarta's traffic. However, it carries a fairly small number of total daily trips at present – 210,00 passengers a day in a city where 7 million people make daily trips (Kompas, 2007b) – and a much more extensive modern public transport system is required to encourage a significant number of the commuters who currently use private vehicles to change their travel behaviour.

Figure 10.7. The TransJakarta Busway – convenient but with low ridership.

Figure 10.8. A busway station.

During and following the Asian economic crisis in the late 1990s, the Indonesian government implemented several emergency anti-poverty programmes for both rural and urban populations. Some of the programmes are place-based, others are people-based. The number of agencies and funding sources involved in poverty alleviation has posed issues of co-ordination during implementation. Firman (1999*a*) has argued that there was no shortage of funds for the social safety net programmes introduced in the Jakarta city-region and in other provinces in Indonesia during the economic crisis. What was lacking was capable and credible management to administer the programmes. For Jakarta, those programmes were at least successful in decreasing the numbers assessed as experiencing poverty from 861,000 in 1998 to 379,000 in 1999. 2008 BPS (Badan Pusat Statistik/Central Board of Statistics) data suggest that Jakarta now has about 400,000 people living below the poverty line (see Firman *et al.*, 2011), but the definition of poverty is set at a low level and it has to be acknowledged that there are many people who live in very poor circumstances in Jakarta – perhaps another 300,000 – even if they do not fall within the technical definition of poverty. Only 52.57 per cent of Jakarta's residents have piped-water (BPS, 2007). Moreover, the recorded unemployment level of 12.1 per cent in Jakarta was much higher than the national average of 7.8 per cent in 2009. The government of Jakarta seeks to make Jakarta a city 'for everyone',[8] but there remains a great challenge in extending the benefits of urban life to the city's lowest income groups. And, as indicated earlier, many of Jakarta's poorest residents live in the areas most susceptible to flooding.

Outward-Looking, Region-Based Strategies

The Master Plan 1965–1985 acknowledged that the problems of Jakarta in the future would require planning that was not limited by existing administrative boundaries and recommended that the 'planning region' for Jakarta should consist of Jakarta city itself and surrounding areas with a direct impact on Jakarta.[9] Since then, a number of region-based strategies have been pursued with the stated intention of protecting the environment and guiding the further development of the Jakarta city-region. One of the more important of these was the Jabotabek Metropolitan Planning Study which led to the Jabotabek Metropolitan Development Plan (JMDP) in 1983. This sought to guide development away from the environmentally sensitive zones of the uplands around Puncak and the northern coastal area. Five zones were identified, based on their environmental characteristics (see figure 10.9). Zone I, the northern coastal area, was described as a low-lying coastal zone best suited for agriculture and fishponds, while Zone V was a steep, mountainous zone with limited vegetation and rapid runoff where natural forests should be preserved and agriculture limited to areas where terraces could be maintained or constructed. Urban development was to be guided primarily to Zone III, higher land with good drainage, rising above the coastal plains and with reasonable soil-bearing capacity for buildings. This zone extended from east to west providing an axis along which decentralization was to be encouraged by the provision of infrastructure under a process termed 'Guided Land Development'.

The West Java Urban Development Project of 1985 proposed the extension of the principles of the JMDP to that portion of Jabotabek under the jurisdiction of the West

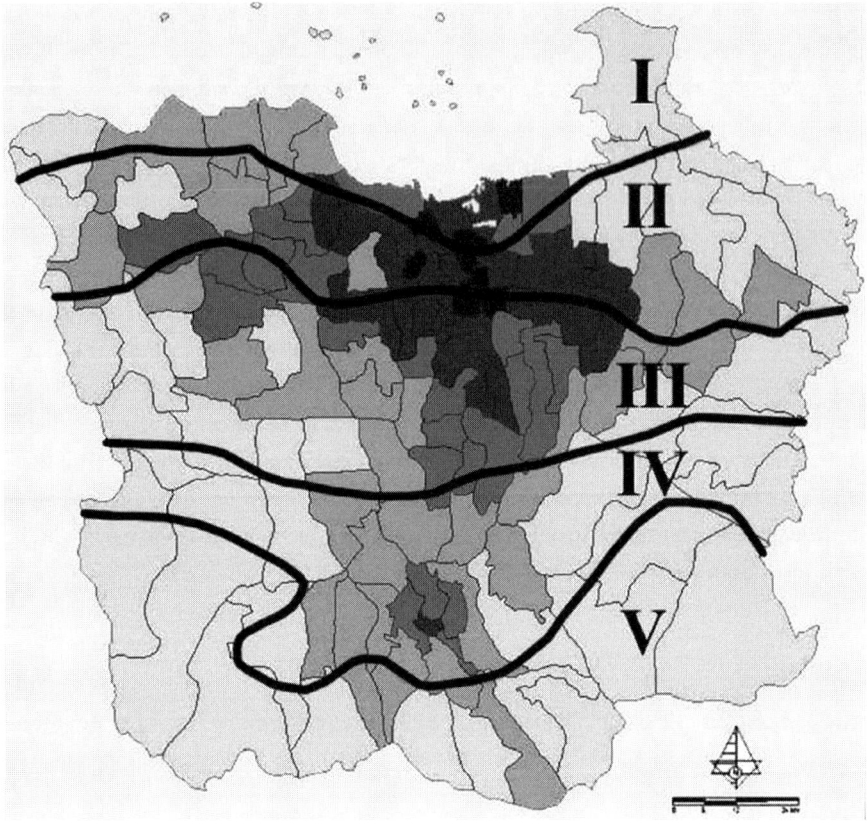

Figure 10.9. Environmental zones in the Jabotabek Metropolitan Development Plan 1983. (*Source*: Adapted from Douglass, 1991, 2010)

Java Province, which, at the time, comprised Bogor, Tangerang, Bekasi, and Puncak-Cianjur. Co-ordination between different levels of government was an important theme at this time and underpinned also the 'Integrated Urban Infrastructure Development Program' (IUIDP), initiated early in the 1980s to provide cross-sectoral city-based planning of land and infrastructure through a collaborative process involving national and city planning staff (Wegelin, 1995).

However, these initiatives were not effective in guiding Jakarta's urban development in the direction of a more environmentally sustainable urban form (Douglass, 1991 and 2010; Hasan, 2003), nor has the desire to achieve better horizontal or vertical co-ordination between government agencies at different levels borne much fruit. Stolte (1995) commented on some of the obvious reasons for this. For example, the JMDP had only advisory status (and the province of West Java did not adopt it); some of its more important recommendations were deferred because of conflicting interests; and enforcement of its policies was weak, with less than half of subsequent development permits issued in the plan area being in accordance with the plan. The overwhelming evidence was that market forces and private interests were able to determine the actual use of the land in most cases, and a plan unsupported by sound implementation measures and political status could do little to stop them.[10] While some development was 'guided' to locations in accordance with the east–west axis concept, there were few controls imposed on developers who wished to locate elsewhere. This was particularly the case in relation to a wave of new town and commercial developments that occurred in the late 1980s and throughout the 1990s including CiPutra's new city of Bumi Serpong Damai, a supposedly self-contained city with a target population initially of 600,000 (see Hasan and Hamnett, 2003); and the Jakarta Waterfront City project in 1996, a proposal to reclaim 28 square kilometres along the northern coast for residential and commercial development (Jellinek, 2000). Following the Jabotabek Metropolitan Development Plan Review (JMDPR) in 1993, a substantial southern growth axis was now added to the earlier east–west axis. Douglass (2010) notes that, by the late 1990s, official government reports were acknowledging that the idea of guiding development away from the environmentally fragile south was not practical, representing 'a broader acknowledgement that, in the new era of private–public partnerships and neoliberal deregulation, environmental sustainability by whatever definition is … becoming more difficult to pursue' (Douglass, 2010, p. 59).

In these latter years of the Soeharto regime, some retired officials at the national level, including former environment minister Emil Salim, and also former Jakarta Governor Ali Sadikin, proposed the development of a new planning region called the Jabodetabekpunjur metropolitan area (an acronym derived from *Ja*karta, *Bo*gor, *De*pok, *Ta*ngerang, *Be*kasi, *Pun*cak and Cian*jur*). This was to be a powerful unitary authority, headed by a cabinet minister. However, the proposal never gained much support from the highly centralized national government of Soeharto. It was considered politically inappropriate to have a minister with authority over a 'country within a country', with an area larger than Brunei and including the seat of the national government.

More recently, during the current reform era, this idea was raised again in 2005 by the then Governor of Jakarta, Sutiyoso, who proposed that the Jabodetabekpunjur metropolitan region be designated as a 'megalopolitan region' under a single development authority, overseen by the Governor of Jakarta, albeit without political authority over the neighbouring regions. This proposal again attracted opposition, especially from the West Java provincial government, which was concerned that its development priorities would be marginalized, and the proposal was finally turned down in its original form by the national legislature. Instead, however, Jabodetabekpunjur was designated as a 'national strategic zone' under Government Regulation No. 26 of 2008 (dealing with national spatial planning) and a spatial strategic plan was enacted under Presidential Regulation No. 54 of 2008. This regulation mentions that the co-ordination of development in the region is the responsibility of a minister but, unfortunately, it omitted to specify which minister has this authority. Law

No. 29 of 2007, dealing with the Governance of Jakarta as a capital city, has a slightly different emphasis, stipulating that all governments should co-operate by agreement under an inter-regional co-operating body. This suggests the reinvigoration of the Co-operation Agency for the Development of the Jakarta Metropolitan Area (BKSP), jointly established by DKI Jakarta and the province of West Java in 1975 and endorsed by ministerial decrees in 1980 and 1984.[11]

Law 29 of 2007 also requires co-operation in planning and the control of spatial development to be co-ordinated by the Ministry of Public Works, the National Development Planning Agency and the Ministry of Home Affairs (Firman, 2008). Thus, there seem to be at least two officially-mandated sets of arrangements for co-ordinating the development of the Jabodetabekpunjur metropolitan region and for ensuring collaboration between its constituent governments. In the light of greater Jakarta's planning history, these arrangements do not inspire enormous confidence. The essential problems are that the planning arrangements for the extended metropolitan region remain collaborative, with no agency or authority having clear powers and responsibility for implementation. This is a particular issue following the post-Soeharto reforms which gave more power to municipalities and districts than hitherto. While there were many good reasons to support the devolution of tasks and responsibilities to local governments, these reforms appear to have compounded the difficulty of agreeing on and implementing a clear strategic direction for the extended Jakarta region.

Conclusion

Jakarta's problems in specific areas like flooding and traffic congestion, as well as the broader challenges of long-term spatial development, have been responded to at various times by the national government, the city administration and the other governments of Jabodetabek, but there is not much to record in terms of significant achievements. The power of private interests in the development process was acknowledged earlier, but it is also manifestly the case that Jakarta is hamstrung by its governance arrangements – multi-level governments; overlapping laws and responsibilities across a very large city-region, for which there is still no clear or comprehensive plan to manage and guide population growth; and no strong instruments for plan implementation and the enforcement of policies. Business and location permits continue to be issued without reference to an overall plan based on the capacity of land to support development. Vehicle sales proceed apace, outstripping the development of traffic infrastructure. Overall, there remains a serious implementation gap between planning aspirations and outcomes.

Some of the examples described earlier are indicative of the governance problems of the Jakarta region. Thus, the serious issue of environmental degradation in Puncak has been recognized for decades. A Presidential Decree was issued in the early 1990s to stop development in that important water catchment area. However, the legislation and institutional arrangements proposed at that time have been ineffective and development in Puncak has continued. The history of urban development around Depok and the delays with the East Flood Canal also show how weak government planning arrangements are in practice. The failures in relation to water catchments in Puncak, the Depok Dam and the East Flood Canal mean that the low-lying city of Jakarta remains highly prone to flooding at a time when the incidence of flooding is becoming more serious and when climate change models suggest that it will only become worse.

Jakarta has been able to continue functioning so far due to the resilience that its citizens have displayed in coping with the decreasing quality of their environment and with economic crisis. However, it is not clear, in the face of the continuing weakness of governance structures and the record of serial and endemic failures to implement important projects, that the city-region can cope with the environmental and other challenges which clearly lie ahead. The leaders of Batavia and Jakarta demonstrated no real commitment in the past to the implementation of their plans and more recent discussions over appropriate governance arrangements for the Jakarta city-region have led to little in practice. The emphasis seems still to lie on posturing by elites and politicians without any actual strategic interventions of significance which might address the chronic problems that the city clearly has.

Thus, a key step towards a more resilient Jakarta is to find an alternative concept for its governance. This should start from an understanding of its characteristics and its network of problems as the basis for identifying strategic measures to deal with growth challenges. This needs to be followed by a fundamental reform that changes the relationship between city administration and residents and between governments and civil society. In a recent publication, Rakodi and Firman (2009) summarized the challenge here as follows:

Planning and management of the metropolitan area has been fragmented and ineffective; developer interests have been prioritized over planning policies, public priorities and the needs of low-income people; development control has been limited and inconsistent; and property rights are weakly defined. It is estimated that only one third of the land is fully titled, one quarter has no official title and the remainder is subject to intermediate forms of title – rights to build or use. Despite attempts to decentralize responsibility for local development and land management, the metropolis still has poorly co-ordinated government, a lack of capacity to implement plan proposals, unsynchronized planning and land laws, inadequate land administration and a dysfunctional development permit system. The 'privatized planning' of new towns, gated communities and shopping malls linked by toll roads continues to provide middle- and upper-income households with protected lifestyles, while most low-income residents have little choice but to seek accommodation in existing or new informal settlements.

In order to address some of these challenges, these authors put forward several proposals for reforming the governance of the Jakarta city-region as follows:

1.   Strengthening the long-standing Co-operation Agency for the Development of the Jakarta Metropolitan Area (BKSP) to enable it to take responsibility for metropolitan governance, with real authority for planning and for the co-ordinated development of major physical infrastructure.

2.   Land rights registration – a quarter of all landowners currently have no officially recorded right of tenure, which makes them vulnerable to land acquisition by developers.

3.   A review of the processes of issuing location or development permits in order to protect community rights to the land and to ensure that permits include provision for land for public infrastructure and open space.

4.   A more strategic, transparent and participatory approach to local planning to increase public awareness.

5.   The use of property taxes as tools for implementing spatial plans and not just as a means of raising revenue.[12]

The most suitable governance model for the Jakarta city-region is perhaps a mixed system that enhances the authority of BKSP, while being supported by legislation based on the principle of subsidiarity. At present BKSP does not have authority for the implementation of any development programmes in the extended Jakarta Metropolitan Area. That lies with each provincial and lower level government. Given the complexity of Jakarta's planning and governance challenges, a first step would be to give BKSP stronger powers in relation to transportation, spatial development, watershed management and solid waste management with the co-ordination of each project led by the relevant national ministry. The provision of finance for development can be shared by all levels of government (Firman, 2008), although a strengthened role for the central government is important because of Jakarta's status as national capital and also because of the huge financial resources required to meet Jakarta's backlog of infrastructure needs.

Jakarta faces great risks to its economic, environmental and social sustainability, some of which relate to global forces beyond the control of its citizens and governments. However, improvements to governance and planning of the sort sketched above have the potential to increase Jakarta's resilience to the major future challenges which it undoubtedly faces. Without them, the prospects are bleak.

Notes

1. The acronym Jabodetabek was first used by BKSP Jakarta (the Jabotabek Development Cooperation Agency) in early 2000 after the establishment of Depok as a municipality.
2. Calculated from the Inter-census Population Survey 2005 (BPS, 2007).
3. *Daerah Khusus Ibukota Jakarta*, usually abbreviated to DKI Jakarta.
4. 'United East India Company'.
5. A formal commitment to a more enlightened interventionist approach on behalf of the indigenous population expressed, amongst other things, in infrastructure which also gave more responsibility to local authorities.
6. Compared to the previously projected 6 cm per year as reported in Hadiwinoto and Leitmann (1994).
7. Central Bureau of Statistics (BPS, Jakarta Office; see also Firman, 1999*a*).
8. The motto of current Governor Fauzi Bowo during the election campaign in 2007.
9. It also recommended an 'administrative region' of a size designated as *Gewest Batavia en Ommelanden* by the Dutch Colonial Government in 1949 (Pemerintah DKI Jakarta, 1977), larger than the current area of DKI Jakarta but smaller than the Jabodetabek region.
10. In the face of continuing pressure of development in the area, the JMDP Review Study was undertaken in the early 1990s and came out with three conceptual alternatives to guide the development of the Jakarta Metropolitan Region: the new towns concept, the 'five fingers' concept, and continued linear development along the east-west axis (the preferred model) (Stolte, 1995). However, the effect of this review was to weaken the emphasis on an east-west axis and to allow development to the south (and, in reality, in all directions).
11. This institution is now headed jointly by the governors of West Java, Banten and Jakarta, with day-to-day operations managed by an executive secretary appointed by each of the three provinces for 5 years in rotation. BKSP membership consists of governors from the provincial administrations, regents (*bupati*) and mayors from authorities within the JMA. The BKSP has now become a forum for coordination among provincial and regional administrations in the JMA.
12. These points are also discussed in detail in Firman (2008).

References

Ali, F. (2010) Jakarta will (not) submerge? *The Jakarta Post*, 9 October.

BPS (2006) *Produk Domestik Regional Bruto Kabupaten/Kota di Indonesia 2001–2005*. Jakarta: Badan Pusat Statistik.

BPS (2007) *Survei Penduduk Antar Sensus 2005*. Jakarta: Badan Pusat Statistik.

BPS DKI Jakarta (2003) *Jakarta in Figures 2003*. Jakarta: Badan Pusat Statistik Provinsi DKI Jakarta.

BPS DKI Jakarta (2007) *Jakarta in Figures 2007*. Jakarta: Badan Pusat Statistik Provinsi DKI Jakarta.

Caljouw, M., Nas, P.J.M. and Pratiwo (2005) Flooding in Jakarta: towards a blue city with improved water management. *Bijdragen tot de Taal-, Land- en Volkenkunde* (Journal of the Humanities and Social Sciences of South-East Asia and Oceania), **161**(4), pp. 454–484.

Dharmapatni, I.A.I. and Firman, T. (1995) Problems and challenges of mega-urban regions in Indonesia: the case of Jabotabek and the Bandung Metropolitan Area, in McGee, T.G. and Robinson, I.M. (eds.) *The Mega-Urban Regions of Southeast Asia*. Vancouver: UBC Press.

Douglass, M. (1991) Planning for environmental sustainability in the extended Jakarta Metropolitan Region, in Ginsburg, N., Koppel, B. and McGee, T.G. (eds.) *The Extended Metropolis: Settlement Transition in Asia*. Honolulu: University of Hawaii Press.

Douglass, M. (2010) Globalization, mega-projects and the environment: urban form and water in Jakarta. *Environment and Urbanization Asia*, **1**(1), pp. 45–65.

Firman, T. (1992) The spatial pattern of urban population growth in Java, 1980–1990. *Bulletin of Indonesian Economic Studies*, **28**(2), pp. 95–109.

Firman, T. (1996) Urban development in Bandung metropolitan region: a transformation to a desa-kota region. *Third World Planning Review*, **18**(1), pp. 1–22.

Firman, T. (1999*a*) A great 'urban crisis' in Southeast Asia. *Cities*, **16**(2), pp. 69–82.

Firman, T. (1999*b*) From 'global city' to 'city of crisis': Jakarta Metropolitan Region under economic turmoil. *Habitat International*, **23**(4), pp. 447–466.

Firman, T. (2004) Demographic and spatial patterns of Indonesia's recent urbanization. *Population, Space and Place*, **10**(6), pp. 421–434.

Firman, T. (2008) In search of a governance institution model for Jakarta Metropolitan Area (JMA) under Indonesia's new decentralisation policy: old problems, new challenges. *Public Administration and Development*, **28**, pp. 1–11.

Firman, T., Subakti, I., Idroes, I. and Simarmata, H. (2011) Potential climate-change related vulnerabilities in Jakarta. *Habitat International*, **35**(2), pp. 372–378.

Forbes, D. (1996) *Asian Metropolis: Urbanization and the Southeast Asian City*. Melbourne: Oxford University Press.

Hadiwinoto, S. and Leitmann, J. (1994) Jakarta. *Cities*, **11**(3), pp. 153–157.

Hasan, M. (2003) Sustainable Development in a Metropolitan Region in a Developing Country: A Case Study of the New Town of Bumi Serpong Damai (BSD), Greater Jakarta, Indonesia. PhD dissertation, University of South Australia.

Hasan, M. and Hamnett, S. (2003) The Relevance of Western Notions of Sustainability to Developing Countries. Paper presented to the 7th Congress of the Asian Planning Schools Association. Hanoi: Hanoi Architectural University, 12–14 September.

Hirose, K., Maruyama, Y., Murdohardono, D., Effendi, A., and Abidin, H. Z. (2001) Land subsidence detection using JERS-1 SAR Interferometry. Paper presented to the 22nd Asian Conference on Remote Sensing, Singapore, 5–9 November.

Jellinek, L. (2000) Jakarta: kampong or consumer culture, in Low, N., Gleeson, B., Elander, I. and Lidskog, R. (eds.) *Consuming Cities*. London: Routledge.

Jones G.W. (2002) Southeast Asian urbanization and the growth of mega-urban regions. *Journal of Population Research*, **19**(2), pp. 119–136.

Kompas (2007*a*) Rugi Akibat Macet Capai Rp. 43 Triliun. *Kompas Newspaper*, 6 November.

Kompas (2007*b*) Penanganan Jakarta Masih Parsial. *Kompas Newspaper*, 5 November.

Laquian, A. (2005) *Beyond Metropolis: The Planning and Governance of Asia's Mega-Urban Regions*. Washington, DC: Woodrow Wilson Center Press.

Mamas, S.G.M. and Komalasari, R. (2008) Jakarta: dynamics of change and livability, in Jones, G. and Douglass, M. (eds.) *The Rise of Mega-Urban Regions in Pacific Asia: Urban Dynamics in the Global Era*. Singapore: Singapore University Press.

Nurbianto, B. (2007) Floods biggest challenge for Fauzi. *The Jakarta Post*, 15 December.

Pemerintah DKI Jakarta (1977) *Rencana Induk Jakarta 1965–1985* (2nd print). Jakarta: Pemerintah Daerah Khusus Ibukota Jakarta.

Pemerintah DKI Jakarta (1987) *Rencana Umum Tata Ruang Jakarta 2005*. Jakarta: Pemerintah Daerah Khusus Ibukota Jakarta.

Pratiwo and Nas, P.J.M. (2005) Jakarta: conflicting directions, in Nas, P.J.M. (ed.) *Directors of Urban Change in Asia*. London: Routledge.

Rakodi, C. and Firman, T. (2009). Planning for an Extended Metropolitan Region in Asia: Jakarta, Indonesia, Case study prepared for Revisiting Urban Planning: Global Report on Human Settlements 2009. Available at http://www.unhabitat.org/grhs/2009. Accessed 22 November 2010

Rustiadi, E. (2007) Spatial Analysis of Development Problems in Jakarta Metropolitan Area. Presentation to Ministry of Public Works. Jakarta, 16 March.

Salim, W. and Kombaitan, B. (2009) Jakarta: the rise and challenge of a capital. *City*, **13**(1), pp. 120–128.

Silver, C. (2011) *Planning the Megacity: Jakarta in the Twentieth Century*. London: Routledge.

Sofian, I. (2009) *Kajian Dasar Akademis* (Scientific Basis)*: Analisis dan Proyeksi Perubahan Iklim di Indonesia* (in Bahasa Indonesia). Jakarta: Bappenas and GTZ.

Stolte, W. (1995) From Jabotabek to Pantura, in Nas, P.M. (ed.) *Issues in Urban Development: Case Studies from Indonesia*. Leiden: Research School CNWS (Centre for Non-Western Studies).

Suroso, D.S.A. and Sofian, I. (2009) Vulnerability of the Northern Coast of Java, Indonesia to Climate Change and The Need of Planning Response. Paper presented to the International Conference on Positioning Planning in the Global Crises, 12–13 November. Bandung: School of Architecture, Planning and Policy Development, Institut Teknologi Bandung.

Susandi, A. (2009) Integration of Adaptive Planning Across Economic Sectors. Paper presented at

NWP Workshop on Technical Approaches to Adaptation Planning, 12–14 October, Bangkok.

Wegelin, E. (1995) IUIDP in a comparative international context, in Suselo, H., Taylor, J.L. and Wegelin, E. (eds.) *Indonesia's Urban Infrastructure Development Experience: Critical Lessons of Good Practice*. Nairobi: United Nations Centre for Human Settlements (HABITAT).

World Bank (2010) *Indonesia Economic Quarterly: Building Momentum*. Washington DC: The World Bank.

Yusuf, A.A. and Francisco, H. (2009) *Climate Change Vulnerability Mapping for Southeast Asia*. Singapore: Economy and Environment Program for Southeast Asia (EEPSEA).

Chapter Eleven

Bangkok:
New Risks, Old Resilience

Douglas Webster and Chuthatip Maneepong[1]

Bangkok is a large, resilient, middle-income Southeast Asian metropolis of 10.3 million people (2010). This figure refers to the core Bangkok Metropolitan Administration (BMA) area and is made up of 5.7 million registered inhabitants and an estimated 4.6 million unregistered migrants. The capital of Thailand (known as Siam until 1939) has been at its present location for about 230 years. Over this period of time it has repeatedly demonstrated the ability to recover quickly from shocks and disturbances.

This chapter focuses on the three leading types of risk affecting contemporary Bangkok – economic, political and 'natural' risk (see figure 11.1). These interact, creating new, difficult to forecast, risk profiles. In relation to each type of risk, a key theme is examined. These are, respectively, risks associated with the rise of Thailand's amenity economy; risks associated with the recent political and social conflict played out on the streets of Bangkok, given its role as capital city; and risks of climate change, given Bangkok's site as an estuary city, near or below sea level.

As with any large metropolitan area, the nature and severity of risks that Bangkok faces continue to change, although some elements have persisted over long periods of time. In the eighteenth century Bangkok faced military risk, primarily from Burma, forcing the capital to move from Ayutthaya to its present location. In the nineteenth century, the city faced risks associated with the threat of colonization by European powers. In the twentieth century, risks were related to the political unrest associated with the demise of the absolute monarchy in the 1930s; the nearby Indo-China wars in the late 1960s and early 1970s; and the impacts of industrialization (large-scale industrial pollution, urban congestion) in the post-war period. Bangkok was also at the heart of the Asian financial crisis in 1997. At the same time, over the last several

Climate Change

• Vulnerability /
Resilience
• Interacting
Risk Factors

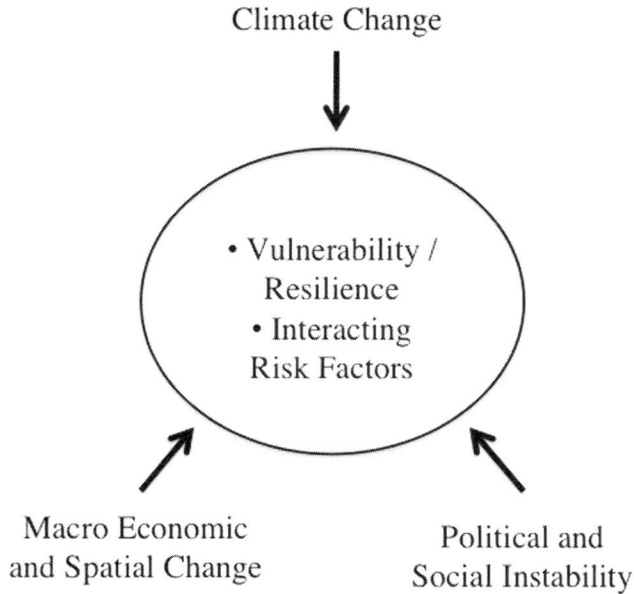

Figure 11.1. Key risk
factors for Bangkok.

Macro Economic
and Spatial Change

Political and
Social Instability

centuries, Bangkok has been blessed with countervailing positives – in particular, food surpluses from its rich agricultural hinterland; relatively low natural hazard risks (Revkin, 2010); high levels of social-cultural coherence and social capital; a stable bureaucracy; and no colonial history.

This chapter attempts not only to assess the nature of contemporary risks facing Bangkok, and the city's resilience to these, but also to emphasize how different types of risk interact. For example, political street conflicts obviously impact negatively on the increasingly important amenity economy – tourism is Thailand's largest source of foreign currency. On the other hand, the rise of post-industrial services has strengthened centripetal spatial forces, driving employment growth and residential densification in the core city, in the form of condominiums. These central areas are much less at risk from sea level rise associated with climate change than suburban areas because of the massive investment that has occurred in flood protection infrastructure.[2]

No one would argue that Bangkok is dull. It is consistently ranked in the top five cities globally as a preferred place to visit by trend-setting publications, such as *Conde Nast* and *Travel and Leisure* In fact, the latter publication ranked Bangkok as the most attractive city in the world to visit in 2010 (albeit on the basis of polling carried out before the street violence in April and May of that year). An open city, it is one of the most cosmopolitan in the world. This is manifest in large numbers of global amenity migrants,[3] foreign investors, international business managers and technicians, tourists, aviation crews, diplomats and international organization personnel. At the other end of the socio-economic scale, but just as critical to the functioning of its globalizing economy, the greater Bangkok region is home to approximately a million undocumented migrants, including refugees from Burma (Myanmar), Cambodia and other neighbouring countries.

The city's rapid emergence as an outward-looking metropolis has been generally welcomed by Bangkok's population. However, the rise of cosmopolitan Bangkok is not necessarily viewed so favourably by residents of the poor northeast of Thailand (known colloquially as *Isan*) and other peripheral regions. The recent street conflicts in Bangkok, which attracted international attention, were between the rural-based 'red shirt' supporters of former Prime Minister Thaksin Shinawatra and the 'yellow shirt' supporters of the current, mainly urban-based, Democrat government.

Sixty years ago, Bangkok existed for its hinterland, exporting rice and teak and supplying other parts of Thailand's hinterland with basic products, either manufactured in the city or imported through its port on the Chao Praya river. Now Bangkok's *raison d'être* is to serve the globe as a major tourist attraction (some 15 million tourists in 2010); a centre for aviation (the sixteenth ranked airport in the world by passenger numbers); an industrial hub (the centre of Southeast Asia's vehicle industry), and a major centre for international organizations in Asia (including the United Nations Economic and Social Commission for Asia and the Pacific).

The BMA forms the inner part of the Bangkok Metropolitan Region (BMR) which

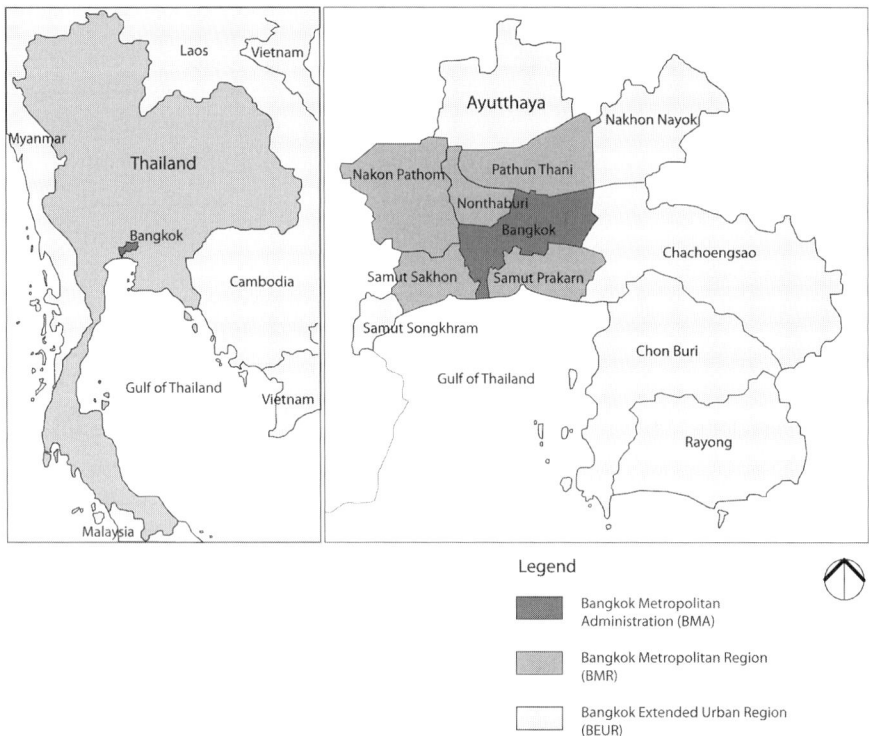

Figure 11.2. Thailand and Bangkok's Extended Urban Region.

has at least 16 million residents. The BMR was established in 1980 by incorporating the urbanizing areas around Bangkok in the provinces of Nakhon Pathom, Nonthaburi, Pathum Thani, Samut Prakan and Samut Sakhon. The continuing growth of Bangkok's periphery has led to the identification of an even larger Bangkok Extended Urban Region (BEUR) which has more than 22 million residents (see figure 11.2).

Bangkok has a large and influential middle class and the BMA had a per capita gross regional product (GRP) of US$22,000 in 2008 (National Economic and Social Development Board, 2008a). Household income overall is fairly modest, however, at US$14,000 per annum (National Statistical Office, 2009). The population of the BMA is growing slowly, at less than 1 per cent per annum between 2001 and 2008, but the suburban and peri-urban areas outside the BMA that constitute the balance of the BMR grew at the rather faster rate of 2.24 per cent over the same period (National Economic and Social Development Board, 2008b).

Bangkok's Changing Role and Form

Historical Context

Bangkok has been the capital of Siam and then Thailand since 1767 when the seat of Royal power was shifted from Ayutthaya, which had served as the national capital for over 400 years. Ayutthaya had been sacked by the Burmese and the new site was thought easier to defend. The new capital was originally situated in Thonburi (see figure 11.3),[4] but in 1782 it was moved across the Chao Praya river by King Rama I to its present site (Smithies, 1986). The Chao Praya river is the key physical feature of Bangkok's flat, near sea level, site on Thailand's central plain.

Thailand has been an open trading nation for much of its history. As early as the fourteenth century China, and states occupying present-day Japan and Vietnam, had significant trading relationships with Ayutthaya. By the seventeenth and eighteenth centuries many Asian and Western countries had embassies at Ayutthaya to represent their commercial interests.

The development of nineteenth and early twentieth century Bangkok was based on outward expansion from three nodes: the Royal/Official city, Chinatown and the International Quarter (see figure 11.3). Bangkok was established as the seat of power of the monarchy, based in the Royal Palace and its surrounding government buildings. As an absolute monarchy, administration of the kingdom from this complex was highly centralized. The official area, known as Rattanakosin, was carefully planned in terms of street layout, parks and monuments. It is currently protected by height restrictions and still contains a large number of inspiring religious, government and royal buildings.

Chinatown sprang up along the river downstream from Rattanakosin. It was populated by large-scale immigration from China over approximately 100 years,

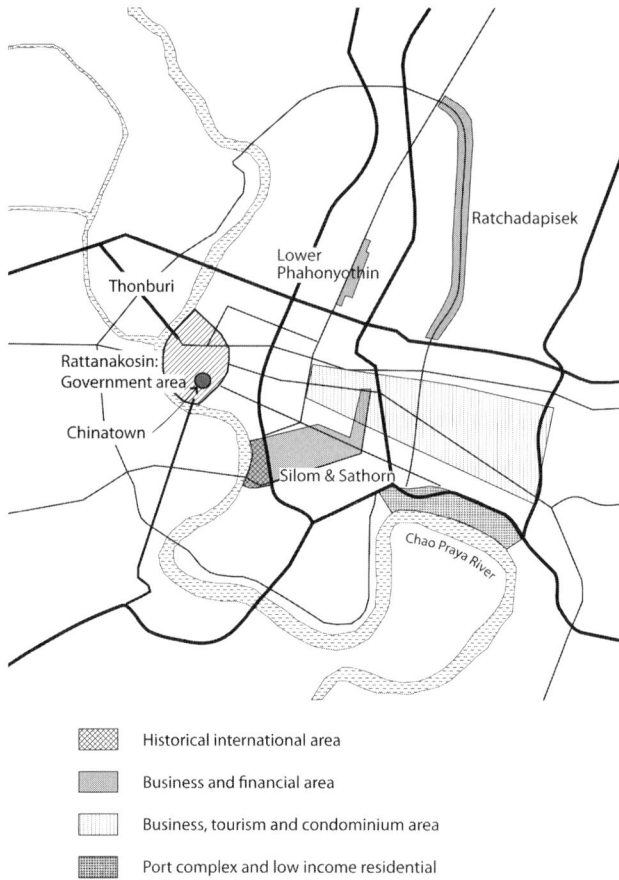

Historical international area

Business and financial area

Business, tourism and condominium area

Port complex and low income residential

Figure 11.3. Bangkok's historical evolution: specialized nodes and corridors.

starting in the mid-1800s (Phongpaichit and Baker, 1995). A large percentage of Bangkok's population can trace its roots back to this migration. Chinatown provided goods and services to the official city and to the greater urban area. It also played an important role in the import and export of goods to and from Bangkok's hinterland and, in particular, the distribution of rice from the agriculturally fertile central plain.

The third early development node was the International Quarter, further downstream again. The International Quarter resulted from the re-opening of Siam to foreign trade in 1855 under the terms of the Bowring Treaty.[5] International trade, particularly with European powers, was conducted in the International Quarter and Bangkok's first road – New Road, now named Charoen Krung – was built here, next to the river. Prior to this, all travel within Bangkok was along the canals or *klongs*. Although Thailand was never colonized, Britain and other Western powers had special rights in the International Quarter, including courts to try their own citizens. There are some parallels with the treaty port cities in China, established in the nineteenth century.

Late Twentieth Century Bangkok

At the end of World War II, Bangkok was a relatively small city, containing no more than a million people. It was the administrative and political capital, and the gateway to a rich agricultural hinterland. However, the Indo-China wars in the 1960s and 1970s unleashed modernization forces that set the city on the path of motorization, a trajectory not checked until the 1990s when the initial sections of the mass rail transit system were opened. Bangkok had been known as the 'Venice of the East' because of its network of canals but, in the 1950s and 1960s, many were filled in to create an arterial road network and motor vehicles became the dominant mode of transportation. Motorization, in turn, drove substantial suburbanization from the 1960s onwards, while traffic jams in the core city became commonplace by the 1970s. In the 1980s, the first of Bangkok's expressways opened, speeding up the suburbanization process. Residential population declined in the older areas of Rattanakosin, Chinatown and the former International Quarter, as residents of these areas pursued more living space and 'modernity' in the suburbs. The expressways also acted as gateways to Bangkok's periphery, supporting rapid peri-urbanization, driven by Foreign Direct Investment (FDI) during the 'golden age of manufacturing' between 1984 and 1997. In the 1990s and early twenty-first century, these expressway systems were extended to the most important peri-urban areas, such as Chonburi on the eastern seaboard and Ayutthaya (the former capital), about 40 kilometres north of Bangkok.

Until 1999, Thailand continued to be highly centralized politically and administratively, with only 8 per cent of public expenditure controlled at the sub-national level. Even after the democratic reforms introduced in the 1997 constitution and the National Decentralization Act of 1999, Bangkok continues to be at the core of a highly-centralized system, with over 70 per cent of public expenditure controlled at the national level in 2010. Most public expenditure within BMA is the responsibility of the national government, either directly or through nationally-controlled state owned enterprises (SOEs).

Prior to 1972, the present area of the BMA had twelve local governments. In 1972, the military junta led by Thanom Kittikachorn abolished local self-government in the metropolitan area and established the BMA (see Ruland and Ladavalya, 1996). In 1974, the governor of the BMA became an elected official and, in 1985, the BMA Act was passed which significantly extended the powers of the BMA government. Since then, the BMA has acquired further powers, particularly related to physical development (see BMA, 2001).

Bangkok Post-1997

The twenty-first century began in Bangkok in 1997. The Asian financial crisis occurred that year and Bangkok was at the epicentre. Virtually overnight, the US dollar value of Bangkok's economy was cut approximately in half. Although Thailand's economy has since surpassed its earlier 1997 highs in real terms, it lost its innocence at that time. The crisis was a major shock to Thailand, and to Bangkok in particular, not just economically but also psychologically and socially. It led to changes to national politics and to the dynamics of growth, including (*i*) a rise in the importance of the amenity economy – tourism, conferences, lifestyle migrants, medical tourism and a creative economy based on elements such as cuisine, design and fashion; (*ii*) the rise of the core city, supported by the expansion of rail rapid transit[6] and contrasting with earlier, manufacturing-driven, peri-urban growth; and (*iii*) growing political tension, manifest in rural versus urban and class-based politics This tension reached its highest levels recently. Bangkok's airport was closed in November–December 2008 for nine days as a result of political action by the 'yellow shirts' and then, in April–May 2010, the 'red shirts' seized the city's commercial core. Widespread arson followed the military action which broke up the red shirts' encampment on 19 May.

Economic Risk to Amenity-Oriented Bangkok

The Rise of Thailand's Amenity Economy

Post-1997, there was a growing consensus among the ruling elites and technocrats that Thailand's comparative and competitive economic advantages now lay in amenity services including: (*i*) fashion and design (jewellery, garments, ceramics, advertising and media; (*ii*) human skills in the application of (high) technology (hospitals, spas); (*iii*) cuisine, including high-value processed foods and specialized restaurants; and (*iv*) high-quality natural and built environments, including retirement communities (domestic and international), beach tourism, second homes and up-market real estate, and other elements of footloose knowledge worker lifestyles. Such an economy places a high premium on the quality and accessibility of place. Part of the argument for amenity as the propulsive sector of the Thai economy is that, under such a scenario, Thai culture and hospitality become assets, increasing the value of goods and services.

The new amenity economy was based on four particular attributes: (*i*) fun, or *sanuk*; (*ii*) beautiful tropical beaches, a geographic feature in short supply in rapidly emerging China to the north (explaining why approximately a million Chinese tourists visit Thailand each year); (*iii*) a population that is socially skilful and excels at tasks involving human contact, such as medicine, spas and tourism; and (*iv*) an open culture which encourages high-quality, fusion products and services, combining elements of the rich local culture and foreign influences.

However, politicians and bureaucrats still understood the importance of the manufacturing economy, even if it was becoming less consistent with Thai cultural and employment preferences. They recognized that Thailand was becoming less competitive in low-end labour intensive products, but that competitive advantage still existed in the case of a few manufacturing clusters – particularly vehicle production (specializing in pickup trucks) and agri-processing, as well as consumer appliances and electronics. Competitive advantage in these economic clusters was largely based on relatively deep supply chains associated with Thailand's 'first mover' advantage in Southeast Asian industrialization. This dated back to the 1970s when import substitution rather than export was the prevailing economic doctrine. Industry has become increasingly concentrated in peri-urban Bangkok – especially the Ayutthaya area and the eastern seaboard (see figure 11.4). Earlier ambitious plans to create large industrial areas remote from Bangkok on the western and southern seaboards have been largely abandoned or downplayed by the national government. In fact, the previously proposed western seaboard industrial area has been renamed as the 'Royal Coast' and is being repositioned as an amenity region.

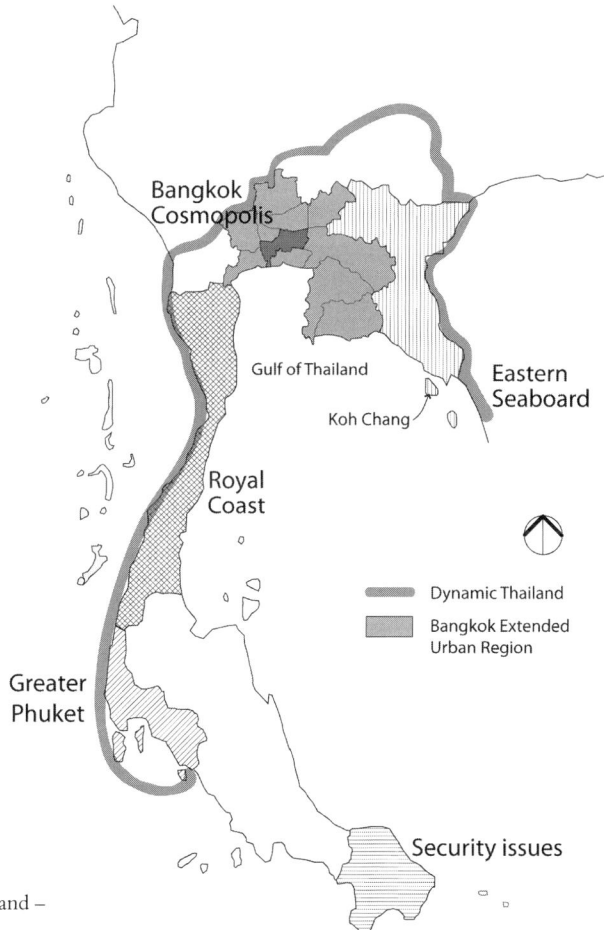

Figure 11.4. Dynamic Thailand – development clusters.

Spatial Implications of Economic Change

Rapid restructuring of Thailand's economy, as described above, is significantly reshaping the country's spatial economy, concentrating wealth and opportunity along the Gulf of Thailand (see figure 11.4). 'Dynamic Thailand' starts at the Cambodian border and hugs the Gulf of Thailand, crossing over to the Andaman Sea and ending in Greater Phuket, which includes Krabi and Phang Nga provinces. In addition to the large-scale amenity regions developing in Greater Phuket and along the Royal Coast, Koh Chang, on the eastern side of the Gulf of Thailand, is now developing as a 'second Phuket'. Together with the peri-urban industrialization of Ayutthaya and the eastern seaboard,[7] these areas support Bangkok's global role. This new spatial economy is supplanting traditional secondary cities performing regional service functions, such as Hat Yai, near the Malaysian border, and Nakhon Ratchasima in the northeast.

So Where's the Risk?

The shift in Thailand's economy described above, driven by both market forces and supportive government policies, seems positive and, from an overall perspective, probably is. As policies evolved, little thought was given to risk. In fact, this was largely perceived to be a risk-reducing strategy – tourists pollute less than heavy industry, while high-end personal services create well-paying jobs for Thai workers.[8] However, in the early years of the twenty-first century, this evolution in the Thai economy and the associated spatial changes have not been a 'free ride'. Rather, they have produced a new set of risks which is discussed below.

Spatial Inequity

As noted earlier, the ongoing changes in the country's spatial economy have caused resentment in northeast Thailand (*Isan*), the poorest and most populous rural region, resulting in the massive 'red shirt' political protest movement. This movement has been catalyzed by the rise and fall of Thaksin Shinawatra, Prime Minister from 2002 until he was overthrown in a military coup in 2006. What is surprising to some is that relative household income gaps between the northeast, the BMA and the nation as a whole decreased during the period 2000–2009. In 2000 household incomes in *Isan* were 31 per cent of the BMA's but, by 2009, they were up to 41 per cent. A similar trend was noted in relation to the nation as a whole, with *Isan* household incomes over the same period rising from 64 per cent to 73 per cent, although the absolute household income gap over the same period between *Isan* and BMA increased from 210,000 to 268,000 Thai Baht (see figure 11.5). What appears to have happened is that the movement of northeast Thailand from a poor economy, close to subsistence, to an emerging middle-

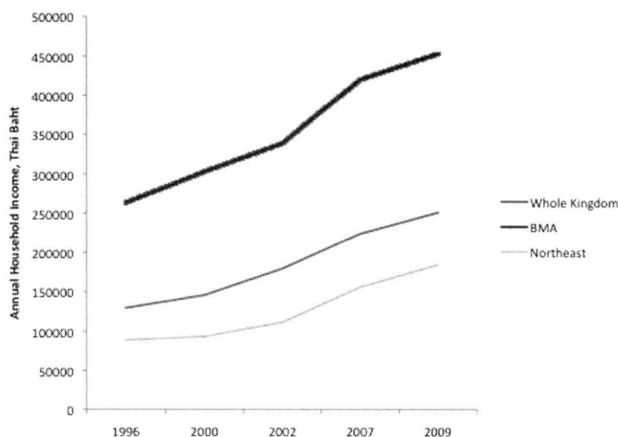

Figure 11.5. Household income 1996–2009: Thailand, BMA and the northeast. (*Source*: National Statistics Office, 2009)

income rural area actually increased expectations and made people more aware of spatial disparities (partially also through increased physical and telecommunication linkages with Bangkok). The resulting 'red shirt' political movement has put the Thai state at risk.

This destabilization has had a significant negative effect on tourist numbers, GDP levels and growth. For example, international tourist visits to Thailand dropped by 442,000 in 2009, the year after the airport was seized by the 'yellow shirts'. The civil disturbances of December 2008 stranded 350,000 tourists, with total losses to the economy exceeding US$1billion (Hunt, 2008). Visitor arrival numbers for April 2010 were down 38 per cent compared to April 2009, and the drop was even sharper in May 2010, directly associated with the street violence in which ninety people died. However, perhaps surprisingly, other components of Thailand's economy, particularly manufacturing and exports, continued to grow at a healthy pace during the violence, with national economic growth forecast in the range of 5–6 per cent for 2010. Recent events have clearly indicated how sensitive the amenity component of the economy is to domestic political conflict, relative to other sectors. Such events affect not only tourist numbers, but also domestic residential property sales, and particularly high-end property (Webster and Maneepong, 2009).

Global Economic Shocks

Global movements of people are sensitive to energy costs. For example, when the crude oil price exceeded US$145 per barrel in July 2008, amenity-oriented activities such as tourism, medical tourism, conventions, exhibitions and second home purchases were negatively impacted as the cost of aviation services rose. The sensitivity of travel to global economic conditions tended to be underestimated until the 2008–2010 global recession impacted on Thailand's amenity economy. Other external economic factors that can produce risk include abrupt exchange rate shifts (a current threat with the ongoing appreciation of the Thai Baht). These can reduce the effective discretionary income of overseas tourists and make Bangkok significantly less attractive to amenity visitors and migrants.

Environmental Deterioration

In the longer run, there is a risk that Bangkok and Thailand's amenity economy could sow the seeds of its own demise. Amenity-based developments, such as beach resorts, international education facilities,[9] and second home purchases by international investors and affluent Thais, depend highly on attractive settings. Deterioration in the environmental quality of cosmopolitan urban zones and beach resort areas would impact negatively on Thailand's emerging amenity economy, therefore. Paradoxically, the more attractive a location becomes, the more likely carrying capacity pressures will degrade the very environmental amenity responsible for that attractiveness.

Urban Terrorism

The current Democrat government of Thailand has officially labelled the street violence in Bangkok of April–May 2010 as 'terrorism'. But a second, perhaps more serious, threat exists in the insurgency raging in the five southernmost provinces of Thailand which has killed over 5,000 people in the last 10 years. Given the importance of tourism to the Thai economy (greater than 13 per cent of GDP broadly measured), an act of urban terrorism, such as a major bombing in Bangkok or a key beach resort like Phuket, could do untold damage to Thailand's amenity economy.

Natural Hazards

Another possible risk of significance to the amenity economy would be a major natural hazard event. The 2004 tsunami killed over 1,000 people in the Greater Phuket area and caused widespread property damage. It showed how a major natural disaster can affect tourism, although the adverse impact on visitor numbers was relatively short-lived in this case.

Human Resources

In Bangkok, there are chronic risks related to human resources. Amenity-based development requires a high level of human skills, in fields like medicine, design, graphics, tourism (management and language skills) and cuisine. Yet Thai post-secondary education in critical areas related to amenity-based development is not keeping pace with demand in either quality or quantity.

Summing up, it is not suggested that an amenity-based development model is particularly risky relative to other possible economic trajectories. All development models have risks. For example, a manufacturing-led model is less sensitive to civil disturbances – international business executives appeared to panic less readily than tourists and continued to stream into Bangkok during the peak of urban violence in April and May 2010 – but can be vulnerable to changes in global demand. An agriculture-based model is sensitive to risk factors such as drought, but, as with amenity-based development, energy costs are also significant, with petroleum costs reflected in the price of fertilizer. The main point is that, given the potential risks to an amenity-based model, particularly in the light of the current national security context, authorities need to pursue this strategy with their eyes wide open.

Political Risk – Bangkok as Capital

As noted earlier, Bangkok's Rattanakosin area plays a key role as the formal centre of national government of a highly centralized nation. It is where all of the key actors (whose relative power can vary considerably over fairly short periods of time) are based: (*i*) the monarchy; (*ii*) the military; (*iii*) the parliament; (*iv*) the prime minister; and (*v*) the powerful Ministry of Interior.

Figure 11.6. Redshirt encampment in downtown Bangkok. (*Photo*: Chuthatip Maneepong)

Rattanakosin has also played another role in the past as the pre-eminent area for the contestation of power. Sanan Luang, the area around the Democracy Monument and Ratcha Damnoen Boulevard were the areas favoured by competing political groups. For example, the violent uprising of 1992, led by former Bangkok Governor Chamlong, which overthrew the military government of the time, was focused here. But this has changed, with the street-based political movements in Thailand now understanding that political action is more effective when it targets strategic economic nodes – the airport or the commercial centre of the city – rather than symbolic political spaces.

Culture Wars: The Rural–Urban Divide

There is a definite Outer Thailand ('the provinces') and a clear rural bias in formal national governance in Thailand. Provincially-based political elites have considerable power. The BMA and the BMR have far fewer members of parliament than might be expected on the basis of their populations. This results partly from the fact that migrants to Bangkok often do not register their presence. They are thus counted as residents of the province from which they migrated and are not allowed to vote in Bangkok. The majority of Bangkok's residents cannot vote. In addition to nearly 5 million unregistered migrants from within Thailand, there are approximately 800,000 low-wage undocumented workers from Cambodia and Burma (working mainly in construction). Also, given the cosmopolitan nature of Bangkok's economy, there are perhaps 500,000 high-salary expatriate workers.

Rural bias is reflected in the tendency for Thai governments to view urbanization cautiously, in contrast to the pro-urbanization policies in other countries, such as China. Until recently, there had never been a Thai prime minister from Bangkok (although this has now changed with the current Prime Minister Abhisit Vejjajiva, and it appears that urban Bangkok's interests prevailed in the outcome of the civil unrest of April–May 2010). There are large net transfers of fiscal revenue from BMA and the BEUR to Outer Thailand, accelerated recently by the populist politics of former Prime Minister Thaksin Shinawatra. For example, BMA generates at least 54 per cent of Thailand's public revenues, but receives much less back (perhaps 20 per cent of national expenditure) in transfers from, and investment by, the national government.

The result of all of this is that Bangkok's residents do not have the power to elect national governments. However through their urban networks, technological sophistication, higher education and physical proximity to power, they have often removed national governments from power by organizing campaigns of dissent and taking to the streets. As the representatives of the rural periphery have also begun to use street power, Bangkok's interests have increasingly been represented through close networking between the military, bureaucracy and business elites. Bangkok benefits from higher levels of educational opportunity and achievement than the population of other parts of the country. Although the quantitative gap in the mean number of years of school completed is narrowing, there is still a qualitative difference in the nature of the education available in Bangkok and Outer Thailand. The result is that most middle-class Bangkok residents tend to prefer technocratic governments and are more likely to make voting decisions on the basis of evidence. There is less vote buying in Bangkok than elsewhere. By contrast, rural Thailand, particularly the northeast, tends to vote for populist policies. For example, the previous Thai Rak Thai government won two elections by significant margins by attracting the rural vote through a series of pro-rural policies, such as village loans, agricultural loan write-offs and '30 baht health care'. Middle-class Bangkok felt threatened by these policies and by the higher taxes required to pay for them. This was a major cause of the 'yellow shirt' street rallies of 2008. It would be misleading to portray Bangkok as politically homogenous, however. Not surprisingly, some of Bangkok's workers, such as its taxi drivers, support the rural populists in street-based political disputes. Nevertheless, recent political alignments in Thailand are threatening national unity through geographically based culture wars.

What Does Being a Capital Mean?

What then are the benefits to Bangkok of being the capital city? One is the direct influence on employment and economy of government activity, with about 9.4 per cent of the BMA's Gross Regional Product (GRP) coming from government functions. The wider impact on GRP and employment is undoubtedly larger, including benefits from firms and lobbyists locating close to the national government to seek to influence decisions. There is also a substantial diplomatic corps – 123 countries have embassies or consulates in Bangkok. Furthermore, the presence of the national government function has resulted in investment in attractive physical spaces and buildings with high cultural value which draw tourists.

Beyond those benefits, however, it is difficult to assess the importance that being the national capital has in determining Bangkok's identity, cosmopolitanism and economic performance. The role of capital was one of the city's main original functions, co-existing with the role of trading entrepôt. Bangkok is likely to continue to thrive as a leading Asian metropolis in future because of its economic importance, but the risks associated with being the capital appear to be increasing relative to the benefits. There is no doubt that the political struggles played out in the streets of Bangkok affect the economy and the incomes of Bangkok's residents. Businesses, investors and household consumers dislike instability more than almost any other condition. Over the past 5 years, Thailand's economy has significantly underperformed against competing Southeast Asian and East Asian states such as Vietnam, Malaysia, Singapore and China, and political instability is certainly part of the explanation for this. In Bangkok 'canary' sectors, such as high-end housing, have been particularly affected, relative to competing cities such as Singapore. People with money are footloose.

Being the national capital is not what puts Bangkok on the world map these days, unlike other capital cities such as Canberra, Ottawa or Washington DC, which are known primarily for their governance function. The world knows, visits, invests in and talks about Bangkok because it is fun (*sanuk*), has great cuisine, a rich culture, offers world class services from health care to silk clothing design, and is a global aviation hub. The question is not whether Bangkok's economy or identity is driven by its role as capital, but whether the risks of playing this role are starting to outweigh the benefits.

Climate Change Risk in the Bangkok Extended Urban Region

Vulnerability to climate change has been assessed in more detail for Bangkok than for most other Asian metropolitan areas. This is a product of Bangkok's relatively advanced economic development, mature bureaucracy and also its high vulnerability to flooding, as a consequence of a considerable portion of the BMA area being below sea level and with high rates of subsidence.[10] A number of technical studies have outlined scenarios of likely impacts (ONEP, 2008; World Bank, 2007, 2008, 2009; Institute of Development Studies, 2007; Greenpeace, 2007; BMA, 2007). The main potential impacts identified in these studies are as follows:

(*i*)   Flood volumes will increase by the same percentage as precipitation, but flood peak discharge will increase more (World Bank, 2009).

(*ii*)   A 1 in 30 year flood in 2050 will inundate an additional 180 square kilometres of the BMA and suburban Samut Prakarn province compared with the equivalent 2008 scenario – a 30 per cent increase in the flooded area (World Bank, 2009).[11]

(*iii*)   Much of the increase in the flood prone area will be in the western areas of the metropolis where protection structures, such as dykes and pumps, are less well-developed (World Bank, 2009). Large-scale flood protection investments, associated with protecting the Suvarnabhumi international airport (opened in 2006), make the eastern side of the metropolis less vulnerable to flooding.

(*iv*)   Mean temperature will increase between 3 and 3.5 per cent by 2100 (Greenpeace, 2007).

Vulnerable Areas and Communities

About a million people would be affected by a 1 in 30 year flood in 2050. One-third would be affected by flooding to a depth of half a metre for at least a week. More than a million buildings would be likely to be damaged by such a flood. However, half of these buildings are yet to be built in potentially impacted peripheral areas, particularly to the west.

Water supply, sanitation, public health, energy and transportation infrastructure would be minimally affected. This is both because much key infrastructure is in protected areas (the core city) and many systems, particularly those built since the 1995 flood, are designed to withstand significant flooding. For example, MRTA subway stations are raised (figure 11.7) and, as noted above, the fairly recent Suvarnabhumi airport was protected, being constructed on land fill which raised its elevation by several metres.

Figure 11.7. Raised subway station entrance to cope with flooding at Huai Khwang. (*Photo*: Andrew Gulbrandson)

Direct economic losses related to business disruption (not including damage to structures) from a 1 in 30 year flood in 2050 would be close to US$1 billion at 2008 prices. However, a wider assessment indicates that land in the BEUR which generates 23 per cent of Thailand's GDP could be flooded, in which case close to one-quarter of the national economy would be at risk. This is because the BEUR dominates the national urban system and economy to a greater extent than the major metropolitan region of any other country in East Asia, accounting for over half the country's economic output. Thailand would have a greater percentage of its GDP at risk from sea level rise than any other major country in East Asia except for Vietnam (World Bank, 2007).

It is likely that there would be significant increases in diseases and accidents associated with flooding and temperature rise. Of special concern is dengue fever, but other risks include salmonellosis, electrocution and drowning.

Looking at the spatial impacts more closely, the upper gulf provinces of Samut Sakhon and Samut Songkran would be most dramatically impacted by sea level rise (BMA, 2007). Long-standing slums, in areas such as Klong Toey (Bangkok's largest substandard housing area) and along rail lines extending northwards from Hua Lampong station, would not be significantly affected by most major potential climate change induced flood events, however, because they are interwoven with the relatively well-protected inner-city area. This finding runs counter to the starting point for many climate change risk studies in developing cities (with stereotypes sometimes substituting for analysis), that poor urban core slums will be most at risk.[12] The core city will be at higher risk, however, from increased temperatures, compared to suburban and peri-urban areas, because of the interaction of heat island and global warming effects. The problem in the core city will be compounded by increased heat and greenhouse gas emissions from air-conditioners as people try to cope with the higher temperatures.

The provinces of Samut Prakarn, and Samut Sakhon are key industrial areas, accounting for large numbers of manufacturing jobs. Vulnerability to major flood events in these provinces is likely to be associated as much with disruption to employment in factories as to residential life and the socio-economic groups most impacted would not be the poorest, but rather lower-middle income, blue-collar industrial workers, including domestic and international migrants.[13] Middle-class suburbanites would also be affected in the coastal district of Bang Khun Thian. This district, with a population of 127,000 in 2007, has attracted attention because its coastline and canal system is already being impacted by sea level rise and increased salinity, affecting locally important orchard, aquaculture and agri-business industries. Middle-class housing could also be at risk in Thonburi. Much of the additional housing proposed, but not yet built, in the western BMA area, which would be at risk from flooding, is intended for the middle-class market.

Vulnerability is a product of exposure to physical risk, the robustness of the built environment and the ability of households and institutions to cope. While there have been several studies of the likely physical impacts of flooding in Bangkok, as detailed above, understanding vulnerability also requires more research into the geography of the urban region's highly dynamic production, housing, employment and household income systems.

Current Adaptation Policies and Plans

To date, the highest level of awareness in regard to climate change has been displayed by the national government, and particularly the Office of Natural Resources and

Environment Policy and Planning (ONEP), within the Ministry of Natural Resources and Environment, which represented the Royal Thai Government in the 2009 Copenhagen negotiations. ONEP has produced a *National Strategy on Climate Change Management* (ONEP, 2008). This is a technically impressive document, analyzing vulnerability and setting out proposed responses in relation to a series of action areas. In its mix of mitigation and adaptation measures, the strategy skilfully relates vulnerability to the structure of the nation's economy, stressing potential damage to economic pillars such as agriculture, agri-processing and tourism. One section of the strategy focuses specifically on potential damage to urban settlements.

At the metropolitan level, the BMA, in conjunction with the World Bank, has undertaken a *Climate Change and Adaptation Study for Bangkok Metropolitan Region* (World Bank, 2009). The motivation for this action stems from increased awareness of the threat of climate change impacts in the BEUR and acknowledgment that more action is needed, much of which will require large amounts of capital. This study advocates the 'mainstreaming' of climate change adaptation in all BMA and national agency operations in the BEUR, and climate-proofing of city buildings. Unsurprisingly, the report advocates less harvesting of ground water (the main cause of subsidence which aggravates flooding), as well as the protection of shorelines, flood-sensitive land-use zoning and the creation of more green retention areas ('monkey cheeks'[14]).

International NGOs are significantly involved in climate change issues in Bangkok. Greenpeace has identified key climate change-related issues facing Bangkok in a study that relates global climate change dynamics to potentially vulnerable eco-systems in Thailand, including mangrove ecosystems (Greenpeace, 2007). Thai NGOs and universities are significantly involved at the local level. This is particularly the case in Bang Khun Thian district where local NGOs, including the Chumchon Thai Foundation and the local chapter of Greenpeace, are working with Rangsit University and the Asian Institute of Technology to devise strategies to tackle both coastal erosion and increasing salinity. The BMA has introduced sand-bagging approaches, but the local NGO–university alliance is proposing alternative adaptation methods, and communities in the most physically vulnerable areas (Bang Khun Thian, Samut Sakhon, Samut Prakarn) have developed a 'network of wetland communities' which shares learning and lobbies governments.

Cities, even large metropolitan areas such as Bangkok, can do little to affect the levels and rates of climate change that they will experience, except by acting collectively through the network of global cities and nation states. Levels of climate change are essentially the result of external global forces. Metropolitan institutions such as the BMA are best advised to focus on adaptation, therefore. The national government is increasingly involved in climate change mitigation and adaptation policy through the ONEP in the Ministry of Natural Resources and Environment although, given the 'silo' nature of Thailand's national government, the ability of this ministry to drive increased climate change awareness, policy formulation and action in other agencies may be limited. NGOs meanwhile focus on local responses where physical impacts are already appearing, as in the Bang Khun Thian case described above.

Actual adaptation measures to date have had a 'hard' bias: large-scale dyke systems, extensive landfill projects to raise the height of critical infrastructure, and sand bags along the Gulf of Thailand (see figure 11.8). Although the important role of capital-intensive physical responses in many cases is acknowledged, effective adaptation is much more likely to be successful if a variety of measures across a wide continuum is taken with the support of local communities. National and BMA laws supportive of adaptation are in place, including BMA's 'Ban on Nine Types of Buildings in Flood Affected Areas' law,[15] but these are only likely to be enforced effectively if lower-level governments are committed to them. Some flexibility to take account of local knowledge may also be appropriate when, for example, determining appropriate sites for buildings. Local communities can also play a valuable role by taking responsibility for climate change-related adaptation initiatives, such as heat shelters and the provision of climate change education in local schools.

Figure 11.8. Barrier construction to combat sea level rise south of Bangkok. (*Photo*: Chuthatip Maneepong)

In the BEUR, many adaptation initiatives have been 'one-off' and tend to be construction projects which are not integrated into some larger strategy. But if, for example, an area is well-protected from flooding, at a cost of tens or even hundreds of millions of dollars, denser development around rapid transit should be encouraged within this protected area by allowing higher floor area ratios and providing fiscal incentives. This would help to maximize the return on the investment made in flood protection. Green flood retention areas, developed as parks, may also attract adjacent up-market housing, creating land value-capture opportunities that can provide capital for further flood-proofing activities. But a strategic approach and institutional change will be required to achieve this sort of level of integrated planning. Important additional initiatives to include in a cost-effective strategy to reduce flood risks from climate change in the BEUR should include the following:

(*i*)   Ground water harvesting should be dramatically reduced immediately.

(*ii*)   Education programmes are required to make communities, local NGOs and relevant private sector actors (including insurance companies) aware of likely impacts and possible responses.

(*iii*)   Stakeholders, private and public, responsible for economic development, should also be educated about the potential of climate change adaptation as a source of competitive advantage and not just as a threat.

(*iv*)   Public land and housing authorities should begin planning relocation areas.

Conclusions and Implications

This chapter has explored three types of risk facing the Bangkok region. All are related to external global drivers, primarily economic or atmospheric in origin. A good deal of official risk mitigation and adaptation planning and action is being undertaken in regard to climate change, probably because it has been *explicitly* defined as a risk area, especially by the global community. However, in terms of broad public understanding of risk, most Bangkok residents would view political risk as the greatest threat to the wellbeing of the city. Handling this risk has, to a large extent, been regarded as a 'law and order' issue, although increasingly it is also viewed popularly as a spatial equity issue. The least awareness, and least explicit action in regard to risk, has been shown in relation to significant, ongoing changes to the national economy and spatial system. This lack of understanding and action may well be related to the abstract nature of macro-economic and spatial system changes.

What is clear, however, is that all of these risk areas are closely inter-related. The growth of the amenity economy and the changing geography of manufacturing industry have led to increased concentration of talent in the BEUR and along the Gulf of Thailand. This has helped to fuel feelings of unfairness in the poor northeast, with consequences seen in widespread political dissent on the streets of Bangkok, despite the fact that relative regional spatial inequalities are diminishing. Increased political and social instability have the potential to slow down macro-economic and spatial change.

The re-emergence of Bangkok's core as an area of economic importance has encouraged investment in flood protection infrastructure and reduced its vulnerability to sea level rise. However, most of Thailand's beaches could be seriously affected by rising sea levels, dealing a severe blow to the amenity sector. A catastrophic climate change-induced event, involving massive flooding in the BEUR area, could cripple Thailand's economy, triggering major social and political instability. The permutations and combinations among these interacting risk factors are enormous and inherently unpredictable. Available forecasting tools, such as probability analysis, may be of limited value (Taleb, 2007).[16] Large extended urban regions, such as the BEUR with its population in excess of 20 million, will always be subject to considerable risk. Risk is something that must be lived with, but what is important is to mitigate it if possible. Where risk cannot be fully mitigated, as in the case of climate change or political and social instability, the challenge is to facilitate adaptive processes that minimize the impacts of shocks on the city's overall wellbeing and performance.

Despite the increased division of society into two political camps at present, Bangkok seems to be a socially resilient city. When things go wrong, as in the 1997 financial crisis, people tend to help each other, through, for example, voluntary job sharing which resulted in a poverty rate much lower than the World Bank had predicted. As a globally open society, Thailand's higher tiers of government benefit by learning from best practice from abroad. As a domestically open society, moreover, Thailand benefits from strong bottom-up action from NGOs, unlike other transitional states in the Asian region, which are more dependent on top-down action to deal with risk. Bangkok's resilience can be demonstrated by the fact that tourism numbers are on track to return to pre-crisis levels by the end of 2010, despite the city having been virtually under siege in April and May 2010. The Greater Phuket area also recovered economically from the Asian tsunami within 18 months, while downturns in Bangkok's economic output in 2008 and 2010 from street conflicts were reversed quickly.

Will Bangkok exist in another thousand years? Who knows? But if it does, it is likely that it will be because its high levels of social capital have enabled it to deal with a myriad of ever-changing risks.

Notes

1. Research assistance in the preparation of this paper by Jon Valentine is gratefully acknowledged.
2. A dam in Lop Buri and dykes to the east, in particular.
3. Foreigners can legally own up to 50 per cent of the units in condominiums.
4. Thonburi is now within the BMA.
5. A treaty signed between the King of Siam and Sir John Bowring, Governor of Hong Kong and Britain's envoy.
6. This rail system now carries about 600,000 people per day.
7. Bangkok's industrially based peri-urban areas are home to close to 8 million people. The Eastern Seaboard's core has over 3 million people spread across the Provinces of Chacheongsao, Chonburi and Rayong.
8. One hospital alone, Bumrungrad International, services 420,000 foreign patients per year (2008).
9. International education facilities cater for both domestic and expatriate markets. Characteristics

of these institutions include tuition in non-local languages, foreign faculty and foreign curricula, e.g., the Baccalaureate programme.

10. At present, the rate of subsidence is 3–5 centimetres a year in the BMR (see Buapeng, 2007).

11. Samut Prakarn is an adjoining industrial province in the BMR, immediately to the south of the BMA.

12. Low-income communities are, of course, at greatest risk in some Southeast Asian metropolitan areas. For example, in Ho Chi Minh City the population of the poorest district (District 6) is most at risk from flooding (see Ho Chi Min City People's Committee, 2010, and also Chapter 12 of this book, on Manila).

13. Large numbers of Burmese migrants work as fishing boat crews, in agri-business (fish processing, etc.) and in chandlery services in Samut Sakhon. A high percentage of factory workers are domestic migrants.

14. Green retention areas in Thailand are called 'monkey cheeks' because they expand when water enters them, in the way a monkey can blow up its cheeks to scare away predators.

15. This law was enacted in 1999, but only recently has it begun to be seriously enforced. It bans nine types of buildings in flood prone areas, including high-rise buildings, warehouses, hotels, shopping centres and factories.

16. For example, days before the 1997 financial crash, the national planning board of Thailand (NESDB), full of PhDs from the world's best universities, was forecasting >8 per cent economic growth for 1998. One foreign advisor, who suggested something was going wrong (based on discussions with business people who were concerned about Thailand's future), was dismissed.

References

BMA (2001) *Bangkok Metropolitan Administration*. Bangkok: Thammasat University Press.

BMA (2007) *Prevention and Solution for Coastal Erosion in Upper Gulf of Thailand Coastal Area*. Bangkok: BMA (for Town and Country Planning Department).

Buapeng, Somkid (2007) *Groundwater Situation and Land Subsidence Mitigation in Bangkok and its Vicinity*. Presentation for the Department of Groundwater Resources, Ministry of Natural Resources and Environment (MONRE). Bangkok: Government of Thailand.

Greenpeace (2007) *Crisis or Opportunity: Climate Change and Thailand*. Bangkok: Greenpeace Southeast Asia.

Ho Chi Minh City People's Committee (2010) *Urban Planning and Development: Responses to Climate Change: Adaptation and Mitigation*. Ho Chi Min City: HCMC People's Committee.

Hunt, L. (2008) Land of disappearing smiles. *Bangkok Post* (Spectrum), **1**(14), pp. 3–5.

Institute of Development Studies (2007) *Governance Screening for Urban Climate Change Resilience-Building and Adaptation Strategies in Asia: Assessment of Ho Chi Minh City, Vietnam*. University of Sussex: Institute for Development Studies.

National Economic and Social Development Board (NESDB) (2008*a*) National Accounting Statistics. Bangkok: Government of Thailand. Available at http://www.nesdb.go.th. Accessed 19 July 2010

National Economic and Social Development Board (NESDB) (2008*b*) National Demographic Data Bangkok: Government of Thailand. Available at http://www.nesdb.go.th. Accessed 19 July 2010.

National Statistical Office (2009) *The 2009 Household Socio-Economic Survey*. Bangkok: Government of Thailand. Available at http://web.nso.go.th/en/pub/eco/531130hse09.htm. Accessed 19 July 2010.

ONEP (Office of National Resources and Environment Policy and Planning) (2008) *Thailand's National Strategy on Climate Change*. Bangkok: Government of Thailand.

Phongpaichit, P. and Baker, C. (1995) *Thailand: Economy and Politics*. Kuala Lumpur: Oxford University Press.

Revkin Andrew (2010) Disaster Awaits Cities in Earthquake Zones. *New York Times*, 25 February.

Ruland, J., and Ladavalya, B. (1996) Managing Metropolitan Bangkok: Power Contest or Public Service, in Ruland, J. *The Dynamics of Metropolitan Management in Southeast Asia*. Singapore: Institute of South East Asian Studies.

Smithies, M. (1986) *Old Bangkok*. Oxford: Oxford University Press.

Taleb, Nassim Nicholas (2007) *The Black Swan: The Impact of the Highly Improbable*. New York: Random House.

Webster, D. and Maneepong, C. (2009) Bangkok: Global actor in a misaligned national governance framework. *Cities*, **13**(1), pp. 80–86.

World Bank (2007) *East Asia Environment Monitor 2007: Adapting to Climate Change.* Washington, DC: World Bank.

World Bank (2008) *Climate Resilient Cities: 2008 Primer – Reducing Vulnerabilities to Climate Change Impacts and Strengthening Disaster Risk Management in East Asian Cities*. Washington, DC: World Bank.

World Bank (2009) *Climate Change Impact and Adaptation Study for Bangkok*. Bangkok: World Bank.

Chapter Twelve

Manila: Metropolitan Vulnerability, Local Resilience

Brian Roberts

Metro Manila[1] is a large polycentric megacity with a population of more than 12 million people, currently growing at around 1.5 per cent per annum, and with a land area of 636 square kilometres. By 2020 the population of the metropolitan area, also known as the National Capital Region (NCR) of the Philippines, is expected to reach 13.4 million (City Mayors, 2009). Metro Manila is located in the southwest of the island of Luzon, on an isthmus bounded by Manila Bay to the west, a large shallow lake (Laguna de Bay) to the southeast and the Sierra Madre Range of hills and volcanoes to the east and northeast. Most of the city lies on a large, wide flood plain, divided by the Pasig River. In reality the metropolitan area spreads beyond the seventeen cities that make up the NCR to include also adjoining urbanized areas in the surrounding provinces of Bulacan (north), Rizal (east) and Cavite and Laguna (south). The population of this Greater Manila area is close to 18 million, making it the seventh largest megacity in Asia (Laquian, 2005).

Metro Manila is a fascinating city of diverse cultures, traditions and religions, as well as being the national centre of government, commerce, education and transportation for the Philippines. Like other large Asian cities, Manila faces many development problems associated with congestion, pollution, infrastructure shortages, weak governance and poverty. It is also a city that is becoming increasingly exposed to the impacts and effects of globalization, terrorism and climate change.

Despite Metro Manila's vulnerability, it remains a dynamic, vibrant and resilient city. It has generally recovered from past crises and destructive events. More often than not, it has been local communities and the private sector which have led the recovery efforts. Although the experience of recovery has often been slow, resilience strategies have been developed by business, the community and government as a result of lessons learned from the management of previous crises.

Few studies have been conducted which examine the dimensions of risk or resilience at a city or regional level in the Philippines (Quisumbing *et al.*, 2008). The research that has been undertaken on urban risk impact and management has also tended to focus on natural and environmental risks, but Manila has a history of other risks besides these which have had severe impacts on its urban systems. This chapter therefore adopts a broad perspective on risk and resilience in Metro Manila.

The chapter begins with a brief discussion of the concept of 'urban resilience' (Resilience Alliance, 2007). This is followed by a brief history of the planning and development of Manila from the Spanish colonial period to the present day. The chapter then describes different types of risk events which have affected Metro Manila in the past and which pose threats for the future. Responses to risk mitigation and disaster management recovery are examined.

Some of Metro Manila's local governments have shown strong leadership in developing risk management strategies and in establishing pathways for recovery. Three case studies are used to illustrate the way the government, business and community sectors have responded to particular disasters.[2] These examples are drawn from three local government areas in Metro Manila – the cities of Marikina, Quezon and Taguig. The lessons identified in these case studies have some relevance to risk management and resilience strategies in other Philippine and Asian cities. The chapter concludes with the message that Manila must improve its resilience to risk through more effective planning at the scale of the metropolitan region.

The Concept of Urban Resilience

Urban systems have a remarkable ability to recover and adapt to shocks, disasters and other crisis events. Indeed, it seems that urban systems, like natural systems, often need shocks to foster change, build immunities, stimulate innovation, and foster creativity to move to a new stage of development. Shocks can be caused by a multitude of factors. They can be sudden or latent, triggered by a cumulative set of seemingly unrelated minor events and factors which build up underlying stresses in social, environmental and economic systems.

The origins of events which cause shocks to urban systems are not always local. The 2008 'Global Financial Crisis', for example, triggered a set of events which began in the United States of America, through sub-prime lending in the housing mortgage and derivatives markets, and subsequently spread to Europe and the rest of the world. Then there are unsustainable urban development practices which are cumulatively contributing to global warming and environmental change. These global influences, along with localized political, structural and cultural systems changes, are making it more difficult for cities to manage risks (UNISDR, 2008).

The phenomenon of recovery in damaged urban systems has given rise to a new focus of research on urban resilience (Walker and Salt, 2006). This is concerned with the degree to which cities are able to adapt to the consequences of catastrophic events, failures or changed environmental situations, often by reorganizing themselves around a new set of structures and processes (Alberti *et al.*, 2003). Urban resilience had not been studied widely until recently (Resilience Alliance, 2007), but it is an area of urban systems science that is likely to become increasingly important as cities become more exposed to global shocks and other events over which they may have little control.

Because the risks faced by cities are becoming increasingly complex, urban administrators need to focus on better ways to address disaster prevention, management and recovery. Although advances have been made in risk management theory, relatively few tools have been developed to handle multi-sector and cumulative risk impacts of disaster or shock events on cities (Roberts, 2003). The responses to risk management and resilience must be multi-sectoral (Roberts and Cohen, 2002; Roberts and Tabart, 2005), since cities are complex systems.

The United Nations International Strategy for Disaster Reduction Regional Task Force on Urban Resilience and Disaster Vulnerability in the Asia-Pacific Region (UNISDR, 2008) notes that there are two particular principles which urban risk managers should focus on. The first is the need to adopt a balanced approach to urban ecosystems – that is, the equilibrium between natural and built environments should be restored and maintained. Second, there is a need to acknowledge the dynamic character of risks which vary in probability, duration and magnitude over time.

An important dimension of urban risk management is that disasters and other shock events can trigger what are known as cumulative causation impacts (Myrdal, 1957). A single event can generate a chain reaction of events, which can have physical, economic and social impacts that extend across a city, region or even several countries. September 11, 2001, triggered a global panic in financial markets, which caused a fundamental change in thinking and practices to make cities and transport systems more secure. The 2003 SARS[3] outbreak in Asia had a devastating and long-lasting impact on urban economies and transport systems in the Asian region. Even local disasters and events can now have international consequences – the so-called 'butterfly effect' of chaos theory.

Cities are likely to become more vulnerable to external economic, environmental and social shocks in the future. Governments and communities will need to become more vigilant and to engage in smart ways to anticipate and manage both known and unknown risks (Roberts, 2006). The latter are almost impossible to prepare for, but systems for resilience management still need to be created to deal with the possibility of unanticipated shock events. This calls for fresh thinking, ideas and strategies to support urban risk management and resilience.

A Framework for Risk and Resilience Assessment

Assessing risk and resilience in cities is tricky as there are no established methodologies for analysis of these factors. By definition, risks have many uncertain dimensions in terms of their nature, scale, time and impact. Much of the analysis conducted on risk and resilience is qualitative, moreover, which makes it difficult to compare the risk exposure of different cities. In order to conduct a more systematic and multi-sector assessment of risk and resilience in researching this chapter on Metro Manila, it was necessary to develop a simple analytical framework (see figure 12.1).

At the first level an analysis was conducted of several types of risk, which affect or have impacted on Metro Manila, and the mitigation and resilience responses to these. The second level of analysis focuses on the way that three local governments in

Figure 12.1 Risk and Resilience Analysis Framework for Metro Manila							
	Planning and Development	Natural	Urban Assets	Environmental	Economic	Social	Political and Legal
Metro Risk Analysis							
Metro Resilience Initiatives and Strategies							
City Case Studies of Risk and Resilience							
Lessons Learned							

Figure 12.1. Risk and resilience analysis framework for Metro Manila.

Metro Manila have developed resilience responses to a selected number of risks, using case studies of three disaster or crisis events. The circumstances leading up to each crisis and the responses are described. Some of these events had causes outside the Philippines. The case studies provide interesting insights into initiatives and responses by stakeholder groups to assist with recovery from crises; rebuild and enhance a city's performance and competiveness; improve local quality of life; and introduce good environmental management and governance practices. The final part of the analysis draws together the lessons derived from the study of Metro Manila and the local government case studies on ways in which governments and communities can reduce their exposure to urban risks and foster resilience.

Development and Planning of Metro Manila

History of Development and Planning

In 1570 the Spanish occupied the Philippines and developed a fortress in the old port area of Manila as their principal centre for administration and trade in the East Indies.

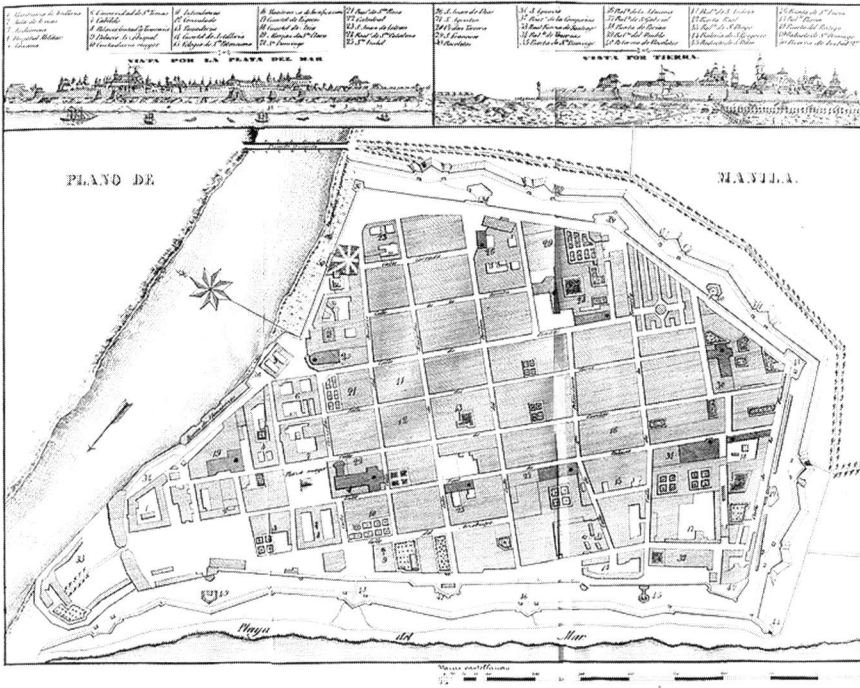

Figure 12.2. Sketch plan of Manila, 1851. (*Source*: http://commons.wikimedia.org/wiki/File:Plano_de_Manila_1851.jpg)

The city gradually expanded beyond the citadel walls as trade with Europe developed. The first detailed plan for Manila was not prepared until 1831, and this was revised in 1842 and 1851 (see figure 12.2). Further expansion plans were prepared in 1863 and 1884.

In 1889 the administration of the country was taken over by the United States, following the defeat of Spain in the Spanish-American Civil War. Little urban planning took place in Manila, however, until 1905 when a detailed plan for the city was prepared by the American planner Daniel Burnham on 'City Beautiful' lines (see figure 12.3). Burnham's plan was never implemented but elements of its design were later incorporated into the plan for Quezon City, proposed as the national capital after Independence in 1946. Although the American colonial administration came to an end at this time, many land management, planning practices and regulations continued to be based on US systems.

During the period from the end of World War II until the fall of the Marcos government in 1986, Manila expanded rapidly as the result of high rates of migration from all parts of the country. This migration contributed significantly to the growth of the national capital, but the urban development that occurred was largely unplanned and unregulated. The city began to sprawl into the surrounding municipalities, giving

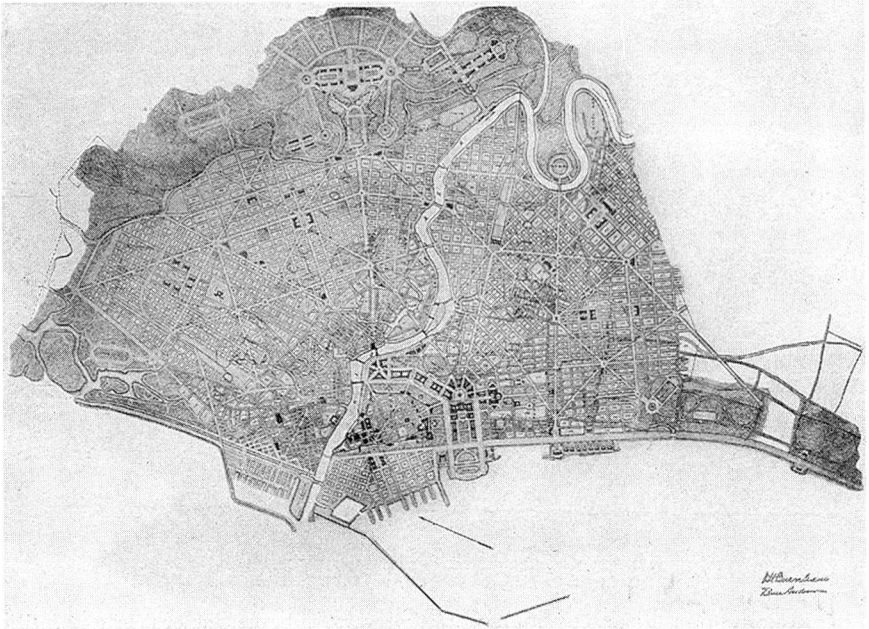

Figure 12.3. Burnham plan of 1905. (*Source*: Burnham and Bennett, 1909)

rise to traffic and environmental issues. While the national government established various organizations and co-ordinating councils to guide urban growth and manage urbanization, by 1972 it realized that it had serious problems. As a result, the first Inter-agency Committee on Metro Manila was established. The aims of this Committee were:

(*i*) to study the system of municipal/city government in Metropolitan Manila; (*ii*) to recommend whatever measures of co-ordination and integration were deemed appropriate; (*iii*) to study the functions and responsibilities of the National Government in the metropolitan area; and (*iv*) to recommend whatever changes in structure and interrelation with municipal governments were deemed appropriate. (Manasan and Mercado, 1999, p. 12)

This led to the recommendation to establish the Metro Manila Commission (MMC) in 1975.

The MMC was primarily responsible for waste collection and disposal, transport and traffic, and the fire service. It was to co-ordinate public and private sector delivery of essential services including water, sewerage, flood control, social welfare, housing and recreation facilities, but unfortunately it was given few resources with which to perform these tasks. It was also charged with developing a comprehensive regional plan to guide the social, economic and physical development of Metro Manila. A master plan for the NCR was developed in 1977, but it was not effectively implemented.

There were improvements reported under the MMC in the co-ordination of squatter settlement development, waste management and transport (Manasan and Mercado, 1999, p. 14), but generally the MMC lacked the power and authority to overrule the self-interest of local governments and private sector service providers. Many local governments began developing their own master plans, ignoring the provisions of the MMC Plan.

The collapse of the Marcos Administration in 1986 and the introduction of a new constitution left the institutional arrangements for the management of Metro Manila in disarray. The incoming Aquino Administration established a Metro Manila Authority (MMA) in an effort to restore business confidence and to manage the urban growth and development of Metro Manila. However, the MMA lacked funds and political authority. Local governments recognized the weakness of the MMA and began to take charge of their own affairs, fostering independently the development of new sub-regional city business districts, shopping and industrial centres. Metro Manila began to develop as a polycentric city, with each city actively competing with others for major development projects.

The ensuing pattern of development caused further traffic congestion which added to the already severe air pollution problems. Moreover, serious shortages of electricity, inadequate water supply and a deterioration of sanitation services plagued the city (see below). At various times the city's urban systems collapsed and Metro Manila developed a reputation as one of the worst cities in Asia for electricity cuts or 'brown-outs'.

Political pressure mounted for a strong central co-ordinating agency to manage the development of Metro Manila. This resulted in the establishment of the Metropolitan Manila Development Authority (MMDA) in 1995. This authority was given powers and responsibilities to administer the development of the cities that make up the Metro Manila Capital Region (see figure 12.4). The legislation, which established the MMDA, required the authority to work closely with non-government organizations (NGOs), community groups and the private sector to cultivate partnerships for development. The MMDA board includes members reflecting the interests of these three groups.

The main functions of the MMDA are planning, monitoring, co-ordination and the exercise of supervisory authority over the delivery of metro-wide services within Metro Manila. These functions are required to be exercised without diminishing the autonomy of lower levels of government in relation to local matters (MMDA, 2009). MMDA has some responsibility for development planning; transport and traffic management; solid waste disposal; flood control and sewerage; urban renewal, zoning and land-use planning; shelter services; health and sanitation; pollution control; and public safety.

Soon after it was established the MMDA prepared 'The Physical Framework Plan for Metropolitan Manila, 1996–2016' (MMDA, 1996) (see figure 12.5). This plan was

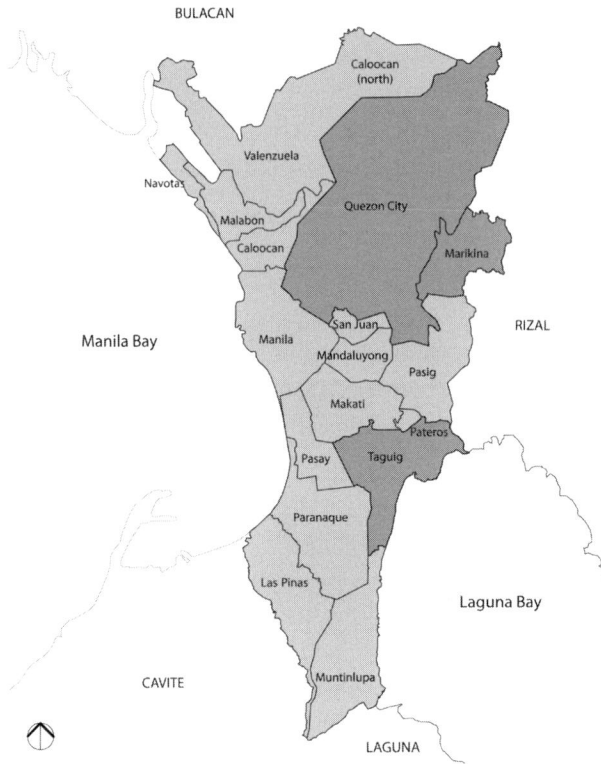

Figure 12.4. Cities of Metropolitan Manila.

to be reviewed every 10 years, although to date no review has occurred. An attempt was made to prepare a new development plan for 2001–2004, but this plan was never completed. The failure to prepare and review the physical framework plan by successive agencies has been a factor adding to the uncertainty of the planning, development and management of Metro Manila.

Planning has been ineffective in Manila for over half a century and the unwillingness of national, metropolitan and local governments to implement plans or enforce regulations has led to an urban development pattern which is not sustainable and which has increased the risk profile of the city. It has also created great uncertainty for the development industry, in land markets and for security of tenure (see below).

The legacy of more than 50 years of failed metropolitan planning will continue to lead to disjointed patterns of urban settlement and development as the city grows. This in turn will reduce opportunities for efficiency gains by promoting greater consolidation of development, better integrated logistics and infrastructure systems, employment clustering and co-ordinated provision of community services. As a consequence, the cost of retrofitting and re-engineering Metro Manila to reduce the city's exposure to risks will be high, running into billions of dollars. These costs could

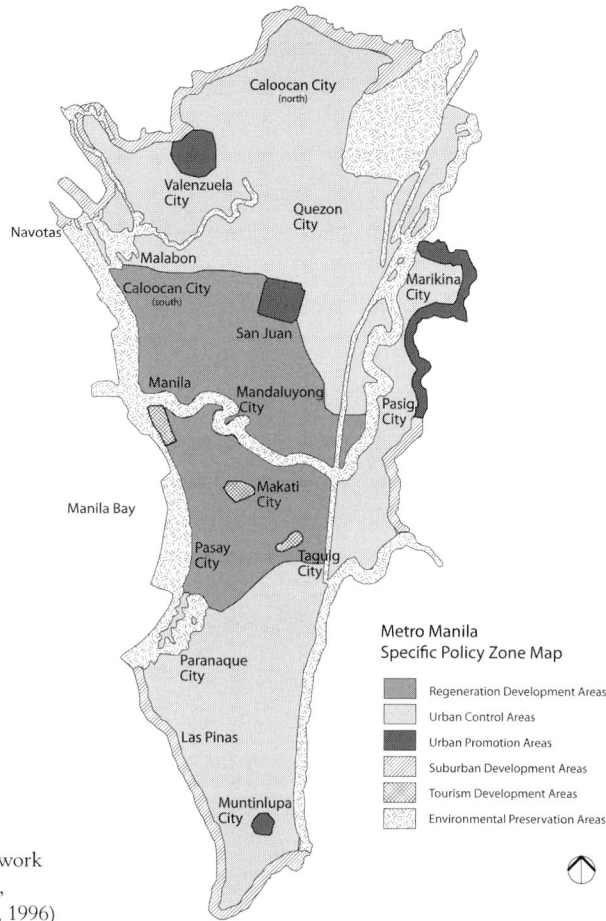

Figure 12.5. Physical Framework
Plan for Metropolitan Manila,
1996–2016. (Source: MMDA, 1996)

have been substantially reduced had the governments of Metro Manila agreed to work collaboratively in the past.

Uncertainty in Urban Development

The *laissez-faire* nature of urban development and investment in Metro Manila has given rise to significant differences in land-use patterns from place to place. The rich live in secure communities around the new city centres, close to large megamall shopping complexes. The middle-income sector is spread through an inner ring of smaller residential estates and in larger urban fringe estates. The city's 4 million urban poor are confined to older inner-city housing, squatter settlements on public land, and flood-prone and land-slip areas. Makati is the wealthiest city in Metro Manila and the main centre for business services, up-market accommodation and diplomatic activities. Ortigas, at the boundary of the cities of Pasig and Mandaluyong, is a major financial, retail and information services centre. Old Manila is a tourist, political and cultural centre; and Quezon City is an educational centre, based around the University of the Philippines, as well as being the home for many national government offices. All the cities of Manila have sought to develop strong commercial centres, in combination with high-rise residential apartments. In recent times, developers have tended to favour locations unencumbered by uncertainties about land tenure, including former government or military sites, such as Fort Bonifacio in Taguig City (see below).

As a result of the high level of competitiveness between cities and ineffectual planning by the MMDA, Metropolitan Manila is now a polycentric megacity linked by a poor transport system. The inadequate transport infrastructure makes Manila one of the most congested cities in Asia, adding to the transaction costs of business and undermining the city's competitiveness when trying to attract investment. Many of Manila's citizens have little choice but to undertake long and frustrating journeys to work.

To address some of these urban development problems and associated environmental risks, the MMDA has supported the decentralization of industries from the older inner-city areas to the outer urban fringe (Kundu, 2009, p. 41). A Special Economic Enterprise Development Zone has been created which incorporates the old US defence bases at Clark airfield and Subic Bay, about 70 kilometres north of the metropolitan area while, to the south, there are several new industrial areas in Taguig and Cavite. The attempt by the MMDA to try to segregate commercial and industrial centres has only compounded the city's transportation problems, however, because of the lack of fast and efficient transport systems connecting the new major employment centres on the periphery to the inner parts of Manila.

Overall, the recent history of urban development in Metro Manila has been one of lost opportunities and uncertainty. Political vacillation, lack of commitment to plan making and poor enforcement of regulations have created a situation in which *ad hoc* development is the norm and corrupt behaviour is widespread. The consequence is a megacity that is exposed to high levels of social, economic and environmental risk. The failure to fund the MMDA and its predecessors adequately has made it extremely difficult for the authority to ensure the efficient delivery of basic services across the whole metro area. Metro Manila has become a collection of fiefdoms, each seeking to secure whatever it can extract from central government or the business sector, rather than working collaboratively to overcome their many common development problems.

Risk and Resilience Analysis of Metro Manila

A recent study ranking the vulnerability of major coastal cities in Asia to climate change issues placed Metro Manila as one of the most 'at risk' cities in Asia, equal second with Jakarta and just below Dhaka (WWF, 2009). While risks to Metro Manila from natural and environmental disasters are significant, governance, planning, unmanaged development, and economic and social issues have also made Manila a highly vulnerable city in which to live and to do business. Some principal areas of vulnerability are discussed below.

Natural Risks

On average the Philippines is hit by more than twenty typhoons a year (International Federation of Red Cross and Red Crescent Societies, 2009), many of which cause severe damage and loss of life. On 30 September 2009, Typhoon Ondoy caused the most destructive floods in living memory. More than 2.5 million people were directly affected by the disaster. A second major typhoon (Parma) hit a week later. These two typhoons left more than 600 people dead and caused damage to infrastructure, buildings and agriculture estimated at 30 billion pesos ($US550 million). Much of the destruction and loss of life occurred in the Marikina Valley and other low-lying areas of Metro Manila.

While winds associated with typhoons cause much destruction, it is the associated flooding which leads to the greatest damage to infrastructure and buildings. The average elevation of Metro Manila's urban area is 10 metres, but more than 1.5 million people live in low-lying areas less than 1 metre above sea level. Flooding in these low-lying areas is becoming increasingly severe as catchment areas become more urbanized, vegetation loss increases and the dumping of solid waste into streams and drainage systems continues. A study by Tharoor (2009) ranked Metro Manila as one of the most flood-prone cities in Southeast Asia.

No one public agency appears to have responsibility for reducing and managing the risks of typhoons and flooding in Metro Manila. While local authorities took action to develop drainage systems and canals to reduce the impacts of flooding during the early development of Manila under Spanish rule, progress since Independence on addressing flood impacts has been slow or totally lacking. In the 1970s it was proposed to build the Paranaque Spillway to allow floodwater to flow from Laguna de Bay directly to Manila Bay, instead of passing along the clogged Pasig River, but this project did not proceed. If it had, it could have saved lives and billions of pesos in subsequent property damage. Lack of attention to flood management and protection works in the Marikina Valley in the 1970s also contributed to the hundreds of deaths from Typhoon Ondoy in 2009. Given Manila's vulnerability to typhoons and flooding, improved flood control measures are amongst the more obvious ways of mitigating future risks.

Metro Manila's vulnerability to earthquakes is a consequence of parts of the city being built across two major fault lines (Simpson *et al*., 2008). Over the past century the Manila region has experienced six severe earthquakes which have caused significant damage to infrastructure and chaos to urban systems (Nelson *et al*., 2000).

Poor enforcement of building regulations and sub-standard construction make Metro Manila one of the highest earthquake risk cities in Asia (Simpson *et al.*, 2008). The Metro Manila Earthquake Impact Reduction Study (MMEIRS) (Japan International Co-operation Agency *et al.*, 2004) suggested that a severe earthquake along the Manila Trench could be expected to destroy more than 5,000 buildings. This could rise to 16,000 buildings in the case of a severe earthquake along the Manila Bay fault. The MMEIRS study stated that:

> Community members will not be assisted by public service immediately, in case of large disasters, such as a major earthquake. Therefore, it is important to maximize the preparedness and disaster response capacity of the community beforehand, through enhancement of social capital by preparing the community and its members for resilient responses. Top priority activities for community preparations are the identification of evacuation route and place, rescue activities, and initial fire extinguishing. In order to establish disaster management mechanisms through enhancement of social capital, the following frameworks support the objectives of (i) enhancing self reliant and mutual help risk management capacity; (ii) inculcating disaster culture in future generations. (Japan International Co-operation Agency *et al.*, 2004, p. 52)

In other words, public agencies acknowledge that they have very limited capacity to deal with an earthquake disaster. There is a need, therefore, to develop the resilience of local communities (Santa Cruz-Makilan, 2004). MMEIRS outlines strategies for this, but lack of training and community education, a paucity of equipment, no city-wide mechanisms for co-ordination or devolution of responsibilities and the lack of emergency powers entrusted to local communities or wards (*barangays*) are likely to result in a chaotic situation if a major earthquake occurs in the near future.

Metro Manila is also located in an active seismic zone known to trigger tsunamis. There are historical records of tidal waves hitting the foreshore at Manila Bay, although the configuration of the bay has a dampening effect on the height of waves. Storm surges resulting from typhoons pose a much greater threat to coastal communities. The storm surges during typhoons Milenya in 2006 and Ondoy in 2009 raised the sea level of Manila Bay by 2.5 metres, impacting on more than 1.2 million people living along the low-lying foreshore areas (National Disaster Co-ordinating Council, 2010). Apart from causing serious risks to life and property, storm surges have a severe impact on local water supplies. Communities along Manila Bay are particularly susceptible to outbreaks of waterborne disease because many houses are built over water and most have no sanitation. These areas will become increasingly vulnerable to environmental and health risks in future as sea levels rise.

Urban Services Risk Management

Metro Manila is a city facing chronic urban management problems and an ongoing crisis in providing urban services and utilities. It has experienced numerous failures to its traffic, energy and water supply systems. The city has tried to manage these recurring crises through changes to governance arrangements, including privatization of water and electricity utilities but most of these have only proved effective in reducing short-term risks.

The private provision of public transport is an interesting case. Metro Manila is renowned for its traffic congestion and pollution problems, yet it has one of the most efficient privately run local public transport systems in Asia. Philippine cities are famous for their *jeepneys* (extended ex-army jeeps). These are a popular, cheap and efficient means of localized public transport. Overall, however, authorities have failed to develop and manage an orderly, hierarchical and integrated road and public transportation system to serve the whole city and this has caused Metro Manila's current chronic traffic, transportation and air pollution problems.

Figure 12.6. Distinctive forms of public transport in Manila. *Left*: a *jeepney*; *above*: a *calesa*.

Lack of compliance with road rules and the love affair of many Manila residents with large cars compound the city's transport problems. When government sought to limit traffic by introducing 'odd and even day' number plates, the wealthier citizens simply bought a second vehicle. The failure of government to invest heavily in public mass transport and to address bad behaviour by drivers has resulted in a series of transport crises since the 1970s, when the city first began suffering high levels of congestion and regular gridlock. The national government's response at that time was to set up a light rail transit system. A 'Monorail' Plan and a Light Rail Transit Authority (LRTA) were established in 1980. The plan proposed a network of rapid transit rail lines along all major corridors of the city, to be constructed within 20 years. Work began on the first Light Rail Transit (LRT1) line in 1981 and was completed in 1984. The Metro Rapid Transit System (MRTS) running along EDSA (Epifanio de los Santos Avenue – the city's main inner ring road) commenced in 1997 and it became fully operational in 2004. After 30 years, however, the metropolitan rail transport network is less than 40 per cent complete. The major cross-city link (LRT2) has not started, nor has the linking of the northern extension of the LRT1 and the MRTS to create a circle line. The interchanges between the metro rail and bus networks along the arterial routes are poorly planned and make it difficult for passengers to transfer from one system to another. Key transport hubs like the domestic and international airports are not linked into the network either. Privatization of the management of the rail networks has not assisted in efficient passenger transfers. In short, as a result of poor planning and management, there is currently no effective integration of public transport systems in Metro Manila.

Manila has become a city that is choking on its traffic. Pollution levels are among the highest in Asia and are having a grave impact on public health (see below). Congestion and health costs add billions of pesos to the costs of production and cause productivity losses. Governments seem incapable or unwilling to take action to invest in solutions or even to confront the problems. The city's transportation network is becoming increasingly stressed, and a paucity of construction and a lack of maintenance have put the system at a high risk of failure. In a city divided by a large river, the failure of one of the five major bridge crossings would be disastrous.

Turning to Manila's power facilities, until the 1970s virtually all of these were operated by private interests.[4] In effect most centres in Luzon, as well as the larger cities in the southern islands of the Visayas and Mindanao, were served by vertically-integrated, investor-owned electricity companies that supplied limited areas. At that time the only significant government interest in the power sector was through the National Power Corporation (NPC), which had responsibility for developing the country's hydro-electric facilities. Shortly after the imposition of martial law under Marcos in 1972, however, the government enunciated a new policy which placed all generation and transmission facilities under the state-owned NPC.

As a by-product of that policy, the government believed that it could implement social pricing policies that kept electricity charges low for residential and other small consumers. The result was chronic underinvestment in power generation as revenues could not cover investment costs and the national budget did not have the capacity to make up the difference. 'Brown-outs' brought businesses in Metro Manila to a halt and some companies transferred manufacturing away from the Philippines to countries where electricity was more reliable and cheaper.

As the power crisis deepened, private ownership came to be viewed once again as the only viable solution for quickly addressing the shortages. In 1987 the government ended the NPC's monopoly on generation facilities and committed to developing a plan for privatizing the power sector. It developed a legal framework, starting with the 'Build Operate and Transfer' (BOT) Law of 1991, to enable private (including foreign) interests to own and operate generating facilities. These measures resulted in investment in some fifty-six generation projects, which progressively reduced shortages. However, because of the rushed public-private partnership programme, implemented at a time when the government was in a very poor bargaining position, the cost of electricity in the Philippines is now amongst the highest in Asia. Thus, while the risk of supply shortages in Metro Manila has been reduced, the cost of electricity through poor contract management has increased the city's economic exposure by undermining its competitiveness. The city will only overcome this by achieving increased competition in its electricity market.

Manila also faces the prospect of one of the most devastating events that could hit any city – the total loss of its water supply system and storage capacity. Several times in the early 1990s, the city almost ran out of water when the Angat Dam reached dangerously low levels. Chronic supply and theft problems meant that only a few areas of the city had access to a continuous water supply. Factories were forced to cut production and high-rise buildings had to bring in water by truck, not only because of limited supplies but also due to low water pressure. Health problems emerged in poor areas of the city as people began using contaminated water for drinking and ablutions. The water supply management authority at the time, the Metropolitan Waterworks and Sewerage System (MWSS), was found to be corrupt, highly inefficient and in debt to the tune of US$800 million (Esguerra, 2003).

By mid-1995, the city was fast running out of water. As an emergency measure to alleviate the supply problem, the government constructed a 13 kilometre long tunnel to divert water from the Umiray river to Angat Dam. However, the US$7.5 billion required to finance new infrastructure and fund repairs to the existing system was beyond the capacity of the government to raise, either locally or through an international development bank loan. As water demand was expected to increase by more than 40 per cent during the decade, the government had little choice but to privatize the city's water supply distribution and introduce other measures to reduce water consumption.

In January 1997, two companies – Maynilad Water Services and the Manila Water Company – were awarded concessions for the west and east zones of Metro Manila, respectively. These companies took over responsibility for managing and rehabilitating facilities to provide water and wastewater services to residents, in exchange for revenues from user fees (Mayong, 2006). They were given a tax break for 7 years, to help reduce investment risk and to maintain a relatively stable price for water to consumers. The privatization led to an immediate fall in water tariffs, by 43.5 per cent in the west zone and 73.6 per cent in the east zone. The threat of catastrophe appeared to have been averted as both companies began to expand water supply and maintenance.

Unfortunately, in June 1997, the Asian Financial Crisis struck. Both consortia had borrowed heavily off-shore and were exposed to foreign exchange risks. Unable to meet their obligations, the companies sought to pass on the costs of servicing their debts to consumers and introduced significant tariff increases within months. This impacted particularly on the poor.

Then, in December 2004, the main supply tunnel servicing the Angat Dam (which supplies 97 per cent of Metro Manila's potable water) collapsed. This occurred after illegally felled logs washed down the catchment as the result of floods, blocking and damaging the tunnel structure. The damage to the tunnel threatened to deprive the city of water within 4 months. Because the dam is also located on an earthquake fault and has serious leakage problems, its potential for failure is very high. In 2005, Manila Water was unable to meet its commitments to the government. Services could not be expanded, and this left many areas without water. To date the supply situation has not improved.

A number of lessons can be learned from Manila's water crises. The most important is the need for government to ensure that urban water supplies are not reliant on one source. If the Angat Dam fails, chaos would ensue and it would take months for the city to develop alternative sources of water to meet demand. Second, while privatization did provide a short-term solution to the city's water supply problems, it came at a cost to consumers. Metro Manila now has some of the highest water costs in Asia, which was not, of course, what the government had intended when it moved to privatization. Water risk management in Metro Manila is likely to become even more challenging in future if predictions of the impact of climate change on water resources become a reality (WWF, 2009). Alternative strategies for water management are vital to provide safe and reliable water supplies for Manila's residents in the future.

Economic Shocks

Manila is the main financial, commercial, transport, government and industrial centre of the Philippines. It is the principal centre in the country for the manufacture of chemicals, steel, textiles, clothing and electronic goods. It is also the wealthiest city in the country, with one study estimating that 32 per cent of the nation's GDP is produced by the Metro Manila area (WWF, 2009). As a megacity, Manila creates its own market but, as noted earlier, because of its polycentric structure and transport system, it also generates diseconomies through the transaction costs of freight and distribution.

About 65 per cent of Metro Manila's GDP is generated by the service sector. Retailing dominates the economy, with Manila having some of the largest shopping malls in Asia. The expanded SM City North Mall on EDSA comprises 425,000 square metres of shops, food courts and supermarkets, and is reputed to be the third largest mall in the world. Metro Manila acts as a super retail and supply centre for much of Luzon. Its malls also contribute to its importance as a major tourist destination, attracting over a million foreign visitors per year. Remarkably, it is projected that, at the current growth rate, Manila's tourism volume will surpass that of Singapore by 2020 (WWF, 2009).

The economic performance of Metro Manila has tended to lag behind that of other large Asian cities. The Philippine economy differs from the economies of other Southeast Asian countries of comparable population size in that it has a higher service industry component in its national GDP. The level of remittances also plays a very prominent role in the economy, as the Philippines has one of the highest per capita rates of remittances in the world. In 2009 more than 9 million overseas Filipino workers sent home an estimated US$20 billion (World Bank, 2010). This has had an effect on patterns of investment and capital flows, which sometimes act in a counter-cyclical manner to regional and global economic trends. The Philippines, being less export dependent, has generally been able to respond better to economic crises than other Asian countries – especially in the case of the Asian Financial Crisis and the Global Financial Crisis. Some reasons why the Manila economy seems to be more resilient to economic shocks are elaborated upon below.

The 1997 Asian Financial Crisis had a substantial impact on several major Asian cities, including Bangkok, Hong Kong, Jakarta and Kuala Lumpur. Property markets were affected most, as many construction projects were funded through foreign debt. The rapid deterioration in exchange rates left local residential and industrial development projects with crippling debt, forcing many into receivership. Metro Manila was less affected by the crisis, as external borrowings to fund property development were at much lower levels than in other Asian countries, with funds often having been secured through contributions from participants in the Filipino diaspora, rather than through the formal banking system. Also, the development industry in the Philippines is dominated by a small number of powerful family companies which secure most of their capital requirements from private sources. The effect of the crisis in Manila, therefore, was that many projects halted but local companies and family businesses absorbed accumulated debt in the expectation that the economy would recover quickly, which it did.

Remittances from overseas also helped greatly to reduce the social impacts of the Asian Financial Crisis, and subsequently the Global Financial Crisis, providing private welfare support for low-income earners, with a high proportion of remittances being used to supplement family incomes and outlays on housing. Migrant workers also tend to return home with their accumulated savings at times of international financial crisis, as in 1997 and 2008.

However, while the Philippines has been able to absorb the effects of economic shocks somewhat better than other Asian countries, the opposite tends to occur during the recovery phase, when significant capital outflows have led to a series of budget crises. The country, and especially Metro Manila, is then deprived of capital for investment in projects that would have helped develop the domestic and export economies. This phenomenon has been a significant factor in explaining the slower rate of economic growth in Manila over recent times as compared to other Asian cities.

Remittances from expatriate workers also play a key role in recovery after natural disasters in the Philippines. There is a consumption 'rebalancing' by households affected by floods or typhoons, which leads to the deployment of savings to repair residences and vehicles, and to buy basic appliances, furniture and fixtures to replace those damaged. An increase in short-term informal loans to family and the community to aid recovery occurs, and this tends to result in a reduction in business activity unrelated to construction. For example, expenditures on non-essential food and consumption of tobacco, beverages and clothing are delayed until such time as private debt or loans to assist recovery are paid back.

Amongst the urban poor, there is a remarkable level of co-operation and trust in sharing resources and labour to support individuals during economic shocks, and in their reconstruction efforts. This indicates the important role that social capital plays in rebuilding communities after a disaster. Middle-income groups, on the other hand, have tended to rely more on insurance, banking and government support to recover from economic and natural crises. Recovery for the middle class tends to be more business driven than community driven.

Environmental Risks

Metro Manila has one of the highest levels of air pollution in Asia (Haq *et al*., 2006). An Asian Development Bank report showed that air pollution in Metro Manila had reached critical proportions in the mid-1990s and that immediate action was required to address the problem (ADB, 1998). The World Bank estimates that 6,000 tonnes of particulates, chemicals and other pollutants are emitted daily by Metro Manila's 3 million vehicles, factories and households (World Bank, 2005). In 2001 the annual average level of particulate emission for Metro Manila was three times the World Health Organization's recommended minimum standard. Air pollution contributes to an estimated 1,900 deaths per annum in Metro Manila, with 8,400 seriously affected by respiratory problems. The cost to the Metro Manila economy was estimated at $US392 million in 2001 (World Bank, 2005).

In 1999, in an attempt to address Manila's serious air pollution problem, the government rushed through the Clean Air Act. However, this has done little to tackle the problem, largely because the regulatory authorities have not had the capacity to implement emission standards and have been unwilling to impose penalties on violators. The failure to address Manila's growing air pollution problem has not only increased risks to public health, but has resulted in an increased incidence of acid rain. In turn, this increases the chance of concrete panels and structures failing and being dislodged during a typhoon or earthquake. With the rapid rise in the number of high-rise apartment and office buildings in metropolitan centres, the risk to public safety of falling material is mounting. These risks are not always obvious but could one day contribute to a catastrophic event, with cumulative causes, resulting in building collapses in the city. Unfortunately, little is being done to reduce the indirect effects of air pollution on the built environment and on public health.

The Intergovernmental Panel on Climate Change (IPCC) has noted that Southeast Asia will experience an 'increase in the frequency of intense precipitation events', and 'extreme rainfall and winds associated with tropical cyclones are likely to increase' (Intergovernmental Panel on Climate Change, 2007, p. 879). Local air pollution is a contributing factor to global climate change, which is expected to impact significantly on Metro Manila. The frequency and patterns of rainfall will influence the quality and quantity of the city's water supply. As noted earlier, climate change will also affect sea levels and increase the propensity to flooding in low-lying areas.

The MMDA has taken some steps to establish programmes to mitigate and adapt to climate change. A Metropolitan Air Quality Action Plan for 2006–2009 was prepared for the Metro Manila Air-Shed Governing Board. The objectives of the plan were to support: (*i*) implementation of urgently needed road rehabilitation works; (*ii*) traffic engine-ering training and management; (*iii*) strengthening of ambient air quality management; (*iv*) public health monitoring; and (*v*) 'anti-smoke belching' programmes for vehicles. Thus far, the measures taken have been disappointing, with political tussles between agencies slowing progress with the implementation of programmes. Several local governments have taken small initiatives to try to reduce air pollution and greenhouse gas emissions but these have proved relatively ineffective in the absence of an overall approach to the management of emissions across the whole metropolitan region.

Nearly 60 per cent of water pollution in Metro Manila is caused by untreated residential sewage being discharged into open waterways or drainage channels (NEDA, 2007). Only about 15 per cent of households are connected to a sewerage system. The Pasig River is effectively an open sewer. Most of the city's smaller rivers and streams are also heavily contaminated from untreated residential and industrial waste. The lack of sanitation facilities, coupled with potential human contact with raw sewage, represents a high risk to health, especially from diarrhoea and cholera, as well as other infections resulting from using contaminated water for ablutions.

The local governments of Metro Manila and the MMDA recognize the seriousness of water pollution but it is a difficult problem to solve as it requires expensive infrastructure, the relocation of people and the acquisition of land, as well as behavioural change. The last is a particularly challenging issue and it will take time to educate the city's population about sustainable waste disposal methods. Measures have recently been taken, through the Pasig River Cleanup Project, to begin the rehabilitation of 24 kilometres of polluted river. The project will involve progressive reduction in the discharge of industrial toxic waste and the use of community-based facilities for the treatment of solid and liquid wastes to reduce the high level of discharge into the river. It is hoped that this will help to clean up the river, but it is likely to take several decades to restore the river's health.

Another compounding factor affecting groundwater quality is that coastal areas of Metro Manila are sinking. Many areas are already below sea level and, as noted earlier, are already at great risk from predicted increases in sea level. The subsidence of these areas, which is the result in part of excessive loss of groundwater caused by over-extraction through bores, has led to saltwater intrusion and contamination of the remaining groundwater. Many wells are close to septic tanks while industrial wastes also seep into the groundwater. The contamination of the freshwater aquifers in these areas has resulted, once again, in greater risks to the urban poor who have nowhere else to go. Those who are able to buy water at high prices from tankers, but in many cases the quality of this water is little better than that drawn from local wells.

Solid waste is also a major environmental problem. Metro Manila generates around 5,500 metric tonnes of solid waste per day, of which only about 70 per cent is collected. The Payatas waste dump, in Quezon City, handles more than half of Metro Manila's waste. The management of waste became a critical problem in Metro Manila and other cities when it was realized that many landfill sites were discharging high levels of contaminants into groundwater systems. This was recognized as a national problem and resulted in government passing the Ecological Solid Waste Management Act in 2000. This Act marked a turning point in national attitudes to improving solid waste management and resource conservation.

Waste management sites have proved problematic in Metro Manila in the past, but solid waste can also provide important opportunities for materials recovery and local employment. The way in which a disaster at Payatas in 2000 led to significant changes to waste management practices is the focus of a case study later in this chapter.

Social Risks

Manila is a city of socio-economic extremes. In a country where the top 30 per cent of income earners account for almost 90 per cent of consumption (Encyclopedia of the Nations, 2007), a very large proportion of the city's population is vulnerable to the traps of poverty, disease and malnutrition. Over 20 per cent of Metro Manila's population fall below the poverty line and 35 per cent live in informal settlements (ABD, 2003). More than 12.5 per cent of the population is jobless. Between 30 per cent and 40 per cent of the working age population is employed in the informal sector, mostly as petty traders or in small-scale service and manufacturing enterprises.

The level of impoverishment in the low-income areas of Metro Manila gives rise to a volatile social system where neighbourhood violence is common. Without security of tenure, an estimated half a million people residing in poor urban communities also live in fear of eviction by the government because they happen to live on land designated for major infrastructure projects. 'The failure of the government's resettlement programs and continued neglect of housing as a basic right has been become a strong focus of NGOs concerned about the plight of these communities' (Perlman, 2007).

More recently, terrorism has become a significant social risk for people living in Metro Manila. In particular, the prolonged civil war in Mindanao presents a threat to the nation's capital, and recent bomb attacks have resulted in deaths. While the government is seeking to negotiate a solution to the Mindanao problem, the peace process is likely to be prolonged. Protection against terrorism and civil crimes is purchased at a high cost by those living in the city's wealthier residential areas. Security also adds greatly to the cost of doing business in Metro Manila when compared to most other Asian cities, where crime rates and terrorism risks are lower. The solution to this problem will involve tackling complex issues, in Mindanao and elsewhere, encompassing poverty, land tenure and economic development. It is highly unlikely that the social risks that confront Metro Manila can be resolved by simply addressing the problems locally and by creating fortress suburbs for the rich. Lateral thinking and integrated strategies are required to manage the city's external risks if Metro Manila is to be a safer city in the future.

Political and Legal Risks

The politics of urban development in Metro Manila are very dynamic, with regular changes of administration at elections, and widespread accusations of corruption and nepotism in local government. Political instability has significantly increased the risk profile of Metro Manila as a city for doing business. The high level of political volatility has tended to drive business and government into supporting projects and activities that generate profits and political gains in the short term. The myopic view of the development process caused by short cycles of local government politics is significantly hindering commitment to the provision of basic infrastructure and services. Short political cycles provide opportunities for rent-seeking and are a contributing factor to the Philippines being ranked as one of the most corrupt countries in Asia (Quimpo, 2009).

The Philippines is also probably one of the most litigious and bureaucratic democracies in Asia. The bureaucratic systems in central and local government are multi-layered, with six tiers of public administration – national, regional, provincial, city, municipality and *barangay*. Metro Manila also has an additional level of metropolitan governance. The effect of these multiple levels and of complex and dated legal systems is that it can take years for development decisions to be made. Consequently, development tends to occur without formal endorsement or approval, especially on land where tenure and ownership are unclear. Authorities are reluctant to prosecute or take action to prevent the development of illegal structures and the private sector uses the legal process as a means of stopping rival development projects from going ahead.

The legal profession is exceedingly profitable in the Philippines, with the country having as many lawyers *per capita* as the United Kingdom and Japan (August, 1992). Few improvements have been introduced, however, to strengthen the legal basis for planning, land and environmental management. This adds greatly to the uncertainty and risks associated with land and property transactions. A step forward was made in 2008 with the approval of a new law on environmental protection and the establishment of a specialist court to oversee the application of this law (Vina *et al.*, 2009). The law relating to planning and development approvals still remains to be streamlined, however, and the slowness of current procedures adds greatly to development costs and delays in the release of land to low-income earners. Judicial and bureaucratic costs and delays not only put the poor at risk, but also provide a further barrier to investment by businesses. An initiative to reduce governance risk by speeding up development processes has been taken by Taguig city and this is presented in a case study later in the chapter.

Risk Management and Resilience Strategies for Metro Manila

A United Nations study in 2009 ranked the Philippines seventh among the ten Association of Southeast Asian Nations (ASEAN) in its capacity to manage disasters (UNISDR, 2009). As earlier sections have shown, there are many risks facing Metro Manila which have the potential seriously to impact on its environment, population and economy. The city is poorly equipped to prevent and manage the impacts of these latent and cumulative risk events. Much more needs to be done to prepare for disasters and to foster resilience in the shock recovery phase. Some measures currently in place to improve disaster management and encourage resilience are discussed below.

Integrated Disaster Management Planning

There has been a long history of attempts by government to prepare and assist communities in the Philippines to manage the impacts of natural disasters. The first attempt was during the Japanese occupation with the creation of the Civilian Protection Service (CPS) (National Disaster Co-ordinating Council, 2008). The CPS was empowered to formulate and implement plans and policies to protect the civilian population during air raids and other national emergencies. In 1954 Civil Defence Councils were established to provide protection and welfare to the civilian population during national emergencies. In the aftermath of Typhoon Sening which caused severe flooding in Metro Manila in 1970, a 'Disaster and Calamities' plan was prepared, leading to the establishment of the National Disaster Control Centre and local Civil Defence Associations. CDAs were primarily intended to deal with disaster management after the event, rather than mitigation and prevention. The responsibility for disaster prevention rested with many other agencies of government, but little co-operation or knowledge sharing occurred between these agencies.

In 1991, the Philippine government made its first attempt to introduce integrated disaster planning and to foster more sustainable urban development in its Medium-Term Philippine Development Plan. All local governments were required to incorporate disaster management into their local development plans (EMI, 2005, p. 6). Few did so, however. Some disaster mitigation and recovery strategies were adopted by local governments in Metro Manila, but there were never sufficient resources available to implement key projects and programs.

In 2007, a Mega-City Disaster Risk Management Plan was prepared for Metro Manila (EMI, 2007). This was the first time a risk management and mitigation plan had been prepared for the whole of the metropolitan area. This plan introduced measures to:

◆   Create a stronger Metro Manila Disaster Coordinating Council.

◆   Promote and adopt disaster management ordinances in each city and municipality.

◆   Promote the revitalization of city/municipal/*barangay* disaster co-ordinating councils.

◆   Enhance lateral and vertical inter-agency and inter-governmental communication and co-ordination.

◆   Institutionalize disaster risk management within local government frameworks and financing.

◆   Promote local government mitigation planning through existing planning tools.

◆   Conduct training needs assessments and develop capacity building programmes.

The Plan has much to commend it but is yet to be fully implemented and, like so many other plans for managing Metro Manila, the resources for implementation are lacking. Several cities have established their own disaster risk management plans recently, however. Makati City, for example, has mapped disaster impact areas that could be affected by typhoons, floods and earthquakes (Seva, 2008). The city has also carried out a series of emergency training programmes and seminars to advise the community on what to do in a crisis, and to train emergency response teams (EMI, 2005). The programme has been extended to deal with mass evacuation in case of a major terrorist attack, health scare or industrial accident. But even Makati City, one of the better-resourced cities in Manila, and also Quezon City, the centre of government, lack the technical capacity and finance to implement fully their disaster management plans. This became apparent during Typhoon Ondoy in 2009. Nevertheless, responses to the typhoon in these two cities were somewhat more effective than in other cities in Metro Manila, which did not have any plans at all. The preparation of local city disaster management plans should be mandatory and their integration into a metropolitan-wide plan is needed urgently to address future events with the potential to cause widespread damage across the metropolitan area.

Case Studies of Community Resilience Initiatives

While strategies to support urban resilience in Manila at city and metropolitan levels still have a long way to go, several local governments and communities have developed and implemented projects which are examples of good practice that could be applied elsewhere. The following three case studies explore different dimensions of economic, social, governance, physical risks and resilience in three cities in Metro Manila. Some of these case studies have had localized impacts, but others have had a more citywide impact and have attracted international attention. Each case study includes some background information on the particular risk event and a description of the impacts. The subsequent response by city governments and communities is then described. Finally, the lessons gained from the resilience strategy and its future relevance to urban and disaster management practices in Metro Manila are summarized.

Fall and Transformation of the Shoe Capital of the Philippines

The city of Marikina, located in the eastern part of Metro Manila, has been the centre of textile, clothing and fashion manufacturing in the Philippines for more than 125 years (Marikina City, 2007). Until the mid-1980s, the shoe industry in Marikina was heavily protected by tariff barriers and other policies designed to support the development of import-substitution manufacturing industries. After the Marcos era, import substitution policies were replaced by more open market economic policies designed to encourage competition and attract new industries to the Philippines.

After the Philippines gained membership of the World Trade Organization (WTO) in 1995, the Marikina shoe industry began to lose its competitiveness to countries like China, Korea and Taiwan (Scott, 2005). By 1998 the industry had fallen into a steep decline. Exports dropped from $US146 million in 1998 to US$25 million in 2006. The local shoe industry was also adversely affected by a high incidence of smuggling, the relocation of local manufacturers to other Southeast Asian countries (Bacalla, 2004, p. 3), and the unwillingness of local consumers to continue purchasing more expensive local shoes (Scott, 2005).

The Philippines now imports most of its shoes. What remains of the Philippines export shoe industry market is mainly mid-range footwear, slippers, sandals and sports shoes, most of which are exported to Mexico, Japan and the USA (ACC, 2008, p. 26).

The majority of these are produced in Marikina. There are about 340 firms in Marikina still involved in the manufacture and export of shoes, with the majority having fewer than ten employees. Most small firms are engaged in manual production, with only a few larger firms using more mechanized processes. It has become the practice to import parts, such as the soles and heels of shoes, and this has led to a significant contraction in the number of local suppliers.

The predicament facing shoe manufacturers triggered efforts to rebuild the industry and to modernize it to become more internationally competitive. The industry recognized that it had not established adequate linkages with international shoe manufacturer supply chains (Scott, 2005, p. 79), nor had it been willing to collaborate to reduce the transaction costs of doing business. The capacity of many firms was underutilized and there was a reluctance to specialize. Marikina retains its skill base in shoemaking, as many of the skilled workers still live in the area, but most have found other forms of employment.

Several initiatives were introduced by the Marikina City government and shoemakers in the early 2000s which have helped to halt the decline in exports and revive the domestic industry, albeit on a much smaller scale than previously. The first was the Philippine Footwear Design Competition ('Sapatero Festival') in 2002, which was supported by the Marikina City government. The objective of the event was to provide an opportunity for shoemakers to market and sell their products and to increase the awareness and visibility of the local industry. In 2002 the Samahan ng Magsasapatos sa Pilipinas (League of Filipino Shoemakers) was established to work together to rebuild the shoe manufacturing industry. The Marikina City government and the Philippine Footwear Federation also lobbied the national House of Representatives successfully to pass the Footwear, Leather Goods and Tannery Industries Development Act, which enabled the shoe industry to secure government assistance to help revitalize the local industry (Scott, 2005, p. 90).

Another important initiative to support the recovery of the industry in Marikina was the establishment of an industry cluster organization, SikapMo Inc, in 2003 (Scott, 2005, pp. 85–88). SikapMo has eighty members (including fifty manufacturers) and is the prime vehicle for shoe manufacturers in Marikina to promote their products in global markets (Valt, 2007). 'Marquina' was established as the common brand name for shoe products by SikapMo members. SikapMo also conducts training programmes and seminars for factory owners and workers to introduce them to more efficient manufacturing and costing systems.

The collaborative efforts of the Marikina City government and local shoe manufacturers have been crucial in reviving the local industry. It is still going through substantial restructuring and transformation, but SikapMo members' sales reached 50 million pesos (US$1 million) in 2006, up from 2 million pesos 3 years earlier, when the cluster was first formed. The partnerships that have been created in the industry provide a strong institutional base to facilitate appropriate programmes and projects to boost production, especially by small and medium-scale operations.

Joint marketing efforts by manufacturers are also helping to establish linkages with global markets. The partnerships facilitated by the local government have enabled a sense of ownership and trust to develop between shoe manufacturers, because their involvement has been collaborative. The programmes and projects established have also been helpful in promoting locally manufactured shoes to Filipino consumers. The shoe industry is transforming and is on the road to recovery, although it is unlikely it will ever regain the dominance it once enjoyed in the times of tariff protection and strong national government support.

Payatas: a Waste Mountain that Moved

The Payatas dumpsite, located in the northeast part of Quezon City, was established in 1993. The area was initially intended to be a 30 hectare upmarket housing estate and, prior to its becoming a dump site, the surrounding areas had already been developed with well-planned streets and homes. The dump currently receives over 6,700 cubic metres of waste a day from Quezon City and Pasig City (Tanchuling *et al.*, 2009). Since it opened, Payatas has received more than two million tons of waste, making it one of the largest dump sites in Asia (see figure 12.7).

Initially the site was intended just for disposal of garbage from the neighbouring Payatas housing development. However, after the closure of Manila's notorious 'Smokey Mountain' landfill site was ordered by the Aquino government in the early 1990s, the volume of waste directed to Payatas increased significantly. In addition,

Figure 12.7. Payatas waste dump, Quezon City.

a number of squatters from Smokey Mountain relocated to Payatas. The Payatas dumpsite was itself scheduled to be shut down in 1998, but the Quezon City government requested that the MMDA extend the use of the site because it lacked the financial capacity to transport waste to an alternative landfill site in San Mateo in Rizal Province, east of Manila (Merry *et al.*, 2005).

In July 2000, Manila was hit by two typhoons, which brought heavy rains over a period of 10 days. The dumpsite became saturated and began to move. A mountain of garbage and mud flowed rapidly downhill, burying hundreds of slum dwellers and causing 278 deaths. In addition, the landslide released methane gas and fires broke out, causing significant damage to property and further loss of life (Merry *et al.*, 2005). The severity of the tragedy was compounded by other factors, including a lack of compaction of garbage; the steepness of slope; poor site management practices; the absence of a barrier fence; and the failure to remove slum dwellings from dangerous areas of the dumpsite.

Five days after the incident the dumpsite was closed. However, it was re-opened in November 2000 – with the exception of the collapsed section – as there was no alternative site available which could take Metro Manila's waste. Ironically, the request to re-open the site was initiated by the residents of Payatas, 8,000 of whom depended on scavenging at the dump to make a living (Gonzales, 2003). Nevertheless, to avoid a repetition of the tragedy, the Quezon City government established a 50 metre wide danger zone at the foot of the waste heap and demolished all squatter dwellings within that zone.

Following the tragedy, the Philippine government established a taskforce for the rehabilitation of the site and the relocation of affected persons. The rehabilitation of Payatas turned out to be a good example of multi-sectoral collaboration between the Quezon City local government, slum dwellers and NGOs. The expertise offered by the University of the Philippines, the local MAPUA Institute of Technology, the University of Singapore and a private group, IPM Environmental Services, was also invaluable in assisting the recovery effort (Gawad Galing Pook, 2008*a*).

Four NGOs worked in partnership with the city government to relocate families residing in the danger zone. Quezon City was prompt in the implementation of the Ecological Solid Waste Management Act of 2001, which mandated the closure of all unsafe dumpsites by February 2006. The conversion of Payatas to a controlled waste management facility began in 2004. This included the reshaping, stabilization and greening of the slopes of the waste mounds, improving the drainage system and the accessibility of the area, and extracting the gas. Several other initiatives have also been undertaken to reduce the risks of a future tragedy (Quezon City, 2009). These include organizing scavengers into formal groups and establishing a used tyre retrieval project in collaboration with a cement company. Those who wish to use the waste management facility are now required to pay an entrance fee. Ten per cent of the fees collected are allocated to the *barangay* council, while the remaining 90 per cent is managed by the city government. The income from the dump is used to pay for road repairs, drainage, security and the provision of local ablution facilities.

The scavenger groups are intended to reduce competition and violence between individuals who derive their living from the Payatas facility. These groups are consulted regularly about the operation and management of the site. Moreover, scavengers, recyclers and junkshop operators now have the opportunity to take advantage of financing, education and skills training which will enable them to earn additional income or to embark on an alternative livelihood if they choose (Sia-Su, 2007).

A 100 kW pilot 'Methane Gas to Power Generation' project was inaugurated in 2004 in collaboration with an Italian company and the Philippine National Oil Company Exploration Corporation (PNOC-EC). This biogas emissions reduction project is registered as a Clean Development Mechanism under the Kyoto Protocol of the United Nations Framework Convention on Climate Change (UNFCC). It has the potential to be expanded to 200kW in future.

The Payatas tragedy provided important lessons for local governments in the Philippines about risks associated with poor waste management. There have been several positive outcomes from this disaster. The initiatives undertaken by Quezon City to revitalize Payatas resulted in it being given the Galing Pook[5] award for innovation and excellence in local governance in the Philippines in 2008 (Gawad Galing Pook, 2008*a*). Payatas has created new local employment opportunities through the recycling and reprocessing of its waste. The biogas project has enabled electricity to be generated from waste, and this has created new employment opportunities, facilitated technology transfer and generated additional financial resources from the sale of carbon credits and electricity produced by the power plant. Finally, the project has created opportunities for innovation, with soil from the old Smokey Mountain waste dump being processed, then bagged and sold for growing vegetables. The recovery from the tragedy in 2000 is a remarkable example of urban resilience.

Taguig City: From Laggard to Leader in Business and Housing Development

Taguig City, on the western shore of Laguna Bay, has an area of 45,382 square kilometres. Formerly an urban municipality, Taguig was declared a city in 2004 and currently has a population of 700,000. It is one of the fastest growing cities in the Philippines.

Prior to achieving city status, Taguig was a very poor area and struggled to attract investment (Gawad Galing Pook, 2008*b*). It was developing as a dormitory settlement within Metro Manila, with a high level of poverty and unemployment and large, uncontrolled squatter settlements (Taguig City, 2009). High crime levels put lives and property at risk. Solutions to the many problems associated with Taguig's rapid urbanization needed to be found. The city had several large areas of relatively unencumbered land available for development, but neither the capacity nor the machinery of government to develop them. The largest of these was the former military area of Fort Bonifacio, in the northwest of the city, adjacent to Makati City. The municipal government recognized that, if the city were to prosper, it would have to improve its attractiveness to potential investors and developers.

In 2001 a newly elected municipal mayor set about enhancing the attractiveness and security of Taguig as a place in which to live and invest. In 2002, a city development strategy was prepared with the support of the World Bank Cities Alliance. The strategy was formulated around three development objectives: to improve delivery of basic services; to widen economic opportunities; and to enhance local participation in development (Cities Alliance, 2005). Subsequently, in 2008, Taguig developed a '2020 Vision' focused on improving health, education, environment and sanitation services; infrastructure and utilities; and housing. The local government had previously implemented an ordinance, encouraging investors and employers to hire 70 per cent of their employees from within Taguig (ADB, 2004, p. 65). More recently, it has adopted a comprehensive land-use plan to manage the development of the city (Taguig City, 2009). Taguig has been very active in fostering public and private sector partnerships in education, social welfare and development to overcome its management capacity problems and to elevate the city to a higher status nationally and internationally.

In 2008, the city also introduced a number of regulatory development reforms. Taguig now tops the list for having the easiest entry requirements and fastest means to start a business in the Philippines. It takes 27 days to complete the application process to start a business in Taguig (World Bank, 2008, p. 7). This takes twice as long in other cities in Metro Manila. The city has introduced a customer service centre and clients are only required to submit their documents to this 'one stop shop' (World Bank, 2008, p. 11). The city government has also introduced Integrated Service Information Software (ISIS), which provides an online system linking the city government and all registered business establishments in the area. This online service makes government-related transactions very convenient for businesses in the city. Taguig also has the advantage of having the fastest process for issuing construction-related permits in Metro Manila.

In September 2006, the Taguig City government launched the Family Townhomes Project to address the squatter housing problem (FAB4, 2008). The desired outcome of the project is the rehousing of squatters in affordable homes. Under the Family Townhomes Project, the city government meets site development costs while NGO partners provide housing subsidies. Low-income residents are given the opportunity to purchase the units with a minimum payment of between 500–1,800 pesos (US$11–38) per month for 30 years. In lieu of a cash deposit, beneficiaries are required to undertake 1,000 hours of voluntary work building houses.

By February 2009, the Family Townhomes Project had provided 535 low-income families with good housing in the area of Pinagsama Village and Western Bicutan. A further 242 housing units were completed in three other *barangay* areas in 2009 and an additional 843 units were scheduled to be built in 2010–2011 (Gawad Galing Pook, 2009). The Family Townhomes Project has been recognized as the country's best housing programme and was a winner in the 2008 Galing Pook Awards (Gawad Galing Pook, 2008*b*).

Faced with the need to manage rapid urbanization, Taguig has achieved significant results by thinking creatively in attracting investment and streamlining business regulations. Civic leadership and entrepreneurship have enabled the city to develop a clear vision of its development priorities, to plan for growth and to provide housing and jobs for the urban poor.

Lessons for a More Resilient Manila

Metro Manila, like many other Asian cities, has become a very large and sprawling city. Its extended mega-region is expected by 2020 to have 20 million people, more than half of whom may well be living in sub-standard accommodation. Metro Manila has experienced many disasters during its history but it has survived and has shown a remarkable ability to recover from past challenges. As Manila continues to grow, however, the risks it faces are likely to become more severe and complex, and the challenges to government and its inhabitants in addressing these risks will become increasingly daunting.

The citizens of Metro Manila have learned to live with natural disasters. With little prospect of employment in the provinces and regional cities, tens of thousands of people migrate to Manila each year in search of employment and a better way of life. The ability of the city to absorb these increasing numbers continues to stretch the capacity of authorities to provide basic services, housing, transport infrastructure and community facilities. New arrivals are being forced more and more to settle on marginal land, susceptible to flooding and subsidence, with insecure tenancy. The vulnerability of citizens of Metro Manila is therefore mounting, and governments, in general, seem unable to reduce their level of risk exposure. The situation is made worse by constant rent-seeking by urban developers and local politicians; weak land administration, management and property tax systems; and high levels of tax avoidance.

Major reforms are necessary to improve land-use and development practices if Metro Manila is to reduce the current level of risk to which its urban systems are exposed. Such reforms have been discussed for decades and many plans and regulations have been put in place but, unfortunately, few of these have been implemented or enforced. Public authorities, for example, were well aware of the dangers of flooding in the Marikina Valley before Typhoon Ondoy struck, but authorities chose not to adopt mitigation measures which had been identified more than 30 years earlier and which would have reduced the impact of this calamity.

The failure of the planning and development control system to manage development risks has resulted in Metro Manila putting itself at even greater risk of future catastrophe. As discussed earlier in the chapter, the city has developed as a polycentric city, and there is enormous competition and rivalry between local governments to attract investment, create jobs, build houses for all and provide basic infrastructure. The failure of the MMDA effectively to plan and co-ordinate development, and thereby to reduce the opportunities for the development industry to play off one city against another, has added to the problems and uncertainty facing the city. Effective planning at the level of the metropolitan region remains an obvious priority, but there is no sense yet in Metro Manila that the constituent cities are interested in collaborating in order to provide better infrastructure and reduce the transaction costs that will enable the metropolitan area as a whole to build some form of competitive advantage to attract business and investment. A similar scenario applies to disaster management planning and mitigation, climate change adaptation and pollution control. These are metro-wide issues requiring metro-wide solutions beyond the resources and capabilities of any one of the local governments that make up Metro Manila.

Despite these enduring issues, the case studies discussed provide some useful and encouraging lessons on risk and resilience. Leadership and regime change is a critical factor in resilience. New leadership at the local level, sometimes as a result of a crisis, resulted in the introduction of fresh ideas, innovative practices, administration reforms and new development and urban management models in these cases.

Secondly, learning to think globally and to act or respond locally remain critical to dealing with external events and shocks. There is currently a low level of awareness by local government and businesses of the importance of linking into global supply chains, and the global economy more generally, through collaboration with international firms and business partners. The shoe industry in Marikina collapsed partly because of a failure in this respect. Such arrangements are critical to protecting and developing markets, reducing business transaction costs and minimizing risks.

Thirdly, collaboration and partnerships are also important in overcoming resource and capacity constraints in local government. They help to spread risks, but they require greater levels of trust, transparency and accountability. This was very apparent in the recovery effort made for the Payatas Waste dump disaster, and also in the Taguig Family Townhome project. And fourthly, communities recover from disasters much better and faster if control over decisions is delegated and there is self-governance and ownership of the recovery project. Communities and businesses in the Philippines often have much greater capacity to mobilize and leverage resources than governments, and there are much closer levels of scrutiny over the way resources are used.

Conclusion

The study of risk and resilience is important in improving the management of cities and reducing their vulnerability. While there is now a substantial research focus on sustainable cities, there remains a generally poor understanding of the processes and factors that make some cities more vulnerable to shocks and disasters than others, and also of why some cities are more resilient in recovery (UNISDR, 2008). There does seem, however, to be persuasive evidence that more resilient cities are those where local communities and citizen groups are actively engaged in local development processes and share in the responsibility for safeguarding and improving their local areas.

This chapter has emphasized the importance of a systematic approach to assessing urban vulnerability and to identifying principles and opportunities for building resilience in urban systems. The Asia Regional Task Force on Urban Risk Reduction notes that 'building resilience is particularly important in areas such as coastlines, cities, agricultural land and industrial zones which are often the most impacted by humans. It is the same areas that people value highly, both economically and aesthetically, and upon which the wellbeing of society often depends' (UNISDR, 2008, p. 2).

Metro Manila is a city which is highly-vulnerable to natural disasters but the vulnerability of many of its citizens is further increased by the city's inadequate and failing infrastructure, and by the large numbers who are obliged to live in informal settlements, with no security of tenure, on flood plains, low-lying coastal land, or sites close to sources of pollution. This chapter has documented some successful examples of local resilience and the lessons of these should be learned. There is an urgent need, however, for Manila to adopt effective strategic planning arrangements for the extended metropolitan region, in order to increase the capacity to address and respond in a co-ordinated way to the many very serious risks to the city's future liveability and sustainability that it clearly faces.

Notes

1. Metro Manila is made up of the cities of Manila, Caloocan, Las Piñas, Makati, Malabon, Mandaluyong, Marikina, Muntinlupa, Navotas, Paranaque, Pasay, Pasig, Pateros, Quezon City, San Juan, Taguig and Valenzuela.
2. The author is indebted to Corella Cabannas for assistance in collecting the background material for the preparation of the case studies.
3. Severe acute respiratory syndrome.
4. Inputs from Dr Michael Lindfield of the Asian Development Bank in relation to this section are gratefully acknowledged.
5. Galing Pook is a leading Philippines foundation continuously promoting innovation and excellence in local governance.

References

ACC (2008) *State of the Sector Report on Philippine Footwear 2006*. Calgary: Agriteam Canada Consulting

Ltd. Available at http://www.pearl2.net/images/ssr/ssr2006/Footwear2006.zip. Accessed 15 May 2010.

ADB (1998) *Philippines: Metro Manila Air Quality Improvement Sector Development Program*. Manila: Asian Development Bank. Available at http://www.adb.org/documents/rrps/phi/rrp-R45506.pdf. Accessed 18 April 2010.

ADB (2003) *Metro Manila Urban Services for the Poor*. Manila: Asian Development Bank. Available at http://www.adb.org/Documents/Profiles/PPTA/31658012.asp. Accessed 14 March 2010.

ADB (2004) *City Development Strategies to Reduce Poverty*. Manila: Asian Development Bank.

Alberti, M., Marzluff, J.M., Shulenberger, E., Bradley, G., Ryan, C. and Zumbrunnen, C. (2003) Integrating humans into ecology: opportunities and challenges for studying urban ecosystems. *BioScience*, **53**(12), pp. 1169–1179.

August, R. (1992) Mythical kingdom of lawyers – America doesn't have 70 percent of the earth's lawyers. *American Bar Association Journal*, **78**, p. 72.

Bacalla, T.B. (2004) Smuggling is killing shoe, garments and textile industries. *Philippine Center for Investigative Journalism Magazine*. Manila. Available at http://www.pcij.org/stories/2004/smuggling3.html. Accessed 26 June 2010.

Burnham, Daniel H. and Bennett, Edward H. (1909) *Plan of Chicago*. Chicago: the Commercial Club.

Cities Alliance (2005) *The Impacts of City Development Strategies*. Washington DC: Cities Alliance/ World Bank. Available at http://www.citiesalliance.org/ca/sites/citiesalliance.org/files/cds-impact-study-final-report-august-11-2005%5B1%5D.pdf. Accessed 14 December 2009.

City Mayors (2009) *The World's Largest Cities and Urban Areas in 2020*. Available at http://www.citymayors.com/statistics/urban_2020_1.html. Accessed 2 February 2010.

EMI (2005) *Metropolitan Manila – The Philippines Disaster Risk Management Profile*. Manila: Earthquake and Megacities Initiative/Pacific Disaster Centre. Available at http://emi.pdc.org/cities/CP-Metro-Manila 0805.pdf. Accessed 21 July 2009.

EMI (2007) *The Disaster Risk Management Plan for Quezon City*. Kihei, HI: Earthquake and Megacities Initiative/Pacific Disaster Centre. Available at http://www.emimegacities.org/upload/3cd_2007_DRMMP_MM_BR0702.pdf. Accessed 2 May 2009.

Encyclopedia of the Nations (2007) *Philippines – Poverty and Wealth*. Available at http://www.nationsencyclopedia.com/economies/Asia-and-the-Pacific/Philippines-POVERTY-AND-WEALTH.html. Accessed 11 December 2010.

Esguerra, J. (2003) *The Corporate Muddle of Manila's Water Concessions: New Rules, New Roles: Does PSP Benefit the Poor?* Manila: WaterAid and Tearfund.

FAB4 (2008) *City of Taguig, Village People*. Taguig: Fab4 Publishing House.

Gawad Galing Pook (2008*a*) *Quezon City: Transforming Payatas Dumpsite*. Manila: Gawad Galing Pook Foundation. Available at http://www.galingpook.org/download/ggpsouvenir/2008_Galing_Pook_Souvenir_Program.pdf. Accessed 14 August 2009.

Gawad Galing Pook (2008*b*) *Condo Living for the Urban Poor*. Manila: Gawad Galing Pook Foundation 24–26. Available at http://www.galingpook.org/main/images/stories/replikit/2008galingpookwinners.pdf. Accessed 14 August 2009.

Gawad Galing Pook (2009) *Paano Tayo Aasenso? Policy Formulation on Local Economic Development*. Paano Tayo Aasenso Policy Forum on Local Economic Development, Asian Institute of Management Conference Center. Makati City: Gawad Galing Pook Foundation.

Gonzales, E. M. (2003) From Wastes to Assets: The Scavengers of Payatas. Paper presented to the International Conference on Natural Assets, Political Economy Research Institute and Centre for Science and the Environment Tagaytay City, Philippines.

Haq, G., Schwela, D., Huizenga, C., Fabian, H., Han, W.J. and Ajero, M. (2006) *Urban Air Pollution in Asian Cities: Status, Challenges and Management*. Lyndhurst, NJ: Barnes and Noble.

Intergovernmental Panel on Climate Change (2007) *Assessment Report* (Chapter 11. Regional Climate Projections). Geneva: Intergovernmental Panel on Climate Change.

International Federation of Red Cross and Red Crescent Societies (2009) *Philippines: Typhoons*. Manila: IFRCRCS. Available at http://www.ifrc.org/docs/appeals/06/MDRPH002fr.pdf. Accessed 8 September 2010.

Japan International Co-operation Agency (JICA), Metropolitan Manila Development Authority (MMDA) and Philippine Institute of Volcanology and Seismology (PHIVOLCS) (2004) *The Metropolitan Manila Earthquake Impact Reduction Study*. Manila: PHIVOLCS.

Kundu, A. (2009) *Urbanization and Migration: An Analysis of Trend, Pattern and Policies in Asia*. New York: United Nations Development Programme.

Laquian, A. (2005) *Beyond Metropolis: The Planning and Governance of Asia's Mega Urban Regions*. Washington, DC: Woodrow Wilson Center Press/Baltimore, MD: Johns Hopkins University Press.

Manasan, R.G. and Mercado, R.G. (1999) *Governance and Urban Development: Case Study of Metro Manila*. Manila: Philippine Institute for Development Studies. Available at http://www3.pids.gov.ph/ris/pdf/pidsdps9903.pdf. Accessed 29 October 2009.

Marikina-City (2007) *Shoe History*. Available at http://www.marikina.gov.ph/PAGES/history2.htm. Accessed 26 June 2010.

Mayong, A. (2006) Drowning in crisis: a look into the privatization of water services. Available at http://akbayanmayong.wordpress.com/2006/09/19/drowning-in-crisis-a-look-into-the-privatization-of-water-services/. Accessed 21 January 2010.

Merry, S.M., Kavazanjian, E. and Fritz, W.U. (2005) Reconnaissance of the July 10, 2000, Payatas landfill failure. *Journal of Performance of Constructed Facilities*, **19**(2), pp. 100–107.

MMDA (1996) *Physical Framework Plan for Metropolitan Manila, 1996–2016*. Manila: Metropolitan Manila Development Authority.

MMDA (2009) *Metropolitan Manila Development Authority Functions and Scope of Services*. Available at http://www.mmda.gov.ph/main.html. Accessed 10 December 2009.

Myrdal, G. (1957) *Economic Theory and Underdeveloped Regions*. New York: Harper and Row.

National Disaster Co-ordinating Council (2008) *History of Disaster Management in the Philippines*. Available at http://ndcc.gov.ph/home/index.php?option=com_content&task=view&id=14&Itemid=28. Accessed 6 January 2010.

National Disaster Co-ordinating Council (2010) *Final Report on Tropical Storm ONDOY and Typhoon PEPENG*. Available at http://www.ndcc.gov.ph/index.php?option=com_content&view=article&id=92. Accessed 14 November 2010.

NEDA (2007) *Project Appraisal Document: Manila Third Sewerage Project*. Manila: National Economic Development Authority. Available at http://www.neda.gov.ph/hgdg/training/WS%20group%203%20_Sewerage%20Case.pdf. Accessed 9 September 2009.

Nelson, A.R., Personius, S.F., Rimando, R.E., Punongbayan, R.S., Tuñgol, N., Mirabueno, H. and Rasdas, A. (2000) Multiple large earthquakes in the last 1500 years on a fault in metropolitan Manila, the Philippines. *Bulletin of Seismological Society of America*, **90**(2), pp. 73–85.

Perlman, J. (2007) *Mega-Cities Project: Innovations for Urban Life*. Manila: Mega-Cities Network. Available at http://www.megacitiesproject.org/about.asp. Accessed 14 June 2009.

Quezon City (2009) *The Payatas Dumpsite: From Tragedy to Triumph*. Available at http://www.quezoncity.gov.ph/index.php?option=com_content&view=article&id=160:the-payatas-dumpsite-from-tragedy-to-triumph&catid=44:special-features&Itemid=38. Accessed 30 November 2009.

Quimpo, N.G. (2009) The Philippines: predatory regime, growing authoritarian features. *The Pacific Review*, **22**, pp. 335–353.

Quisumbing, A.R., McNiven, S. and Godquin, M. (2008) *Shocks, Groups, and Networks in Bukidnon, Philippines*. Washington, DC: Consultative Group on International Agricultural Research (CGIAR) System-wide Program on Collective Action and Property Rights.

Resilience Alliance (2007) *A Research Prospectus for Urban Resilience: A Resilience Alliance Initiative for Transitioning Urban Systems towards Sustainable Futures*. Available at http://www.resalliance.org/files/1172764197_urbanresilienceresearchprospectusv7feb07.pdf. Accessed 9 May 2009.

Roberts, B.H. (2003) Regional risk management and economic development. *Australia New Zealand Regional Science Association*, **9**, pp. 67–96.

Roberts, B.H. (2006) Analysing and managing local economic development risk: a study of two Australian cities. *Built Environment*, **32**(4), pp. 413–433.

Roberts, B.H. and Cohen, M.E. (2002) Enhancing sustainable development by triple value-adding to the core business of government. *Economics Quarterly*, **16**, pp.127–137.

Roberts, B.H. and Tabart, C. (2005) *Managing the Risks: An investigation of risks affecting the ACT Economy*. Canberra: University of Canberra, Centre for Developing Cities.

Santa Cruz-Makilan, A. (2004) Disaster execs admit they can't handle big quake. Available at http://bulatlat.com/news/4-27/4-27-disaster.html. Accessed 10 June 2010.

Scott, A.J. (2005) The shoe industry of Marikina City, Philippines: a developing-country cluster in crisis. *Kasarinlan: Philippine Journal of Third World Studies*, **20**, pp.76–99.

Seva, V.S. (2008) *Disaster Risk Management Plan: The Makati Way.* Available at http://www.scribd.com/doc/5567977/DRM-the-MAKATI-WAY. Accessed 10 June 2010.

Sia-Su, G.L. (2007) Determinants of economic dependency on garbage: the case of Payatas, Philippines. *Asia-Pacific Social Science Review*, **7**, pp. 77–85.

Simpson, A., Cummins, P., Dhu, T., Griffin, J. and Schneider, J. (2008) Assessing natural disaster risk in the Asia-Pacific region. *AusGeo News*, June, pp. 1–6. Available at http://www.ga.gov.au/ausgeonews/ausgeonews200806/disaster.jsp. Accessed 21 November 2009.

Taguig City (2009) *Taguig City Comprehensive Land Use and Zoning Plan.* Taguig: Taguig City.

Tanchuling, M.A., Seedao, C., Acosta, L., Cariaga, J., Cruz, E. and Takemura, J. (2009) Assessing the effect of the Quezon City Controlled Dumping Facility (QCCDF) on soil and water quality, in *Proceedings of the Sixth Regional Symposium on Infrastructure Development.* Bangkok: Kasetsart University with Tokyo Institute of Technology and University of the Philippines.

Tharoor, I. (2009) The Manila floods: why wasn't the city prepared? *Time Magazine*, 29 September.

UNISDR (United Nations International Strategy for Disaster) (2008) *Urban Resilience and Disaster Vulnerability in the Asia-Pacific Region: Concept Note for Regional Urban Task Force.* Geneva: UNISDR Secretariat. Available at http://www.adrc.asia/events/RTFmeeting20080130/PDF_Presentations/01_Concept-Note-forRegional-Urban-Task-Force.pdf. Accessed 1 January 2010.

UNISDR (2009) *Global Assessment Report on Disaster Risk Reduction.* Geneva: UNISDR Secretariat. Available at http://www.preventionweb.net/english/hyogo/gar/report/index.php?id=9413. Accessed 1 January 2010.

Valt, E. (2007) Marikina shoes get a new shine, *The Good News Magazine* Available at http://www.inquirer.net/specialfeatures/thegoodnews/view.php?db=1&article=20070923-90234. Accessed 14 July 2009.

Vina, D.T.L., Agsaoay-Sano, E., Dulce, J. and Jambalos, J. (2009) Never again Ondoy and Pepeng. Note 4: Addressing the Roots of the Havoc in urban areas. *Good News Magazine*. Available at http://gator366.hostgator.com/~ateneo/index.php?option=com_content&view=article&id=160:never-again-ondoy-and-pepeng-note-no-4-addressing-the-roots-of-the-havoc-in-urban-areas&catid=23:news-highlights&Itemid=45. Accessed 10 January 2010.

Walker, B. and Salt, D. (2006) *Resilience Thinking: Sustaining Ecosystems and People in a Changing World.* Washington, DC: Island Press.

World Bank (2005) *Wanted a Breath of Fresh Air!* Manila: World Bank. Available at http://siteresources.worldbank.org/INTPHILIPPINES/Resources/TOC-Preface-Abbrev.pdf. Accessed 17 May 2009.

World Bank (2008) *Doing Business in the Philippines: Comparing Regulations in 21 Cities and 178 Economies.* Washington, DC: World Bank/International Bank for Reconstruction and Development. Available at http://www.doingbusiness.org/documents/subnational/DB08_Subnational_Report_Philippines.pdf. Accessed 12 December 2010.

World Bank (2010) *Migration and Remittances.* Washington DC: World Bank Migration and Remittances Team/Global Development Finance Database.

WWF (2009) *Mega-Stress for Mega-Cities: A Climate Vulnerability Ranking of Major Coastal Cities in Asia.* Gland: WWF International.

Index